LECTURE NOTES AND SUPPLEMENTS IN PHYSICS

Elements of Nuclei

Many-Body Physics with the Strong Interaction

Philip John Siemens
Department of Physics
University of Tennessee
Texas A & M University
United States of America

Aksel Stenholm Jensen
Institute of Physics
University of Aarhus
Denmark

Addison-Wesley Publishing Company, Inc.
The Advanced Book Program
Redwood City, California • Menlo Park, California • Reading, Massachusetts
New York • Amsterdam • Don Mills, Ontario • Sydney
Bonn • Madrid • Singapore • Tokyo • Madrid • Bogotá • Santiago
San Juan • Wokingham, United Kingdom

Publisher: Allan Wylde

Cover Photo Fifty years of nuclear reactions. The drawing on the left was used in 1936 by theorist Niels Bohr to illustrate his concept of the statistical model of compound-nuclear reactions (*Nature* **137** *(1936) 351*). The photo on the right was taken in 1986 by experimenter Kevin Wolf; it shows the traces left by the fragments of a barium nucleus upon being hit by an argon ion (left) emerging at one-fourth the speed of light from the Bevalac accelerator at Lawrence Berkeley Laboratory. These reactions are discussed in Chapters 10 and 11, respectively.

This book was typeset in TEX using a VAX 782 computer. Camera-ready output was produced from an Imagen laser printer.

ISBN 0-201-15572-9
ABCDEFGHIJ-AL-8987

Lecture Notes and Supplements in Physics
Edited by John David Jackson and David Pines

Series #	Code	Author	Title
1	34900	Jackson	Mathematics for Quantum Mechanics (1962)
2	31400	Brouwer	Matrix Methods in Optical Instrument Design (1964)
3	33601	Hagedorn	Relativistic Kinematics (1964; 5th Printing 1980)
4	35600	Knox/Gold	Symmetry in the Solid State (1964)
5	37912	Pines	Elementary Excitations in Solids (1963; 3rd Printing 1978)
6	30590	Barton	Introduction to Dispersion Techniques in Field Theory (1965)
7	31001	Bohm	The Special Theory of Relativity (1965; 2nd Printing 1979)
8	37750	Park	Introduction to Strong Interactions (1966)
9	30750	Bethe/Jackiw	Intermediate Quantum Mechanics (1964; 5th Printing 1982)
10	30667	Baym	Lectures On Quantum Mechanics (1969; 9th Printing 1981)
11	37397	Nishijima	Fields and Particles (1960; 4th Printing 1980)
12	10471	Brandsen	Atomic Collision Theory (1970)
13	38490	Sard	Relativistic Mechanics (1970)
14	32602	Frauenfelder	Nuclear and Particle Physics (1975)
15	39856	Wyld	Mathematical Methods for Physics (1976; 3rd Printing 1981)
16	30203	Amrein, et al.	Scattering Theory in Quantum Mechanics (1977)
17	31181	Brandsen	Atomic Collision Theory, Second Edition (1983)
18	30757	Bethe/Jackiw	Intermediate Quantum Mechanics, Third Edition (1985)
19	31505	Capri	Non-Relativistic Quantum Mechanics, Second Edition (1985)
20	38754	Ichimaru	Plasma Physics: An Introduction to Statistical Physics of Charged Particles (1986)
21	15572	Siemens/Jensen	Elements of Nuclei (1987)

Editor's Foreword

Everyone concerned with the teaching of physics at the advanced undergraduate or graduate level is aware of the continuing need for a modernization and reorganization of the basic course material. Despite the existence today of many good textbooks in these areas, there is always an appreciable time-lag in the incorporation of new viewpoints and techniques which result from the most recent developments in physics research. Typically these changes in concepts and material take place first in the personal lecture notes of some of those who teach graduate courses. Eventually, printed notes may appear, and some fraction of such notes evolve into textbooks or monographs. But much of this fresh material remains available only to a very limited audience, to the detriment of all. Our series aims to fill this gap in the literature of physics by presenting occasional volumes with a contemporary approach to the classical topics of physics at the advanced undergraduate and graduate level. Clarity and soundness of treatment will, we hope, mark these volumes, as well as the freshness of the approach.

Another area in which the series hopes to make a contribution is by presenting useful supplementing material of well-defined scope. This may take the form of a survey of relevant mathematical principles, or a collection of reprints of basic papers in a field. Here the aim is to provide the instructor with added flexibility through the use of supplements at relatively low cost.

The scope of both the Lecture Notes and Supplements is somewhat different from the Frontiers in Physics Series. In spite of wide variations from institution to institution as to what comprises the basic graduate course program, there is a widely accepted group of "bread and butter" courses that deal with the classic topics in physics. These include: mathematical methods of physics, electromagnetic theory, advanced dynamics, quantum mechanics, statistical mechanics, and frequently nuclear physics and/or solid-state physics. It is chiefly these areas that will be covered by the present series. The listing is perhaps best described as including all advanced undergraduate and graduate courses which are at a level below seminar courses dealing entirely with current research topics.

The above words were written in 1962 in collaboration with David Jackson who served as co-editor of this Series during its first decade. They serve equally well as a Foreword for the present volume. During the past two decades the study of the nucleus as a many-body system of interacting neutrons and protons has not only led to a significant improvement in our physical understanding of nuclear properties but has made evident the close connection between nuclei and atoms, and between nuclear matter and other quantum liquids, such as the electron and the helium liquids which play such an important role in condensed matter physics. Nuclear physics has thus both contributed to, and benefited from, the advances in understanding many-body problems which have characterized the work of this period.

In the present volume, Philip J. Siemens, who has been a seminal contributor to our understanding of the nucleus as a many-body system, and his able collaborator, Aksel S. Jensen, introduce graduate students and colleagues in other fields to the basic concepts of nuclear physics in a way which connects clearly the methods of nuclear physics with those of condensed matter, atomic, and particle physics. Their book thus provides a lucid introduction to the key facts and concepts of nuclei, including many of the most recent developments, while emphazing the similarities and the differences between the behavior of nuclei, atoms, elementary particles, and condensed matter. It should thus prove useful, not only as a text for an introductory graduate course in nuclear physics, but as a reference book for all scientists interested in a unified picture of our understanding of physical phenomena associated with many-body systems.

David Pines
Aspen, Colorado
July 1987

Preface

This book is intended to serve as a text for a first course in nuclear physics, for graduate students in both experimental and theoretical physics. We assume that the student has a good knowledge of quantum mechanics, equivalent to a year's course of graduate study at the level of, e.g., Schiff's **Quantum Mechanics** or Messiah's two-volume work of the same name.

Many good books are available for study at this level, e.g., Bohr and Mottelson, **Nuclear Structure** (2 volumes); Preston and Bhaduri, **Structure of the Nucleus**; de Shalit and Feshbach, **Theoretical Nuclear Physics**. However, these books are mainly intended for students who wish to specialize in nuclear physics, and consequently require at least a full year's course of study. The present work is meant instead as an overview of the field for those who have not yet decided whether to concentrate on nuclear physics, or who have chosen another specialty but wish to broaden their knowledge with a survey of this important field. Therefore, we keep our exposition brief enough to be studied in a single semester or two quarters.

To achieve the desired brevity, we focus our attention on key facts about nuclei and the basic concepts that explain them. Although we purposefully omit many detailed facts, we expound the concepts at a fairly advanced level of abstraction while emphasizing their foundation in experiment. As we demonstrate the similarities and differences between the nucleus and other many-body systems, we try to show the connections between the methods of nuclear physics and other branches of our science. We expect that this perspective may prove helpful in encounters with other disciplines of physics. We hope that these notes will stimulate the reader's curiosity to learn more about nuclei, and provide the student with a sound framework for understanding them.

Contents

Introduction

These notes expound a picture of the nucleus as a many-body system of neutrons and protons interacting by strong forces of mostly indefinite origin. We see the nucleus as a meeting ground between the many-body physics of quantum liquids and atoms, and the interacting fields of particle physics.

The composite structure of the nucleon in terms of quarks and glue is invoked only when other, more generic explanations are inadequate. We anticipate that current advances in strong-interaction particle physics will soon lead to a much clearer picture of the fundamental nature of hadrons and their interactions, of which only the longest-range force from pion exchange is presently on a firm footing. The rapid progress of strong-interaction physics, due to the recent recognition that the vacuum is a strongly interacting many-body system, will surely produce a revolution in our understanding of nuclei.

Meanwhile, it is remarkable that we can come so far in relating observations of the nuclear system at different levels without a more detailed insight into the structure of the constituents or into the origins of the forces among them. This view of the nucleus provides a challenging application of the concepts and techniques of many-body physics which have been so successful for understanding atoms and condensed matter.

Our exposition begins with a short description of the astrophysical origins of nuclei. We then discuss the most important features of the force between nucleons, introducing for later use the concepts of the effective-range expansion and of effective forces. Next, we relate how the scattering of nucleons from nuclei can be described by the phenomenological optical model, and introduce two characteristic features of nuclei: the saturation density of nuclear matter, and the long mean free paths of nucleons in nuclei. We show how the picture of nucleons moving independently in nuclei can be used to explain reactions in which nucleons are deposited in or removed from a nucleus, and observe that the picture fails to account for the binding energies of nuclei. We obtain a better understanding of these features by way of the mean-field picture with effective interactions to account for short-range correlations and many-body effects.

The mean-field picture implies the spontaneous breaking of translational symmetry which we demonstrate to be an artifact of the mean-field approximation. We then show how the mean field also leads to broken rotational symmetry. The effects of correlations beyond the mean field are discussed in terms of the BCS picture of time-reversed pairs. This allows us to understand the properties of nuclear ground states: angular momentum and magnetic moments as well as binding energies. The isospin symmetry between neutrons and protons is shown to lead to characteristic relationships among the excited states of different nuclei.

The collective vibrations of nuclei are introduced in a linear-response picture by allowing the mean field to depend on time, leading to the RPA eigenvalue equation for the frequencies of collective modes. Many normal modes of nuclear motion are described within this simple framework, and the groundwork is laid for equations governing irreversible large-amplitude collective motion. Rotational motion is shown to require a viewpoint more sophisticated than the mean field, because of its intimate relation to the breaking of rotational symmetry.

The decay of nuclear excited states by emission of nucleons, nucleon clusters, photons and electrons as well as by fission is summarily described. For this purpose the models developed previously are extended into the domain of the statistical mechanics of small systems. The field theory of the electroweak interaction is applied to nuclear decays, and we tell how the weak decay of nuclei led to the discovery that parity is a broken symmetry.

We conclude with an overview of the phenomena arising when complex nuclei collide. Here we see many macroscopic traits in our tiny systems: friction, a liquid-gas phase transition, and hydrodynamic flow are among the most fascinating phenomena observed so far, at temperatures from 10^{10} to 10^{12} Kelvin. Speculations about an additional phase transition in which nucleons dissolve into quarks and gluons point to future experiments.

We use quantum mechanics freely, with detailed explanations only for those aspects, such as effective-range theory or effective interactions, that may not appear in every graduate quantum-mechanics course. We sometimes use the second-quantized formalism of creation and annihilation operators, and often assume knowledge of results of angular-momentum algebra and of the three-dimensional harmonic oscillator as well as Dirac and Pauli spinors. Our conventions of notation on these and other basics are summarized concisely in an appendix. Measured quantities are mostly given in units of MeV (energy) and fm (length); the appendix also contains a brief collection of physical constants in these units.

A list of references is included at the end of the book. These are mentioned in the text by author and, when necessary, year of publication. We have not attempted to give references for every point, nor to find the original, latest, or most authoritative sources; rather we intend the references as an entryway for further study of subjects introduced in the text. When experimental information is given without an explicit source, it is either our own judgment from a varied collection of data, or is to be found in the periodic summaries of the Particle Physics Data Group in Berkeley or the Nuclear Data Sheets from Oak Ridge.

Many of the illustrations have been reproduced from other sources. These sources are acknowledged in the figure captions. We are grateful to the following for kind permission to reproduce their illustrations: Aage Bohr and Ben Mottelson, Macmillan Publishers Ltd. (for reproductions from *Nature*), The North-Holland Publishing Company, F. Ajzenberg-Selove, F. Becchetti, F. Bertrand, J. de Boer, S. Fernbach, B. Frois, C. Goodman, J. Harris, D. Goren, J. Huizenga, C. Mahaux, L. McLerran, S. Nagamiya, O. Nathan, J. Negele, H. Pauli, F. Plasil, J. Rasmussen, R. Reid, J. Schiffer, R. Silbar, J. Speth, R. Vandenbosch, V. Weisskopf, G. Westfall, Aa. Winther, C.S. Wu. In addition, several of the illustrations contain previously unpublished results, thanks to the generosity of Fred Bertrand, Bernard Frois, Charles Goodman, Dan Horen, Ray Nix, J.J. Simpson, Michael Strayer, V. Volkov, Janusz Wilczynski, and Kevin Wolf.

We thank Helmut Hofmann for original ideas used in our treatment of collective motion; Abel Miranda for frequent enlightenment; Scott Pratt and Gary White for useful comments; our patient students for many corrections and suggestions; Shalom Shlomo and Scott Pratt for help in computing numbers used in the figures; Tove Asmussen and Erik Mortensen for drawing the original figures; Cheryl Wurzbacher for assistance in producing the cover and format; and Mary Anderson, Shannon Bays, Janice Epstein, Rie Vangkilde, and especially Philip Bingham for the technical preparation of the manuscript. Finally, we thank our lucky stars for the tutelage of our teachers Hans Bethe, Aage Bohr, Gerry Brown, Ben Mottelson and David Pines, whose deep insights have inspired us to attempt the following synthesis.

1

Origins of Nuclei

More than 99% of the identifiable mass in the universe is in the form of atomic nuclei. About three-fourths of these are **protons**, the nuclei of the lightest isotope of hydrogen, which is the lightest element. Nearly all the rest are helium nuclei, with small amounts (a percent or less) of deuterium (heavy hydrogen), carbon, and other elements. Although for every proton there are at least 10^8 photons (quanta of electromagnetic radiation) and a comparable number of neutrinos, the energy of these light particles is negligible compared to the rest energy of the protons. We can say that the known universe consists mostly of empty space and nuclei.

1.1   THE FIRST NANOSECOND

It was not always thus. Apparently, the universe originated in the explosion of a tiny but enormously heavy speck of matter. The explosion is called the **Big Bang**. The story of the Big Bang is related very beautifully in Weinberg's book, *The First Three Minutes*. We do not know what the matter was like at the very beginning, but by the end of the first nanosecond, it probably consisted of **quarks** and **antiquarks**, **gluons**, **electroweak bosons**, and **leptons**, the smallest constituents of matter that can be detected in today's particle-physics laboratories. By this time there were more quarks than antiquarks; earlier still, there may have been equal numbers of quarks and antiquarks, with the antiquarks decaying by processes which have been suggested theoretically but not yet verified by experiments. If and when the radioactive decay of the proton is observed, these theories may be verified and we will be able with confidence to look still further back towards the dawn of creation. Here, we will briefly describe the universe's more recent history, shown schematically in fig. 1.1.

1.2   THE FIRST MICROSECOND

The existence of nuclei is due to the predominance of quarks over antiquarks in the early universe. The enormous density of the matter at early times was accompanied

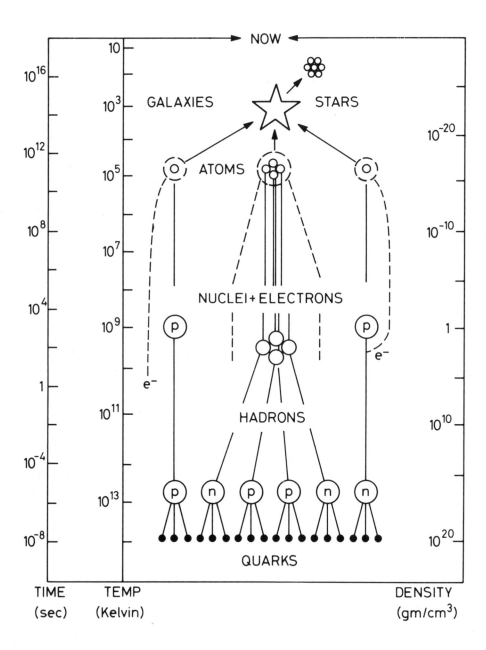

TIME   TEMP   DENSITY
(sec)   (Kelvin)   (gm/cm³)

Figure 1.1   History of nuclei

by an enormous temperature; the pressure of the hot, dense matter caused it to expand rapidly, cooling adiabatically as it expanded. By the end of the first few microseconds, the matter had cooled enough that the forces among the quarks and antiquarks could hold them together in small (about 1 fm radius) clusters called **hadrons**, containing three quarks (**baryons**) or a quark and an antiquark (**mesons**). The forces among the leptons and electroweak bosons are much weaker than the forces among the quarks and had little effect in this epoch. As the universe continued to expand, the hadrons separated from each other; the hadrons, bosons, and leptons collided with each other, or sometimes decayed into other hadrons or leptons, until by the end of the first microsecond only the most stable species were left. Among the electroweak bosons and leptons, the electrons, neutrinos, and photons survived; of the hadrons, only the **nucleons** remained. There are two types of nucleons: **protons**, which have a positive electric charge, and **neutrons**, their electrically-neutral twins.

1.3   THE FIRST SECOND

After the first few microseconds, not much happened for the rest of the first second. The universe continued to expand and cool. The relative numbers of neutrons, protons, electrons, positrons, neutrinos, and photons were determined by the temperature according to the laws of chemical equilibrium. At the end of the first second, the temperature had fallen to about 10^{10} Kelvin, and there were nearly equal numbers of neutrons and protons.

1.4   THE FIRST QUARTER HOUR

As the temperature continued to fall, the nucleons began to stick together in small clusters, mainly deuterons (one proton and one neutron) and alpha particles (helium nuclei with two protons and two neutrons). This is the epoch of the "primordial nucleosynthesis", the first formation of elements. During this epoch the energy of thermal motion $\approx k_B T$ became less than the energy of binding holding the nuclei together. The cooler temperature favored the formation of nuclei, especially the tightly-bound alpha particles.

As the cooling progressed, all the nucleons would have been bound in alpha particles, had it not been for the fact that the rest energy of a neutron is somewhat greater than that of a proton. This enables a neutron to undergo a radioactive **beta decay**, turning into a proton, an electron, and an antineutrino after an average time of about 1000 seconds. For reasons we shall study in Chapter 2, protons are more attracted to neutrons than to each other; a pair of protons cannot stick together. Thus the conversion of neutrons into protons via beta decay caused a surplus of protons, which survived as hydrogen nuclei. Only the neutrons which were bound in primordial deuterons and alpha particles survived. To turn into a proton, they would have had to release enough energy not only to make an electron and a neutrino, but also to compensate for the loss of the attractive interaction energy within the nuclei. But the rest energy of the neutrons is not big enough to make up for the binding forces, so the nuclei preserved the neutrons bound inside them. Most of the deuterium and helium in our sun and other similar stars was formed in this way. In fact, detailed computations of the competition between nucleosynthesis and beta decay are able to explain quantitatively the observed abundances of deuterium and helium. The quantitative success of these computations provides evidence for the "Big Bang" picture of the origin of the universe.

1.5   THE FIRST MILLION YEARS

After the era of primordial nucleosynthesis, the universe continued to expand uneventfully for almost a million years, a mixture of nuclei, electrons, neutrinos, and photons. Gradually, many of the electrons and nuclei stuck together to form atoms and molecules, as the temperature grew cool enough to avoid disrupting these relatively fragile structures, which are held together by electric attraction. As this happened, the universe became transparent to light, since photons are much less disturbed by neutral atoms than by charged electrons. The photons which were present then have never been absorbed, and still exist as the "microwave background radiation" of millimeter wavelength, which forms a significant part of the background noise in microwave communications and radar. The observation of these fossil photons by Penzias and Wilson in 1965 was the first direct experimental evidence of the Big Bang.

1.6   THE FIRST BILLION YEARS

Gradually, many of the atoms and molecules formed in the big bang clustered together due to the mutual attraction of their gravity. Eventually, these clusters gathered enough mass to become stars. The gravitational attraction causes a star to compress together and heat up, reversing the expansion-cooling cycle of the Big Bang. As the stars heat up, a sequence of nuclear reactions allows the protons to combine into helium and other, heavier elements. These reactions are too slow to have played a role in the primordial nucleosynthesis, but can go to completion in stars, which last millions or, usually, billions of years. Most elements lighter than cobalt have been formed by **stellar nucleosynthesis**, which is still going on.

When these reactions use up all the hydrogen and light elements, the star ceases to produce energy. The pressure of the radiation and thermal motion, fueled by to the exothermic nuclear reactions, is no longer there to support the star's weight against the crush of gravity. The star collapses (sometimes the collapse occurs earlier in the star's evolution, if it swallows up a neighboring star). If it is a relatively small star like our sun, the collapse stops when the pressure of the electrons' quantal zero-point motion balances gravity, and the star becomes a "white dwarf". If the star is bigger, or if it swallows another star after it has stopped burning, the weight of its mass may be too great for the electrons to resist, and it continues to collapse rapidly (implode) until the nuclei touch each other. When this happens, if the star is not too massive, the implosion is reversed and becomes a supernova explosion. (Otherwise, it becomes a black hole.) During this process of implosion/explosion, endothermic nuclear reactions can occur due to the driving force of the motion of the matter. These reactions lead to the formation of the elements heavier than nickel. The force of the explosion scatters the heavy elements into space, where they may eventually become part of another star or its planets. All the heavy elements on earth were formed in supernovas; it is estimated that the earth contains portions of the remnants of several thousand different supernovas.

1.7   CONSERVATION LAWS AND NUCLEAR FORCES

We have seen that the matter of today's universe has been formed in a succession of processes, the outcome of which is determined by fundamental laws of nature and by detailed properties of nuclei and nucleons. It is our aim in this course to see how the properties of nuclei are related to the properties of their constituents, the neutrons and protons, and to the fundamental laws of physics. We have already alluded to some of these laws and properties. A list of some important properties is shown in table 1.1, for a few important particles.

The first columns represent conserved quantities: the total energy, baryon number, electric charge, and angular momentum remain unchanged by every reaction. It is believed that experiments currently underway may disprove the conservation of baryon number, but the reactions involved would have played no role after the first picosecond. The next columns, isospin and parity, are approximately conserved quantities: they are conserved by the strong and, in the case of parity, electromagnetic interactions, but not by the weak force. The last entries, the intrinsic magnetic moment and lifetime, are included for completeness.

As an example of the workings of the conservation laws, consider the beta decay of the neutron:

$$n \rightarrow p + e + \bar{\nu}_e.$$

(The bar indicates an antineutrino; antiparticles have the same properties as the corresponding particles, except that their charge and baryon number have the opposite signs). Because the neutron has baryon number 1, the final state has to include a proton, which is the only baryon with a smaller rest mass. That is why the proton can't decay: it couldn't conserve both baryon number and energy. To conserve electric charge, the final state has to include an electron too; it is the only charged particle light enough to be made from the remaining rest energy of the neutron, after the proton's rest energy has been taken care of. That makes a final state

Symbol	Particle	Rest Energy mc² (MeV)	Baryon Number	Electric Charge (e)	Angular Momentum (ℏ)	Isospin	Parity	Magnetic Moment (eℏ/2mc)	Lifetime (sec)
hadrons									
p	proton	938.28	1	1	$\frac{1}{2}$	$\frac{1}{2}$	+	2.793	stable
n	neutron	939.57	1	0	$\frac{1}{2}$	$\frac{1}{2}$	+	-1.913	900
Δ	delta	1230 to 1234	1	2,1,0,-1	$\frac{3}{2}$	$\frac{3}{2}$	+	not measured	6×10^{-24}
π	pi meson	134.96 / 139.57	0	0 / ±1	0	1	-	0	8×10^{-17} / 7×10^{-19}
ρ	rho meson	769	0	0 / ±1	1	1	-	not measured	4×10^{-24}
ω	omega meson	782.6	0	0	1	0	-	"	2×10^{-22}
leptons									
e	electron / positron	.511	0	-1 / +1	$\frac{1}{2}$	$\frac{1}{2}$	+	1.001	Stable
μ	muon	105.66	0	±1	$\frac{1}{2}$	$\frac{1}{2}$	+	1.001	2×10^{-6}
ν	neutrino	0 (< 0.00006)	0	0	$\frac{1}{2}$	$\frac{1}{2}$	+	not measured	Stable
γ	photon	0 ($< 6 \times 10^{-22}$)	0	0	1	-	-	0	Stable

Table 1.1 Some of the lightest particles

with two half-integer angular momenta; they add together to make integral angular momentum, while the neutron was half-integer. So we need a neutral fermion too: the neutrino. (It turns out that each charged lepton has its own kind of neutrino, which carries the same kind of "lepton number," another conserved quantity.) The left-over energy is distributed among the products in the form of kinetic energy.

Other, more complicated arguments lead to the other conclusions cited in this chapter. We shall see many of these arguments in the course of our study. For example, the binding of a neutron and a proton into a deuterium nucleus is a central topic of the next chapter.

PROBLEMS

1.1   Use the fact that the reaction $d \rightarrow p + p + e^- + \bar{\nu}$ is energetically forbidden to place a bound on the binding energy of the deuteron (d).

1.2   What final states can a π^0 meson decay into? a π^+ meson?

1.3   Ever since the photons in the universe decoupled from the electrons and nuclei, they have been a non-interacting gas with a constant entropy. As the universe expanded, the gas of photons expanded too, cooling by adiabatic expansion. Their temperature now is about $3°K$, observed as the microwave radiation. When they decoupled, the temperature was somewhat less than the ionization energy of hydrogen atoms. By what factor has the universe expanded in the interim?

2

Nuclear Forces

The simplest nucleus is the deuteron, or heavy hydrogen, which is a bound state of just one neutron and one proton. It has only one bound state: all its excited states are continuum or scattering states. In this chapter, we study this simple system to learn what we can about the forces between nucleons, in the hope that this knowledge will be useful in studying more complicated nuclear systems. Our study of the nuclear force will reveal that the lightest mesons, the pions, play a dominant role when the nucleons are far apart. Therefore we will want to extend our study to include the scattering of pions by nucleons. At short distances the internal structure of the nucleon becomes important; we learn about this from the scattering of electrons by nucleons.

2.1 THE GROUND STATE OF THE DEUTERON

The only bound state of two nucleons is the deuteron. Its binding energy,

$$B = (m_p + m_n) c^2 - E_{g.s.} = 2.23 \text{ MeV} \tag{2.1.1}$$

is barely 0.1% of its rest mass, in contrast to most heavier nuclei, whose binding energies are about 0.8% of their masses. The deuteron's size is also anomalous: from its interactions with atomic electrons, and from electron-scattering probes, we learn that its r.m.s. radius

$$R_{r.m.s.} = \langle r^2 \rangle^{1/2} \tag{2.1.2}$$

is 2.8 fm, about the same as that of neon, which has ten times as many neutrons and protons. To get an idea of the magnitude and range of the forces, we can consider the bound state of a square-well potential,

$$V(\vec{r}) = -V_0 \, \Theta(|\vec{r}| < R) \tag{2.1.3}$$

where Θ, the truth function, takes on the value 1 if its argument is true and zero if it is false. The Schrödinger equation for the wave function's dependence on the relative coordinate $\vec{r}$ between the neutron and proton is

$$E\psi(\vec{r}) = -\frac{\hbar^2}{2\mu}\vec{\nabla}^2\psi(\vec{r}) + V(\vec{r})\psi(\vec{r}) \tag{2.1.4}$$

9

where $\psi(\vec{r})$ is the wave function in its coordinate representation, and

$$\mu = \frac{m_n m_p}{m_n + m_p} \tag{2.1.5}$$

is the reduced mass. Since we have assumed a central potential,

$$V(\vec{r}) = V(|\vec{r}|) \tag{2.1.6}$$

the eigenfunctions $\psi(\vec{r})$ will be proportional to the spherical harmonics $Y_L^M(\theta_r, \phi_r)$:

$$\psi(\vec{r}) = \frac{u_L(|\vec{r}|)}{|\vec{r}|} Y_L^M(\theta_r, \phi_r) \tag{2.1.7}$$

where $u_L(r)$ satisfies the radial Schrödinger equation

$$-\frac{\hbar^2}{2\mu} \frac{d^2}{dr^2} u_L(r) + \frac{\hbar^2}{2\mu} \frac{L(L+1)}{r^2} u_L(r) + V(r) u_L(r) = E\ u_L(r). \tag{2.1.8}$$

The state of lowest energy will be found for $L = 0$, because other angular momenta have a repulsive centrifugal potential in (2.1.8).

For the square-well potential (2.1.3), the ground-state radial wave function u is given by

$$u = \begin{cases} A \sin(\sqrt{2\mu(E + V_0)}r/\hbar) & \text{for } r < R \\ A \sin(\sqrt{2\mu(E + V_0)}R/\hbar) \exp\left[-(r - R)\sqrt{-2\mu E}/\hbar\right] & \text{for } r > R \end{cases} \tag{2.1.9}$$

and the eigenvalue $E = -B$ solves the characteristic equation

$$\cot\left(\frac{R\sqrt{2\mu(E + V_0)}}{\hbar}\right) = -\sqrt{\frac{-E}{E + V_0}}. \tag{2.1.10}$$

Like E, the r.m.s. radius depends on both R and V_0; we expect it to be an increasing function of R, for fixed E. It is easy to show that, if R is much less than $R_{\text{r.m.s.}}$, then $R_{\text{r.m.s.}}$ is given by

$$R_{\text{r.m.s.}} = \frac{\hbar}{2}(-\mu E)^{-1/2} = 3.0\,\text{fm}. \tag{2.1.11}$$

which is close to the observed value. Thus we can only conclude that the range of the force is considerably less than 2.8 fm, and the nucleons spend most of their time outside the range of the force. Consequently, the depth of the potential must be much greater than the observed binding energy.

2.2   CONTINUUM STATES. EFFECTIVE RANGE THEORY

Disappointed by the paucity of information about nuclear forces revealed by the deuteron, we turn to the continuum or scattering states. Surely the scattering cross sections contain more information, since the scattering can be measured at various energies and angles.

Scattering experiments show that the neutron-proton scattering cross section $d\sigma/d\Omega$ is nearly isotropic for laboratory kinetic energies up to about 50 Mev (see fig. 2.1).

Since isotropic scattering is characteristic of s-wave ($\ell=0$) scattering, we may guess that the absence of p-wave scattering is due to the short range of the potential. At 50 MeV laboratory energy ($= 25$ MeV in the center of mass), the relative momentum is $p = \sqrt{2\mu E} = 153$ MeV/c, so that two non-interacting particles with relative angular momentum $\ell = 1\hbar$ have a classical turning point at $r = \sqrt{1(1+1)}\hbar/p = 1.8$ fm (using $\hbar c = 197$ MeV fm). It is not inconsistent with our knowledge of the deuteron to suppose that the forces are still small at this distance. Thus we will

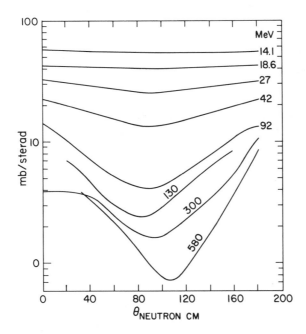

Figure 2.1   Neutron-proton scattering cross sections $d\sigma/d\Omega$, as a function of the center-of-mass angle of the deflected neutron, for various laboratory energies of the incident neutron (MeV).

concentrate on the $\ell = 0$ partial wave.

Still assuming a local interaction V(r) (i.e. diagonal in the relative distance r between the nucleons) we may write the Schrödinger equation for the $\ell = 0$ radial wave function u at center-of-mass scattering energy E as

$$\frac{d^2u}{dr^2} + \left(k^2 - \mathcal{V}\right) u = 0 \qquad (2.2.1)$$

where $k = \sqrt{2\mu E/\hbar^2}$ and $\mathcal{V} = V \cdot 2\mu/\hbar^2$. The radial wave function u(r) approaches 0 at the origin; for large distances r outside the range of the potential, V(r) goes to 0 and

$$u(r) \rightarrow C \sin(kr + \delta) \qquad (2.2.2)$$

Solving the radial Schrödinger equation determines δ, which in turn determines the cross section

$$\frac{d\sigma}{d\Omega} = \frac{1}{k^2} \sin^2 \delta. \qquad (2.2.3)$$

By trial and error, it was soon found that many different potentials V(r), with disparate radial dependences, could be made to fit the observed phase-shifts for $E_{lab} < 50$ MeV.

The reason that the potentials cannot be determined in detail from low-energy nucleon scattering may be found in the effective range theory [Messiah vol.1], a sort of pertubation theory which is useful when the potential energy dominates the kinetic energy (the opposite limit compared to the Born Approximation). Consider two radial wave functions, u_1 and u_2, corresponding to energies E_1, E_2 and wave numbers k_1, k_2. At the same time, consider two free-scattering wave functions $\tilde{u}_1, \tilde{u}_2$ corresponding to the same energies but without interactions (V = 0), and whose asymptotic behavior for large r is the same as the u's, i.e.

$$\tilde{u}_i = C_i \sin(k_i r + \delta_i). \qquad (2.2.4)$$

Multiplying the wave equations for u_1, u_2, $\tilde{u}_1$, and $\tilde{u}_2$ by u_2, u_1, $\tilde{u}_2$, and $\tilde{u}_1$ respectively, we obtain

$$u_2\frac{d^2u_1}{dr^2} + \left(k_1^2 - \mathcal{V}\right) u_2 u_1 = 0 \qquad (2.2.5a)$$

$$u_1\frac{d^2u_2}{dr^2} + \left(k_2^2 - \mathcal{V}\right) u_2 u_1 = 0 \qquad (2.2.5b)$$

$$\tilde{u}_2\frac{d^2\tilde{u}_1}{dr^2} + k_1^2\tilde{u}_2\tilde{u}_1 = 0 \qquad (2.2.5c)$$

$$\tilde{u}_1\frac{d^2\tilde{u}_2}{dr^2} + k_2^2\tilde{u}_1\tilde{u}_2 = 0 \qquad (2.2.5d)$$

If we now add (2.2.5d) to (2.2.5a), subtract (2.2.5b) and (2.2.5c), and integrate from r=0 to a radius R_{max} beyond the range of the force, we find

$$\left(\tilde{u}_1 \frac{d\tilde{u}_2}{dr} - \tilde{u}_2 \frac{d\tilde{u}_1}{dr} - u_1 \frac{du_2}{dr} + u_2 \frac{du_1}{dr}\right)_{r=0} = (k_2^2 - k_1^2) \int_0^{R_{max}} (\tilde{u}_1 \tilde{u}_2 - u_1 u_2) dr. \quad (2.2.6)$$

Choosing for convenience the normalization

$$C_i = \frac{1}{\sin \delta_i} \quad (2.2.7)$$

leads to the result

$$k_2 \cot \delta_2 - k_1 \cot \delta_1 = (k_2^2 - k_1^2) \int_0^{R_{max}} (\tilde{u}_1 \tilde{u}_2 - u_1 u_2) dr. \quad (2.2.8)$$

As a special case, take $k_1 = 0$, $k_2 = k$; denoting the corresponding $\tilde{u}$'s and u's by $\tilde{u}_0$, u_0, $\tilde{u}_k$, u_k, we find

$$k \cot \delta(k) = \frac{1}{a} + \frac{1}{2} k^2 R(k) \quad (2.2.9)$$

where

$$\frac{1}{a} = \lim_{k \to 0} k \cot \delta(k) \quad (2.2.10)$$

$$R(k) = 2 \int_0^\infty dr (\tilde{u}_0 \tilde{u}_k - u_0 u_k). \quad (2.2.11)$$

a is called the scattering length, and R(k) the effective range. Note that the opposite sign convention is often used for a.

We are now in a position to understand why low-energy scattering is similar for different potentials. In (2.2.11) the integrand vanishes outside the range of the potential, since there the u's and $\tilde{u}$'s are equal. If the potential is strong, u_k does not differ very much from u_0 as long as the energy E is less than the depth of the potential. Then $\tilde{u}_k$ and $\tilde{u}_0$ will also be similar, provided $kR_{max} \ll 1$. We conclude that

$$R(k) \approx R(0) \quad (2.2.12)$$

for energies less than the depth of the potential. For such energies, the scattering is determined by only two parameters, a and R(0). Comparing the observed phase shifts to approximation (2.2.12), we estimate the depth of the potential at 50 to 100 MeV. (See problem 2.3).

2.3   SPIN AND ISOSPIN

Until now, we have ignored the fact that neutrons and protons have an intrinsic angular momentum, or spin, equal to $\frac{1}{2}\hbar$. Thus it is not sufficient to specify their wave functions' radial dependence: the wave functions also have to be specified in the space of their intrinsic spins. For example, a scattering experiment may involve a beam of polarized neutrons, whose spins are aligned parallel to an arbitrarily-chosen z-axis. We denote the spin wave functions (spinors) by $|\uparrow\rangle_n$ and $|\downarrow\rangle_n$ depending on whether the alignment is parallel or antiparallel. Similarly, the protons in the target may be polarized. Thus the spinor will be given by

$$\langle\vec{r}|\psi\rangle = \psi^{\uparrow\uparrow}(\vec{r})|\uparrow\rangle_n|\uparrow\rangle_p + \psi^{\downarrow\downarrow}(\vec{r})|\downarrow\rangle_n|\downarrow\rangle_p$$
$$+ \psi^{\uparrow\downarrow}(\vec{r})|\uparrow\rangle_n|\downarrow\rangle_p + \psi^{\downarrow\uparrow}(\vec{r})|\downarrow\rangle_n|\uparrow\rangle_p. \tag{2.3.1}$$

Alternately, the wave function may be specified in the basis of eigenfunctions of the total spin $\vec{S} = \vec{s}_1 + \vec{s}_2$, or more precisely, its square and its z-component.

$$\vec{S}^2|SS_z\rangle = \hbar^2 S(S+1)|SS_z\rangle \tag{2.3.2a}$$
$$S_z|SS_z\rangle = \hbar S_z|SS_z\rangle \tag{2.3.2b}$$

which are related to the individual spinors by

$$|1\,1\rangle = |\uparrow\rangle_1|\uparrow\rangle_2$$
$$|1-1\rangle = |\downarrow\rangle_1|\downarrow\rangle_2$$
$$|1\,0\rangle = \frac{1}{\sqrt{2}}\left(|\uparrow\rangle_1|\downarrow\rangle_2 + |\downarrow\rangle_1|\uparrow\rangle_2\right) \tag{2.3.3}$$
$$|0\,0\rangle = \frac{1}{\sqrt{2}}\left(|\uparrow\rangle_1|\downarrow\rangle_2 - |\downarrow\rangle_1|\uparrow\rangle_2\right).$$

In either case, there are four spin basis vectors.

There is nothing to prevent the target and projectile nucleons from exchanging angular momentum, or even trading spin for orbital angular momentum $\vec{L}$. The total angular momentum,

$$\vec{J} = \vec{L} + \vec{S} \tag{2.3.4}$$

has to be conserved, though. If the orbital motion is in an s-state, then $\vec{L} = 0$ and $\vec{J} = \vec{S}$; thus S^2 and S_z will be conserved. Furthermore, the forces must be independent of J_z and therefore, for $L = 0$, of S_z. Thus the basis $|SS_z\rangle$ is convenient: for $L = 0$, the scattering matrix must be diagonal and independent of S_z. It may, however, depend on S.

In fact, the low-energy cross sections for scattering of parallel and antiparallel spins are not the same: we need different phase shifts for $S = 1$ (the triplet spin state) and $S = 0$ (singlet). The singlet and triplet scattering lengths and effective ranges are different [Bohr and Mottelson vol.1]:

$$\text{singlet}\quad a_s = 23.7 \text{ fm}, \quad R_s = 2.7 \text{ fm}$$
$$\text{triplet}\quad a_t = -5.39 \text{ fm}, \quad R_t = 1.70 \text{ fm}. \tag{2.3.5}$$

The deuteron has $S = J = 1$: there is no bound state with $S = J = 0$. The lack of a singlet bound state is consistent with the measured scattering lengths: a negative scattering length indicates that the potential is strong enough to make the low-energy wave functions turn over inside the well, which is the criterion for the existence of a bound state (fig. 2.2).

The simple picture of the deuteron as an $L = 0$, $S = 1$ bound state implies that its distribution of electric charge is spherically symmetric. Observations of energy levels of deuterium atoms, on the other hand, lead to the conclusion that the deuteron's charge is not quite symmetrically distributed, but is correlated with its angular momentum. Thus the deuteron's wave function must contain a component with $L > 0$. Since the parity is a constant of the motion, the orbital angular momentum must be even; thus the non-spherical admixture has $L \geq 2$. Since the total angular momentum J is a constant of the motion, and since L has to add to s_1 and s_2 (each $1/2$) to give $J = 1$, L cannot be greater than 2; thus the admixture is $L = 2$. The quadrupole moment

$$Q = \left\langle x^2 + y^2 - 2z^2 \right\rangle \tag{2.3.6}$$

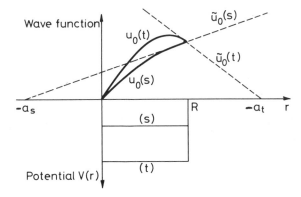

Figure 2.2   Singlet and triplet scattering lengths, a_s and a_t, for a square-well potential of radius R with different depths for singlet (s) and triplet (t). The wave functions u_0 and $\tilde{u}_0$ are shown for each case; note that $u_0 = \tilde{u}_0$ for $r > R$ (see text, eqs. (2.2.2) through (2.2.11)). There is a bound state for the triplet potential but not for the singlet.

allows a rough estimate of the probability that the orbital motion is in a d-state, but also depends on the radial distribution of the d-state component; estimates range from 4% to 8% probability that L = 2.

We conclude that the nuclear force does not conserve orbital and spin angular momenta separately, but only their sum. Thus there must be a component of the nuclear interaction which does not commute with $\vec{L}^2$ and $\vec{S}^2$ separately, but does commute with $\vec{J}$. The simplest operator of this sort is the tensor operator

$$S_{12} = \frac{12}{r^2} \left(\vec{s}_1 \cdot \vec{r} \right) \left(\vec{s}_2 \cdot \vec{r} \right) - 4\vec{s}_1 \cdot \vec{s}_2$$
$$= \frac{6}{r^2} \left(\vec{S} \cdot \vec{r} \right)^2 - 2S^2. \tag{2.3.7}$$

We shall see in Sect. 2.5 that an interaction proportional to S_{12} is indeed expected to result from the exchange of pi mesons between nucleons.

While the nuclear forces have a strong spin dependence, they are remarkably independent of the charge of the nucleons. Indeed, it is the spin dependence of the forces that is responsible for the lack of a di-neutron bound state: the antisymmetry of the wave function for two like fermions means that, for two neutrons to have an even-parity orbital wave function (e.g. L = 0), their spin wave function must be odd under the interchange of the particles spins, implying S = 0 (see (2.3.3)). The S = 0, even-L phase shifts are practically the same (within about 1%) for neutron-neutron and proton-neutron scattering, and the differences in proton-proton scattering can be traced to the Coulomb interaction (again within about 1%). Recalling that the neutron and proton rest masses are also nearly equal, we can see that the nuclear forces are nearly charge-independent: for the strong interactions, neutrons and protons might as well be the same particle. The significant differences in the np or pp systems arise from the fermions' antisymmetry.

A convenient formulation, which exploits the charge independence of nuclear forces, consists of introducing the concept of **isospin**, an abstract, inner degree of freedom, somewhat analogous to spin, which each nucleon possesses. The nucleons are in a subspace of this new degree of freedom, a Hilbert space with two basis states analogous to "up" and "down" states of spin-1/2. We say that the nucleon has isospin t = 1/2. The neutron and proton are eigenstates of the operator t_3, analogous to s_z :

$$t_3|\text{proton}\rangle = +\frac{1}{2}|\text{proton}\rangle$$
$$t_3|\text{neutron}\rangle = -\frac{1}{2}|\text{neutron}\rangle. \tag{2.3.8}$$

The isospin raising and lowering operators, $t_\pm = t_1 \pm it_2$ change neutrons into protons and vice versa. Two-nucleon states can have isospin T = 1 or T = 0, whose isospin wave functions are respectively symmetric or antisymmetric. A two-neutron or two-proton state has $T_3 = t_3(1) + t_3(2) = \pm 1$ which implies T = 1; a neutron-proton state has $T_3 = 0$ and thus may be either T = 0 or 1. The antisymmetry for identical particles is replaced by a **generalized antisymmetry** for all multifermion states: the total wave function, including the isospin part, must be antisymmetric under interchange of two fermion coordinates. Since nn or pp can only have the symmetric isospinor T = 1, the remaining part of the wave function must be antisymmetric, as was required by the rule about identical fermions. For np, the isospinor may be T = 1 or T = 0, so the rest of the wave function may be either antisymmetric or symmetric.

The charge independence of the nucleonic Hamiltonian may be expressed by saying that it is invariant under rotations in "isospace"; for example, a 180 degree rotation about the 1- or 2- axis in isospace reverses neutrons and protons. The strong nuclear forces are apparently invariant under such a rotation; to date, all experiments are consistent with the isospin symmetry of the strong interactions. The electromagnetic interaction, of course, does not possess isospin invariance: we say that it breaks the isospin symmetry. Since nn and pp are unbound systems, in contrast to the deuteron, we conclude that the deuteron is a T = 0 np state.

The isospin of the nucleon, and isospin symmetry of the nuclear forces, may be traced to a corresponding symmetry in the quarks of which the nucleons are made. The nucleons are made primarily out of the light quarks known as u and d , for "up" and "down" projections of the isospin. These quarks, whose rest masses are only a few MeV, are equivalent with respect to the strong interactions in the same sense that the neutron and proton are: while the forces depend on spin (the quarks also have spin 1/2), they do not depend on their isospin. Like the nucleons, the u and d quarks have different properties with respect to the electroweak interactions: their electric charges are +2/3 and -1/3 respectively.

In most contexts, we will find it unnecessary to distinguish between the neutron and proton, and simply refer to them as nucleons. Usually the mass difference of the neutron and proton will be insignificant; in these cases we will denote the nucleon's mass by m_N.

2.4 EFFECTIVE INTERACTIONS AND THE $\mathfrak{S}$ MATRIX

We have seen that the scattering of nucleons of moderate velocities (a kinetic energy of 50 Mev corresponds to a velocity of 0.3c) can, for a given spin state, be accurately described by only two parameters, the scattering length a and the effective range R(0). The description of the nucleons' interaction by a potential V(r), which determines the scattering by way of the Schrödinger equation, seems a rather circuitous (and laborious) procedure. Obviously, it is much more efficient to parametrize the phase shifts in terms of the relative momentum. On the other hand, it is convenient to formulate the results of the scattering experiments in terms of a quantum-mechanical operator, so that we can use the language and results of quantum mechanics to relate the scattering information to properties of more complex nuclear systems. The phase shifts are not an especially suitable choice, since they are not easily expressed in terms of operators in Hilbert space.

A convenient way to describe the measurements of scattering amplitudes is by the effective interaction, an operator $\Im$ in the Hilbert space of two particles which has the property of **giving, in the Born Approximation, the same scattering** as the true interaction V gives when the Schrödinger equation with the true interaction is solved exactly [Messiah vol.2]. To see how $\Im$ is related to the cross sections $d\sigma/d\Omega$, we recall that they may be expressed in terms of the scattering amplitude $f_{\vec{k}}(\Omega)$ (neglecting spin, for the moment):

$$\frac{d\sigma}{d\Omega} = \left| f_{\vec{k}}(\Omega) \right|^2 \tag{2.4.1}$$

where $\Omega = \hat{r}$ is the scattering angle, and $f_{\vec{k}}(\Omega)$ describes the asymptotic form of the scattering wave function for the relative motion:

$$\psi_{\vec{k}}(\vec{r}) \rightarrow e^{i\vec{k}\cdot\vec{r}} + f_{\vec{k}}(\Omega)\frac{e^{ikr}}{r} \quad \text{for } r \rightarrow \infty \tag{2.4.2}$$

where $\hbar\vec{k}$ is the relative momentum before the scattering.* At all $\vec{r}$, $\psi_{\vec{k}}$ satisfies the Schrödinger equation

$$\left(-\frac{\hbar^2}{2\mu}\vec{\nabla}^2 + V(\vec{r}) \right) \psi_{\vec{k}}(\vec{r}) = E\psi_{\vec{k}}(\vec{r}). \tag{2.4.3}$$

By comparison, a plane wave $\Upsilon_{\vec{k}'}(\vec{r})$ of the **same energy** but directed along another arbitrary direction $\vec{k}'$ satisfies the Schrödinger equation

$$-\frac{\hbar^2}{2\mu}\vec{\nabla}^2 \Upsilon_{\vec{k}'}(\vec{r}) = E\ \Upsilon_{\vec{k}'}(\vec{r}) \tag{2.4.4}$$

which is the same as (2.4.3) for large $\vec{r}$, but differs inside the range of the force. Multiplying eq. (2.4.3) by $\Upsilon_{\vec{k}'}^*(\vec{r})$, subtracting $\psi_{\vec{k}}(\vec{r})$ times the complex conjugate of eq. (2.4.4), and integrating over a sphere of radius $R \rightarrow \infty$, we obtain

$$\langle \Upsilon_{\vec{k}'} | V | \psi_{\vec{k}} \rangle = \lim_{R \rightarrow \infty} \int_{|\vec{r}|<R} d^3\vec{r}\ \Upsilon_{\vec{k}'}^*(\vec{r}) V(\vec{r}) \psi_{\vec{k}}(\vec{r})$$

$$= \lim_{R \rightarrow \infty} \int_{|\vec{r}|<R} d^3\vec{r}\ \frac{\hbar^2}{2\mu} \left[\Upsilon_{\vec{k}'}^*(\vec{r}) \vec{\nabla}^2 \psi_{\vec{k}}(\vec{r}) - \psi_{\vec{k}}(\vec{r}) \vec{\nabla}^2 \Upsilon_{\vec{k}'}^*(\vec{r}) \right]. \tag{2.4.5}$$

The right-hand side of eq. (2.4.5) may be evaluated using Green's theorem

$$\underset{\text{volume}}{\int d^3\vec{r} \left(f\vec{\nabla}^2 g - g\vec{\nabla}^2 f \right)} = \underset{\text{surface}}{\int R^2 d\Omega \left(f\vec{\nabla}g - g\vec{\nabla}f \right) \cdot \hat{n}} \tag{2.4.6}$$

---

* $\psi_{\vec{k}}$ and $f_{\vec{k}}$ are often denoted $\psi_{\vec{k}}^{(+)}$ and $f_{\vec{k}}^{(+)}$, respectively, to emphasize the outgoing boundary condition.

where $\hat{n} = \hat{r}$ is a unit vector normal to the surface. Inserting (2.4.6) and (2.4.2) in eq. (2.4.5), we find

$$
\begin{aligned}
\langle \Upsilon_{\vec{k}'} | V | \psi_{\vec{k}} \rangle &= \frac{\hbar^2}{2\mu} \lim_{R \to \infty} R^2 \int d\Omega \left[e^{-i\vec{k}' \cdot \vec{r}} \frac{\partial}{\partial r} \left(e^{i\vec{k} \cdot \vec{r}} + \frac{f_{\vec{k}}(\Omega)}{r} e^{ikr} \right) \right. \\
&\quad \left. - \left(e^{i\vec{k} \cdot \vec{r}} + \frac{f_{\vec{k}}(\Omega)}{r} e^{ikr} \right) \frac{\partial}{\partial r} e^{-i\vec{k}' \cdot \vec{r}} \right] \\
&= -\frac{\hbar^2}{2\mu} f_{\vec{k}}(\Omega') \cdot 4\pi
\end{aligned}
\tag{2.4.7}
$$

where Ω' is the angle corresponding to $\vec{k}'$.

The Born approximation consists of approximating $\psi_{\vec{k}}$ by an unperturbed plane wave $\Upsilon_{\vec{k}}$. Thus we are led to identify the matrix element of the effective interaction $\Im$ (the transition matrix) as

$$
\langle \Upsilon_{\vec{k}'} | \Im | \Upsilon_{\vec{k}} \rangle \equiv \langle \Upsilon_{\vec{k}'} | V | \psi_{\vec{k}} \rangle
\tag{2.4.8}
$$

According to (2.4.7), it is related to the scattering amplitude by

$$
\langle \Upsilon_{\vec{k}'} | \Im | \Upsilon_{\vec{k}} \rangle = -\frac{\hbar^2}{2\mu} f_k(\Omega) \cdot 4\pi
\tag{2.4.9}
$$

f depends only on the magnitude of $\vec{k}$ and on the angle Ω between $\vec{k}$ and $\vec{k}'$.

The requirement that the effective interaction $\Im$ produce the correct scattering amplitude in the Born approximation is not sufficient to completely specify the operator $\Im$, since it only tells us its matrix elements between states of the same energy. The obvious generalization is to try to use (2.4.8) to define the matrix elements of $\Im$ for all $\vec{k}'$ and $\vec{k}$. To see more directly the relationship between the effective interaction $\Im$ thus defined and the true interaction V, it is helpful to find an equation equivalent to (2.4.8), in which the wave function $\psi_{\vec{k}}$ does not appear explicitly.

To this end, consider the integral equation satisfied by $\psi_{\vec{k}}$:

$$
\psi_{\vec{k}}(\vec{r}) = e^{i\vec{k} \cdot \vec{r}} + \int G_E^{(+)}(\vec{r}, \vec{r}') V(\vec{r}') \psi_{\vec{k}}(\vec{r}') d^3 \vec{r}'
\tag{2.4.10}
$$

where the Green function $G_E^{(+)}(\vec{r}, \vec{r}')$ satisfies the defining relation

$$
\delta(\vec{r} - \vec{r}') = (E - H_0) G_E^{(+)}(\vec{r}, \vec{r}') = \left(\frac{\hbar^2}{2\mu} \vec{\nabla}_r^2 + E \right) G_E^{(+)}(\vec{r}, \vec{r}')
\tag{2.4.11}
$$

and is given explicitly by

$$
G_E^{(+)}(\vec{r}, \vec{r}') = -\frac{\mu}{2\pi\hbar^2} \frac{e^{i\sqrt{2\mu E/\hbar^2}|\vec{r} - \vec{r}'|}}{|\vec{r} - \vec{r}'|}
\tag{2.4.12}
$$

where, by convention, the (+) denotes the outgoing boundary condition. It is easy to verify that (2.4.10) guarantees that $\psi_{\vec{k}}$ solves the Schrödinger equation (2.4.3); it also has the asymptotic form (2.4.2), provided $V(\vec{r}') \neq 0$ only for small $\vec{r}'$, and provided $E = \hbar^2 k^2/2\mu$.

We may use (2.4.10) to evaluate the matrix elements on the right hand side of (2.4.8), for $\left|\vec{k}'\right| \neq \left|\vec{k}\right|$:

$$\langle \Upsilon_{\vec{k}'} |V| \psi_{\vec{k}} \rangle = \langle \Upsilon_{\vec{k}'} |V| \Upsilon_{\vec{k}} \rangle + \int d^3\vec{r} \int d^3\vec{r}' \Upsilon_{\vec{k}'}^*(\vec{r}) V(\vec{r}) G_E^{(+)}(\vec{r},\vec{r}') V(\vec{r}') \psi_{\vec{k}}(\vec{r}')$$

$$= \langle \Upsilon_{\vec{k}'} |V| \Upsilon_{\vec{k}} \rangle + \int d^3\vec{r} \int d^3\vec{r}' \langle \Upsilon_{\vec{k}'} |V| \vec{r} \rangle \langle \vec{r} |G_E^{(+)}| \vec{r}' \rangle \langle \vec{r}' |V| \psi_{\vec{k}} \rangle$$

$$(2.4.13)$$

or, introducing the complete set of basis states $|\Upsilon_{\vec{k}''}\rangle$ and $|\Upsilon_{\vec{k}'''}\rangle$ instead of $|\vec{r}\rangle$ and $|\vec{r}'\rangle$,

$$\langle \Upsilon_{\vec{k}'} |V| \psi_{\vec{k}} \rangle = \langle \Upsilon_{\vec{k}'} |V| \Upsilon_{\vec{k}} \rangle + (2\pi\hbar)^{-6} \int d^3\vec{k}'' \int d^3\vec{k}''' \langle \Upsilon_{\vec{k}'} |V| \Upsilon_{\vec{k}''} \rangle \times$$

$$\langle \Upsilon_{\vec{k}''} |G_E^{(+)}| \Upsilon_{\vec{k}'''} \rangle \langle \Upsilon_{\vec{k}'''} |V| \psi_{\vec{k}} \rangle \qquad (2.4.14)$$

The matrix elements of $G_E^{(+)}$ between plane waves are easy to evaluate: using (2.4.12) or directly from the definition (2.4.11), we find

$$\left\langle \Upsilon_{\vec{k}''} \left|G_E^{(+)}\right| \Upsilon_{\vec{k}'''} \right\rangle = \lim_{\epsilon \to 0} \left\langle \Upsilon_{\vec{k}''} \left| \frac{1}{E - H_0 + i\epsilon} \right| \Upsilon_{\vec{k}'''} \right\rangle \qquad (2.4.15)$$

which is diagonal in the plane-wave basis, since H_0 is. Using (2.4.15), (2.4.8), and the completeness relation

$$(2\pi\hbar)^{-3} \int d^3\vec{k} |\Upsilon_{\vec{k}}\rangle \langle \Upsilon_{\vec{k}}| = 1 \quad \text{(the unit operator)} \qquad (2.4.16)$$

(2.4.14) simplifies to

$$\langle \Upsilon_{\vec{k}'} |\Im| \Upsilon_{\vec{k}} \rangle = \langle \Upsilon_{\vec{k}'} |V| \Upsilon_{\vec{k}} \rangle + \left\langle \Upsilon_{\vec{k}'} \left|V \, G_E^{(+)}\Im\right| \Upsilon_{\vec{k}} \right\rangle$$

or, simply,

$$\Im_E = V + V \, G_E^{(+)}\Im_E$$

$$= V + \lim_{\epsilon \to 0} V \frac{1}{E - H_0 + i\epsilon}\Im_E. \qquad (2.4.17)$$

The relation (2.4.17) between $\Im$ and V involves the energy E as a parameter, since E appears in the Green function $G_E^{(+)}$. Thus (2.4.17) actually defines a whole family of operators $\Im_E$. The relation to the scattering amplitude f, (2.4.9), only holds for the effective interaction $\Im_E$ whose energy E is equal to the experimental energy, $E = \hbar^2 k^2/2\mu$. Eq. (2.4.17) can also be used as a definition of the $\Im$ matrix, from which its other properties follow. This definition is not restricted to local potentials, or indeed to non-relativistic quantum mechanics. This is the view we will adopt for the rest of this book. It may be shown that eq. (2.4.17) implies eq. (2.4.9) for the scattering amplitude, and therefore also eq. (2.4.8), even in the most general case of relativistic field theory, provided the mass is taken as the full relativistic energy.

2.5 PION EXCHANGE FORCE

Neutrons and protons are only two of many particles that interact via the strong interactions that bind nuclei together. Such strongly-interacting particles are called hadrons. Most hadrons have masses comparable to the nucleons', except for the pi mesons, or pions, which are lighter. There are three sorts of pions, with positive, negative and neutral electric charges (of the same magnitude as the electron's or proton's charge). Their rest masses are similar: 135 MeV and 139 MeV for neutral and charged pions respectively. The pions are members of an isospin triplet: each has $t = 1$, and $t_3 = \pm 1$ or 0. They have negative intrinsic parity. Since they have no intrinsic angular momentum (spin = 0), they obey Bose statistics.

The pions play a special role in the forces between nucleons: the longest-range part of this force arises from the nucleons' exchange of a single pion, just as the long-range Coulomb interaction between electrically charged particles is due to the exchange of photons.

This idea is due to Yukawa (1935), but we will follow a more modern development similar to that of Brown and Jackson. To see how the exchange of pions produces a force between nucleons, we begin by supposing that there is a term $V_{\pi N}$ in the Hamiltonian or Lagrangian that permits a nucleon to emit a pion. $V_{\pi N}$ has matrix elements between a state $|a\rangle$ with a single (non-relativistic) nucleon, represented by a 4-component wave function (for the 4 possible combinations of spin and isospin projections) Φ_a; and a state $|b\rangle$ with a nucleon and a pion, whose wavefunction $\Phi_b \phi_b$ is the product of a nucleon spinor-isospinor Φ_b, and a pion isospinor ϕ_b (a 3-component wave function containing amplitudes for π^+, π^- and π^0). If we assume a **local** interaction, i.e. that the emitted pion and the final-state nucleon materialize at the same point where the initial-state nucleon disappeared, then the simplest possible form for the matrix element of $V_{\pi N}$ between states $|a\rangle$ and $|b\rangle$ is

$$\langle b|V_{\pi N}|a\rangle = f_\pi c \sqrt{4\pi} \cdot 4 \left(\frac{\hbar}{m_\pi c}\right)^{3/2} \sum_{\alpha=1}^{3} \int d^3\vec{r}\ \Phi_b^+(\vec{r})\,\vec{s}\cdot\left(\vec{\nabla}\phi_b^\alpha(\vec{r})\right)^* t_\alpha \Phi_a(\vec{r}) \quad (2.5.1)$$

where α is an isospin index. $\vec{t}$ and $\vec{s}$ are the isospin and spin operators for the nucleon. The scalar product $\Sigma \phi_b^\alpha t_\alpha$ assures isospin symmetry, and may be written in terms of the charge eigenstates $\phi^0 = \phi^3$ and $\phi^\pm = (\phi^1 \pm i\phi^2)/\sqrt{2}$ as the sum of 3 terms: $\phi_b^+ t_-/\sqrt{2}$, corresponding to emitting a π^+ from a proton leaving a neutron;

$\phi_b^- t_+/\sqrt{2}$, emitting a π^- from a neutron leaving a proton; and $\phi_b^0 t_3$, emitting a π^0 leaving the nucleon unchanged. Because of the pion's negative intrinsic parity, V has to include an operator with negative parity; but because the pion is rotationally invariant (intrinsic angular momentum $J = 0$) this negative-parity operator also has to be rotationally invariant; the simplest possibility is to take the scalar product $\vec{s} \cdot \vec{p}$ of the nucleon's spin with the pion's momentum $\vec{p}$, which has been used in (2.5.1). The coupling constant f_π expresses the strength of the interaction. The factors make f_π dimensionless; its value is about $2/7$. Often f_π is quoted in units of the pion rest mass, which gives $f_\pi \approx 39$ MeV/$m_\pi c^2$.

The effect of $V_{\pi N}$ on nucleon scattering may be found by way of eq. (2.4.17) for the effective interaction,

$$\Im_E = V + V\, G_E^{(+)} \Im_E, \tag{2.4.17}$$

which we now generalize to a larger Hilbert space including pion states. This equation may be expanded in the Born series in powers of the interaction by iteration: substituting the entire right-hand side into the last term, we get

$$
\begin{aligned}
\Im_E &= V + V\, G_E^{(+)} \left(V + V\, G_E^{(+)} \Im_E \right) \\
&= V + V\, G_E^{(+)} V + V\, G_E^{(+)} V\, G_E^{(+)} V + \dots
\end{aligned}
\tag{2.5.2}
$$

To describe the elastic scattering of 2 nucleons at energies below the threshhold for pion production, we have to take matrix elements of $\Im_E$ between 2-nucleon wave functions of energy E. For $V = V_{\pi N}$, the first term on the right-hand side of (2.5.2) vanishes, because $V_{\pi N}$ only connects states with differing numbers of pions. The next term $V G_E^{(+)} V$ is not zero, provided the Hilbert space includes states with two nucleons and one pion. Since the energies of these states are always greater than the energy E of the scattering states, $G_E^{(+)}$ will have no singularities and we can drop the $+$ sign (there are no asymptotic states of two nucleons and a pion at these energies.)

In a shorthand notation leaving out spin and isospin, we introduce unit-normalized 2-nucleon states $|\vec{p}_1 \vec{p}_2\rangle$ writing only the momenta of the nucleons; similarly a state with 2 nucleons and a pion is denoted $|\vec{p}_1' \vec{p}_2' \vec{q}\rangle$ where $\vec{q}$ is the momentum of the pion. $V_{\pi N}$ connects these states, leaving one of the nucleons undisturbed ($V_{\pi N} = \sum_{i=1}^2 V_{\pi N_i}$):

$$
\begin{aligned}
\langle \vec{p}_1' \vec{p}_2' \vec{q} | V_{\pi N} | \vec{p}_1 \vec{p}_2 \rangle = {}& \langle \vec{p}_1' | \vec{p}_1 \rangle \langle \vec{p}_2' \vec{q} | V_{\pi N} | \vec{p}_2 \rangle \\
& + \langle \vec{p}_2' | \vec{p}_2 \rangle \langle \vec{p}_1' \vec{q} | V_{\pi N} | \vec{p}_1 \rangle
\end{aligned}
\tag{2.5.3}
$$

Inserting (2.5.3) into (2.5.2), writing out the operator product $V G_E V$ using a complete set of plane-wave intermediate states with 2 nucleons and a pion, and recalling that $G_E = (E - H_0)^{-1}$ is diagonal in this basis, we find that the pion in the intermediate state has a momentum $\vec{q}$ equal to the momentum transferred between the nucleons: to second order in V,

$$\langle \vec{p}_3\vec{p}_4 |\Im_E| \vec{p}_1\vec{p}_2 \rangle = \sum \int \frac{d^3\vec{q}}{(2\pi\hbar)^3} \left\{ \frac{\langle \vec{p}_4|V_{\pi N}|\vec{p}_2, \vec{q}\rangle \langle \vec{p}_3, \vec{q}|V_{\pi N}|\vec{p}_1\rangle}{E - [\varepsilon_N(\vec{p}_3) + \varepsilon_N(\vec{p}_2) + \varepsilon_\pi(\vec{q})]} \right.$$
$$\left. + \frac{\langle \vec{p}_3|V_{\pi N}|\vec{p}_1, -\vec{q}\rangle \langle \vec{p}_4, -\vec{q}|V_{\pi N}|\vec{p}_2\rangle}{E - [\varepsilon_N(\vec{p}_4) + \varepsilon_N(\vec{p}_1) + \varepsilon_\pi(-\vec{q})]} \right\}$$

(2.5.4)

where the sum is over isospin projection of the pion and $\varepsilon_N(\vec{p})$ is the energy of the nucleon with momentum $\vec{p}$. Since the matrix elements of $V_{\pi N}$ conserve momentum, we see that $\vec{q} = \vec{p}_1 - \vec{p}_3 = \vec{p}_4 - \vec{p}_2$. For the scattering process, $E = \varepsilon_N(\vec{p}_1) + \varepsilon_N(\vec{p}_2)$ and the energy denominators are nearly equal to the pion energy (relativistically, since although $|\vec{q}| \ll m_N c$, it may be comparable to $m_\pi c$)

$$\varepsilon_\pi(\vec{q}) = \sqrt{m_\pi^2 c^4 + \vec{q}^2 c^2}$$

(2.5.5)

which is much larger than the energy difference imparted to the recoiling nucleon.

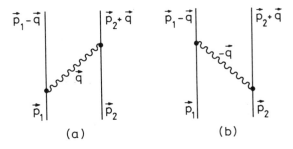

(a)   (b)

Figure 2.3 Graphical representation of eq. (2.5.4). The story told by the graphs proceeds upward with increasing time, and is obtained by reading the formulas from right to left. Straight lines indicate nucleons, wavy lines pions.

We can understand this result with the help of a simple picture, fig. 2.3. Reading the first term of the right-hand side of eq. (2.5.4) from right to left, we see that nucleon 1, with momentum $\vec{p}_1$, emits a pion of momentum $\vec{q}$, leaving the nucleon with momentum $\vec{p}_1 - \vec{q}$. The pion travels to the other nucleon, with the denominator representing the energies of the two nucleons and the pion, which are said to be "off the energy shell" because their energies add up to more than the E of the original nucleons. Reaching nucleon 2, which still has its original momentum $\vec{p}_2$, the pion is absorbed, imparting its momentum $\vec{q}$ to the nucleon which thereby assumes its final state of motion. This process is depicted in fig. 2.3a. Fig. 2.3b depicts the process represented by the other term in eq. (2.5.4), in which a pion of momentum $-\vec{q}$ is emitted by nucleon 2 and absorbed by nucleon 1. The pion cannot escape because there isn't enough energy; we may think of it as owing its ephemeral existence to the time-energy uncertainty principle, according to which it may exist for a time given by $\hbar$ divided by the missing energy, of the order of 10^{-23} seconds.

Since we view the $\Im$-matrix as an effective interaction, it is natural to ask how it depends on the distance between nucleons. Recalling that the relativistic wavefunction for the pion (compare Sect. 10.3.A for a discussion of relativistic boson fields) is

$$\phi_{\vec{q}}(\vec{r}) = \left(\frac{m_\pi c^2}{2\varepsilon_\pi}\right)^{1/2} e^{i\vec{q}\cdot\vec{r}/\hbar}, \tag{2.5.6}$$

we can evaluate the matrix elements using (2.5.1) and (2.5.4) to find

$$\langle \vec{p}_3\vec{p}_4|\Im_E|\vec{p}_1\vec{p}_2\rangle \approx \frac{-64\pi\hbar^3 c}{m_\pi^2}\vec{t}_1\cdot\vec{t}_2\frac{(\vec{s}_1\cdot\vec{q})(\vec{s}_2\cdot\vec{q})}{\varepsilon_\pi(\vec{q})^2}f_\pi^2\delta(\vec{p}_1+\vec{p}_2-\vec{p}_3-\vec{p}_4)(2\pi\hbar)^3, \tag{2.5.7}$$

where $(\vec{s}_i, \vec{t}_i)$ are the spin and isospin operators for the ith nucleon. The expectation value in (2.5.7) does not include spin and isospin variables. This is just the matrix element (between plane waves) of a Yukawa potential in the nucleons' separation $\vec{r}$

$$\Im_{\text{OPEP}}(r) = 16f_\pi^2\frac{\hbar}{m_\pi^2 c}\vec{t}_1\cdot\vec{t}_2(\vec{s}_1\cdot\vec{\nabla})(\vec{s}_2\cdot\vec{\nabla})\frac{e^{-m_\pi rc/\hbar}}{r} \tag{2.5.8}$$

This effective one-pion-exchange potential (OPEP) may also be written, after carrying out the derivatives in (2.5.8) (which act only on the last factor):

$$\begin{aligned}
\Im_{\text{OPEP}} = 4f_\pi^2\frac{m_\pi c^2}{\hbar^2}\vec{t}_1\cdot\vec{t}_2\Bigg\{ & S_{12}\left[\left(\frac{\hbar}{m_\pi rc}\right)^3 + \left(\frac{\hbar}{m_\pi rc}\right)^2 + \frac{\hbar}{3m_\pi rc}\right]e^{-m_\pi rc/\hbar} \\
& + \frac{4}{3}\vec{s}_1\cdot\vec{s}_2\left[\frac{e^{-m_\pi rc/\hbar}}{m_\pi rc/\hbar} - \frac{4\pi\hbar^3}{m_\pi^3 c^3}\delta(\vec{r})\right]\Bigg\}
\end{aligned} \tag{2.5.9}$$

The range of the effective interaction is

$$\frac{\hbar}{m_\pi c} \approx 1.4\text{fm}. \tag{2.5.10}$$

A similar computation could be carried out for other mesons. The most important is the ρ meson, which has angular momentum $\hbar$ like the photon and rest mass $m_\rho c^2 = 769$ MeV. It leads to a repulsive interaction, analogous to the Coulomb interaction (which is mediated by the massless photon) but with a range $\hbar/m_\rho c \approx 0.3$ fm. Attractive effective interactions with shorter ranges also arise from intermediate states with two or more mesons, or with more massive baryons, such as the delta, instead of one or both of the nucleons. A detailed description of nucleon scattering by way of meson exchanges has not been practical, but the meson-exchange mechanism at least allows us to understand quantitatively the longest-range part of the force, and to predict that there will also be a repulsive force at short distances.

2.6   PHENOMENOLOGICAL FORCES BETWEEN NUCLEONS

The incompleteness of the meson-exchange model of nuclear forces is partly due to the computational complexity of describing multiple meson exchange. More fundamentally, though, the meson-exchange picture must break down when the distance between the nucleons becomes smaller than the distance between the quarks of which the baryons and mesons are composed. Suppose, for example, that the hadrons are made up of quarks distributed at a distance r from their centers of mass. Then the nucleons must be separated by a distance 4r if a meson is to be able to fit between them: 2r for the radii of the nucleons, and 2r for the diameter of the meson in between. Estimates of r range from 0.3 to 0.8 fm; thus the meson-exchange picture ought to begin to break down before the nucleons move within 1 fm of each other. It is, indeed, remarkable that the meson-exchange model has had apparent success well inside the range of the one-pion exchange potential.

While the masses of the mesons are fixed from independent measurements, the coupling constants f_π, f_ρ, etc. have to be fit to the two-nucleon scattering data. Additional constants must be introduced to describe, for example, the process in which a nucleon turns into a delta baryon while emitting or absorbing a meson. To allow for the fact that the meson-exchange model must break down at short distances, the coupling constants are made to depend on the momentum transfer in such a way that they become small for large momentum; this cut-off introduces additional parameters. All these parameters have to be chosen to fit the data on two-nucleon scattering.

The forces computed in the meson-exchange model depend not only on spin, isospin, and the distance between the nucleons, but also on the nucleons' relative momentum. The momentum dependence disappeared from our computation of the pion-exchange potential when, in passing from eq. (2.5.4) to eq. (2.5.7), we neglected the change of energy of the recoiling nucleon in evaluating the denominator. When the momentum dependence is taken into account, the meson-exchange model is able to account for all the data on the scattering of nucleons up to energies where meson production becomes the dominant process (about 600 MeV laboratory energy).

Two major efforts to determine the parameters of the meson-exchange model by fitting two-nucleon data have produced phenomenological forces known as the "Paris" and "Bonn" potentials, respectively (see Vinh Mau et al, Holinde). These so-called potentials are actually a kind of effective interaction like $\mathfrak{S}_{OPEP}$. They are intended to represent the effects of meson exchange, but they are not the final effective interaction $\mathfrak{S}$: instead, they have to be used like a potential in the Schrödinger or Dirac equations for the nucleons or, better, in the coupled equations for nucleons and deltas. Unfortunately, these phenomenological forces are so complicated that they have seldom been used in computations of nuclear structure.

A simpler and therefore more popular approximation for the phenomenological force is the static potential developed by Reid in the 1960's. This static potential depends only on spin, isospin, orbital angular momentum, and the radial separation between the nucleons. It was obtained by fitting two-nucleon scattering data to the phase shifts and tensor-coupling parameters obtained from a Schrödinger equation in which the potential is taken as the sum of terms similar to eq. (2.5.8), but with various spin-isospin forms (e.g. spin-orbit) and ranges, and with coupling constants adjusted to fit the data (the bound-state properties of the deuteron were also used in Reid's fit).

Fig. 2.4 shows the Reid potential for nucleons with zero orbital angular momen-

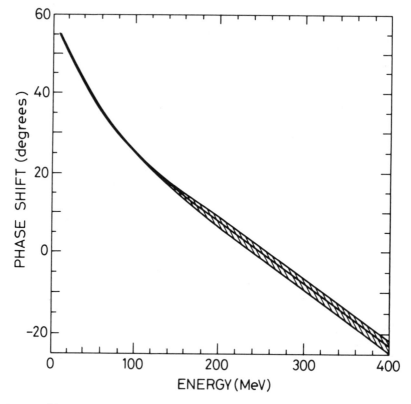

Figure 2.4a   Phenomenology of nucleon-nucleon scattering for L = S = 0.

Measured phase shift (degrees) as a function of laboratory energy. The shaded band indicates the experimental uncertainty. [*From M. MacGregor, R. Arndt and R. Wright, Phys. Rev.* **169** *(1968) 1128.*]

tum in a spin-singlet state, together with the measured phase shifts from which it is computed. The phase-shifts, and therefore the Reid potential, are well-determined by data for orbital angular momentum L≤2, but information for higher partial waves is incomplete. Of course, the Reid potential is not meant to be used when meson production is an important part of the two-nucleon cross section.

2.7   SCATTERING OF PIONS BY NUCLEONS

We have seen that the pions play an important role as the origin of the longest-range force between nucleons. To learn more about the relation between pions and nucleons, we study the scattering of pions by nucleons. The information we learn

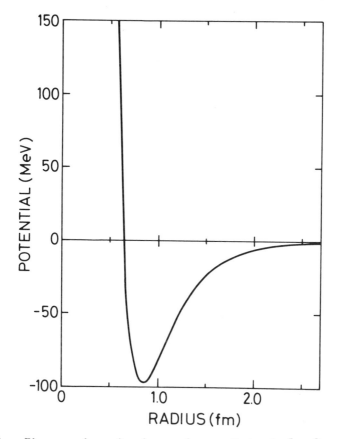

Figure 2.4b   Phenomenology of nucleon-nucleon scattering for L = S = 0.

Reid's phenomenological potential (MeV) as a function of the nucleons' separation (fm). [*From R. V. Reid*]

from pion-nucleon scattering can help to give a perspective on the phenomenological forces introduced in the preceding section. An important part of current research into nuclear structure is concerned with integrating the information about the pion-nucleon system into our picture of the nucleus.

2.7 A. Characterization of Pion-Nucleon Scattering

The effective interaction $\Im_E^{\pi N}$ for pion-nucleon scattering, like the one for nucleon-nucleon scattering, has to be a matrix in the spin and isospin degrees of freedom, as well as being an operator in the variables describing the relative motion. While the beams and detectors of the scattering experiments necessarily observe pions which are eigenstates of the pion's isospin projection t_3^π, this component is not conserved. Instead, the charge of the pion may change in the scattering, for example $\pi^- p \rightarrow \pi^0 n$. The scattering amplitude $f_k^{\pi N}(\theta)$, related to $\Im_E^{\pi N}$ by eq.(2.4.9), will also be a matrix operator in the spin-isospin variables. It is convenient (see Gasiorowicz) to represent $f_k^{\pi N}(\theta)$ in terms of partial-wave amplitudes $f_{LJT}^{\pi N}(E)$ characterized by the conserved quantities: the total isospin $\vec{T} = \vec{t}^\pi + \vec{t}^N$, and the total angular momentum $\vec{J} = \vec{L} + \vec{s}$:

$$f_k^{\pi N}(\theta) = \sum_{LJT} \Pi_{LJ} \Pi_T (2L + 1) f_{LJT}^{\pi N}(E) P_L(\cos\theta) \tag{2.7.1}$$

where Π_{LJ} and Π_T are projection operators for total angular momentum $J = L \pm \frac{1}{2}$ and total isospin $T = \frac{1}{2}$ or $\frac{3}{2}$. They are given by

$$\Pi_{LJ} = \frac{J + \frac{1}{2} + 4(J - L)\vec{s} \cdot \vec{L}/\hbar^2}{2L + 1} \tag{2.7.2a}$$

$$\Pi_T = \frac{1}{3}(T + 1 + 4(T - 1)\vec{t}^\pi \cdot \vec{t}^N) \tag{2.7.2b}$$

which may be easily verified using the identities

$$J(J + 1) = L(L + 1) + \frac{3}{4} + 2\vec{L} \cdot \vec{s}/\hbar^2 \tag{2.7.3a}$$

$$T(T + 1) = \frac{11}{4} + 2\vec{t}^\pi \cdot \vec{t}^N \tag{2.7.3b}$$

The action of the operator $\vec{s} \cdot \vec{L}$ on the angular dependence in $P_L(\cos\theta)$ may be found by expressing $\vec{L} = -\vec{p} \times \vec{r}$ in momentum representation; introducing unit vectors $\hat{k}$ and $\hat{k}'$ for the direction of the relative momentum before and after the scattering we have

$$\hbar^{-1}\vec{s} \cdot \vec{L} P_L(\cos\theta) = -\vec{s} \cdot \hat{k}' \times i\vec{\nabla}_{k'} P_L(\hat{k}' \cdot \hat{k}) = -i\vec{s} \cdot \hat{k}' \times \hat{k} \, P_L'(\cos\theta) \tag{2.7.4}$$

where $P'_L(x) = dP_L(x)/dx$. Using eqs. (2.7.4) and (2.7.2a) in eq. (2.7.1) we obtain

$$f_k^{\pi N}(\theta) = \sum_{L=0}^{\infty} \sum_{T=\frac{1}{2}}^{3/2} \Pi_T \Big\{ \big[L f_{L,L-1/2,T}^{\pi N}(E) + (L+1) f_{L,L+1/2,T}^{\pi N}(E) \big] P_L(\cos\theta)$$

$$- \frac{2i}{\hbar} \vec{s} \cdot \hat{k} \times \hat{k}' \big[f_{L,L-1/2,T}^{\pi N}(E) - f_{L,L+1/2,T}^{\pi N}(E) \big] P'_L(\cos\theta) \Big\} \quad (2.7.5)$$

for the πN scattering amplitude as an operator in spin-isospin space.

We will be primarily interested in πN scattering at fairly low energies where inelastic processes, such as production of an additional pion, are negligible. In that energy region, the partial-wave amplitudes may be parametrized by real phase shifts $\delta_{LJT}^{\pi N}(E)$,

$$f_{LJT}^{\pi N}(E) = k^{-1} e^{i\delta_{LJT}^{\pi N}(E)} \sin\delta_{LJT}^{\pi N}(E). \quad (2.7.6)$$

For these low relative momenta, only L=0 and L=1 are likely to be important. Thus the scattering amplitude is given by

$$f_k^{\pi N}(\theta) \approx \sum_{T=\frac{1}{2}}^{3/2} \Pi_T \Big\{ f_{0\frac{1}{2}T}^{\pi N}(E) + \big[f_{1\frac{1}{2}T}^{\pi N}(E) + 2 f_{1\frac{3}{2}T}^{\pi N}(E) \big] \cos\theta$$

$$- \frac{2i}{\hbar} \vec{s} \cdot \hat{n} \big[f_{1\frac{1}{2}T}^{\pi N}(E) - f_{1\frac{3}{2}T}^{\pi N}(E) \big] \sin\theta \Big\} \quad (2.7.7)$$

where $\hat{n}$ is a unit vector in the direction of $\vec{k} \times \vec{k}'$, i.e. normal to the reaction plane.

2.7 B. Observed Pion-Nucleon Scattering and Reactions

The measured S-wave (L=0) phase shifts are roughly linear in the relative momentum for relative momenta $\hbar k$ up to $m_\pi c$ and are characterized by the scattering lengths $a_1^{\pi N}$ and $a_3^{\pi N}$ for $T = \frac{1}{2}$ and $T = \frac{3}{2}$ respectively:

$$a_1^{\pi N} = \lim_{K \to 0} \delta_{0\frac{1}{2}\frac{1}{2}}^{\pi N}(E)/k(E) = 0.24 \text{fm} \quad (2.7.8a)$$

$$a_3^{\pi N} = \lim_{K \to 0} \delta_{0\frac{1}{2}\frac{3}{2}}^{\pi N}(E)/k(E) = -0.15 \text{fm} \quad (2.7.8b)$$

Above this energy range, the observed scattering is predominantly $T = \frac{3}{2}$, $J = \frac{3}{2}$, L=1 with the phase shift $\delta_{1\frac{3}{2}\frac{3}{2}}^{\pi N}$ passing through $\frac{\pi}{2}$ for pions of laboratory kinetic energy 190 MeV, corresponding to a center-of-mass resonance energy of 1232 MeV. The elastic cross section for $\pi^+ p$ shows (see fig. 2.5) a maximum at this energy with a full width at half maximum of about 115 MeV. Throughout the region of this peak, the spin-averaged angular dependence of both elastic scattering and charge-exchange ($\pi^- p \to \pi^0 n$ etc.) reactions has an angular distribution proportional to $1 + 3\cos^2\theta$, as predicted by eq.(2.7.7) for the case when only $f_{1\frac{3}{2}\frac{3}{2}}^{\pi N}$ is non-vanishing:

$$\frac{d\sigma}{d\Omega}\left(L = 1, J = \frac{3}{2}, T = \frac{3}{2}\right) = \left| \langle f | \Pi_{\frac{3}{2}} | i \rangle \right|^2 \left| f_{1\frac{3}{2}\frac{3}{2}}^{\pi N}(E) \right|^2 (1 + 3\cos^2\theta) \quad (2.7.9)$$

where $|i\rangle$ and $|f\rangle$ are the isospin state vectors of the initial and final states respectively. The matrix elements of $\Pi_{3/2}$ may be obtained from eq. (2.7.2b), or even more readily by observing that $\Pi_{3/2} = \sum_{t_3} |\frac{3}{2}t_3\rangle\langle t_3 \frac{3}{2}|$ so that the isospin factor is just the product of the squares of Clebsch-Gordon coefficients times a charge-conserving Kronecker delta function:

$$|\langle f|\Pi_{3/2}|i\rangle|^2 = \delta_{t_3(i),t_3(f)}|\langle\frac{3}{2},t_3(i)|1,t_3^\pi(i),\frac{1}{2},t_3^N(i)\rangle|^2$$
$$\cdot |\langle\frac{3}{2},t_3(f)|1,t_3^\pi(f),\frac{1}{2},t_3^N(f)\rangle|^2 \qquad (2.7.9a)$$

Thus the cross sections contain a factor 1/3 for each initial or final state of π^+n or π^-p, a factor 2/3 for π^0p or π^0n, and a factor 1 for π^-n or π^+p so that for example the ratio of π^+p to π^-p elastic scattering is 1/9, as is the ratio of π^-n to π^-p in the neighborhood of the resonance. Outside the resonance region, all four amplitudes interfere coherently to produce the total scattering or charge-exchange cross sections.

2.7 C.   Origins of Pion-Nucleon Scattering

To interpret the observed pion-nucleon scattering, it is instructive to compute the consequences of the interaction $V_{\pi N}$ (eq. 2.5.1) used to derive the one-pion-exchange potential for nucleon-nucleon scattering. As in that case, $V_{\pi N}$ gives no contribution in first order (see eq. 2.5.2) because it changes the number of pions. In second order, two contributions to $\mathfrak{S}^{\pi N}$ are represented diagrammatically in fig. 2.6, and are given,

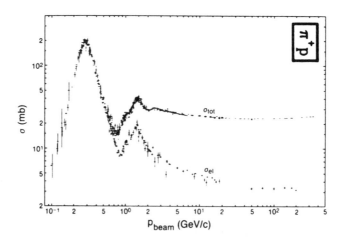

Figure 2.5 π^+p total and elastic cross section vs laboratory beam momentum. [*From Review of Particle Properties", Reviews of Modern Physics* **56**(1984)S1.]

analogous to eq. (2.5.4), by

$$\langle \vec{p}_2\vec{q}_2|\Im_E^{\pi N}|\vec{p}_1\vec{q}_1\rangle = \sum \int \frac{d^3\vec{p}}{(2\pi\hbar)^3} \left\{ \frac{\langle \vec{p}_2\vec{q}_2|V_{\pi N}|\vec{p}\rangle\langle \vec{p}|V_{\pi N}|\vec{p}_1\vec{q}_1\rangle}{E - \varepsilon_N(\vec{p})} \right.$$

$$\left. + \frac{\langle \vec{p}_1|V_{\pi N}|\vec{p}\vec{q}_2\rangle\langle \vec{p}\vec{q}_1|V_{\pi N}|\vec{p}_2\rangle}{E - (\varepsilon_N(\vec{p}) + \varepsilon_\pi(\vec{q}_1) + \varepsilon_\pi(\vec{q}_2))} \right\} \qquad (2.7.10)$$

The matrix elements are evaluated as in Sect. 2.5. Using the identities $\Pi_{1/2} = \sum_{t_3} |\frac{1}{2}t_3\rangle\langle\frac{1}{2}t_3|$ and $(\vec{s}\cdot\vec{q}_2)(\vec{s}\cdot\vec{q}_1) = \hbar^2\vec{q}_2\cdot\vec{q}_1/4 + i\hbar\vec{s}\cdot\vec{q}_2\times\vec{q}_1/2$, we find in the center of mass for $E = \varepsilon_\pi(\vec{q}_1) + \varepsilon_N(\vec{p}_1) = \varepsilon_\pi(\vec{q}_2) + \varepsilon_N(\vec{p}_2) = \varepsilon_\pi(\vec{q}) + \varepsilon_N(\vec{q})$ that the effective interaction is given for $q^2/2m_N \ll \varepsilon_\pi(\vec{q})$ by

$$\langle \vec{p}_2\vec{q}_2|\Im_E^{\pi N}|\vec{p}_1\vec{q}_1\rangle_{\text{direct}} = -(2\pi\hbar)^3\delta(\vec{p}_1 + \vec{q}_1 - \vec{p}_2 - \vec{q}_2)\frac{8\pi cf_\pi^2}{m_\pi^2}\frac{\hbar^3}{\varepsilon_\pi^2(\vec{q})}$$

$$\cdot (\vec{q}_1\cdot\vec{q}_2 - \frac{2i}{\hbar}\vec{s}\cdot\vec{q}_1\times\vec{q}_2)\Pi_{T=\frac{1}{2}}. \qquad (2.7.11a)$$

Comparing to eq. (2.7.7), we see that the "direct" scattering amplitude is of the form expected for scattering in the state $J = T = \frac{1}{2}, L = 1$. The scattering amplitude vanishes like q^2 for small pion momentum q. The "crossed diagram" of fig. 2.6b, corresponding to the second term of eq.(2.7.10), may be computed in a similar way. The main differences are the spin and isospin traces. The result is

$$\langle \vec{p}_2\vec{q}_2|\Im_E^{\pi N}|\vec{p}_1\vec{q}_1\rangle_{\text{crossed}} = +(2\pi\hbar)^3\delta(\vec{p}_1 + \vec{q}_1 - \vec{p}_2 - \vec{q}_2)\frac{8\pi\hbar^3 cf_\pi^2}{\varepsilon_\pi(\vec{q})^2 m_\pi^2} \qquad (2.7.11b)$$

$$\cdot (\vec{q}_1\cdot\vec{q}_2 + \frac{2i}{\hbar}\vec{s}\cdot\vec{q}_1\times\vec{q}_2)(\frac{2}{3}\Pi_{T=\frac{3}{2}} - \frac{1}{3}\Pi_{T=\frac{1}{2}})$$

Comparison with eqs.(2.7.7) and (2.7.11a) shows that the crossed diagram is a linear combination of $J = \frac{1}{2}$ and $J = \frac{3}{2}$ amplitudes, just as (2.7.11b) gives a combination

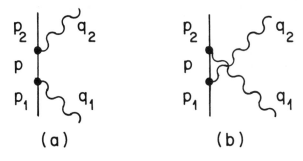

(a)   (b)

Figure 2.6   Lowest-order Born contributions of $V_{\pi N}$ to pion-nucleon scattering. (a) "direct" term; (b) "exchange" term

of $T= \frac{1}{2}$ and $T= \frac{3}{2}$: using eq.(2.4.7) with the reduced mass μ replaced by the relativistic energy $\varepsilon_\pi(\vec{q})/c^2$ we have

$$f^{\pi N}_{1 \frac{3}{2} \frac{3}{2}} = \frac{f^2_\pi}{m^2_\pi} \frac{4\hbar^2 \vec{q}^2}{\varepsilon_\pi(\vec{q})c} \cdot (-\frac{4}{9})_{\text{crossed}} \tag{2.7.12a}$$

$$f^{\pi N}_{1 \frac{1}{2} \frac{1}{2}} = \frac{f^2_\pi}{m^2_\pi} \frac{4\hbar^2 \vec{q}^2}{\varepsilon_\pi(\vec{q})c} \cdot \left[(1)_{\text{direct}} - (\frac{1}{9})_{\text{crossed}}\right] \tag{2.7.12b}$$

$$f^{\pi N}_{1 \frac{1}{2} \frac{3}{2}} = f^{\pi N}_{1 \frac{3}{2} \frac{1}{2}} = \frac{f^2_\pi}{m^2_\pi} \frac{4\hbar^2 \vec{q}^2}{\varepsilon_\pi(\vec{q})c} \cdot (\frac{2}{9})_{\text{crossed}} \tag{2.7.12c}$$

We see that $V_{\pi N}$ leads mainly to scattering in the $L=1$, $J=T= \frac{1}{2}$ state. Clearly the processes related to $V_{\pi N}$, fig. 2.6, are a relatively small part of the pion-nucleon scattering.

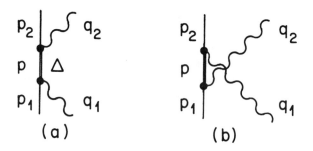

Figure 2.7   Lowest-order Born contributions of Δ isobars to pion-nucleon scattering. (a) "direct" term; (b) "exchange" term

The large scattering in the state $J = T = \frac{3}{2}, L = 1$ is due to the existence of the delta baryon Δ, which has the quantum numbers $J = T = \frac{3}{2}$. Like the nucleon, the Δ is made primarily of three u and d quarks, each of spin and isospin 1/2; in the Δ the quarks' spins and isospins are coupled to $(J,T)= (\frac{3}{2}, \frac{3}{2})$ instead of the $(\frac{1}{2}, \frac{1}{2})$ of the nucleon. When a pion and nucleon interact in the $(J,T)= (\frac{3}{2}, \frac{3}{2})$ state, they can combine to form a delta, with an antiquark from the pion annihilating one of the nucleon's quarks. Conversely the delta can produce a quark-antiquark pair and decay into a nucleon plus a pion. Since the simple three-quark delta is so strongly coupled to the pion-nucleon states, the physical eigenstates are mixtures with components of both types. The simplest contributions to scattering are shown in fig. 2.7. It is easy to see that the process of fig. 2.7a would lead to a large scattering cross section at $E_\pi \approx m_\Delta c^2 - m_N c^2$ since the denominator corresponding to the first term of eq.(2.7.10) would vanish there. The computation of the matrix elements and spin traces are complicated for a spin$-\frac{3}{2}$ particle, but we know from the analysis of Sect. 2.7.A what the resulting angle, spin, and isospin dependence must be. Since the scattering amplitude is large near the resonance, we would not believe the Born approximation anyway.

Neither the coupling $V_{\pi N}$ nor the corresponding coupling to the Δ could explain the S-wave πN scattering, which dominates at low energy. In a meson-exchange type of model, S-wave scattering would be largely due to processes involving the creation of antinucleons or other massive objects. Such processes would be expected to occur only at short distances. Of course, at these distances the quark structure of both the pion and nucleon would be expected to become important. The small values of the S-wave scattering lengths are an expression of the short range of the forces between pions and nucleons without relative orbital angular momentum.

2.8 SCATTERING OF ELECTRONS BY NUCLEONS

Our attempts to understand and describe the forces between nucleons, as well as the pion-nucleon interaction, inevitably raise the question of the internal structure of the nucleon. The clearest picture of this structure is obtained by scattering electrons from nucleons.

The elastic scattering of electrons from nucleons yields directly information about the distribution of electric charge and current inside the nucleon. Since we wish to probe distances of less than 1 fm, we know from the uncertainty principle that we will need electrons of momentum greater than $\hbar/1\text{fm} \approx 200\text{MeV}/c$, so the electrons' energies must be much larger than their rest mass of 0.511 MeV. Thus we have to understand the scattering of relativistic spin-$\frac{1}{2}$ particles. These are described by 4-component Dirac spinors $\Phi(\vec{r})$, representing spin up and down, particle and antiparticle. The effective potential that scatters the electrons is an operator in this spinor space, represented by a 4×4 matrix

$$\Im^e(\vec{r}) = \frac{1}{c} \sum_\mu \mathcal{J}_{e\mu} A^\mu(\vec{r}) = \frac{1}{c} \sum_\mu \mathcal{J}_e^\mu g_{\mu\nu} A^\nu(\vec{r}) \qquad (2.8.1)$$

where $g_{\mu\nu}$ is the metric tensor, $A^\mu(\vec{r}) = (A^\circ(\vec{r}), \vec{A}(\vec{r})) = (\phi(\vec{r}), \vec{A}(\vec{r}))$ is the 4-component vector potential of the time-independent electromagnetic field from which the electron is scattered, and the electric current operator $\mathcal{J}_e^\mu$ of the electron is given in terms of the charge density ϱ_e and current $\vec{\mathcal{J}}_e$ by

$$\mathcal{J}_e^\mu = (\mathcal{J}_e^\circ, \vec{\mathcal{J}}_e) = (c\varrho_e, \vec{\mathcal{J}}_e)$$
$$= ec\gamma^\circ\gamma^\mu. \qquad (2.8.2)$$

The γ^μ are 4×4 Dirac matrices whose properties are summarized in the Appendix at the end of this volume. The factor γ° is $+1$ for particles and -1 for antiparticles, and occurs because they contribute with opposite signs to the probability current in the final state. There is an additional factor γ° in the charge density to account for the opposite electric charges of the electron and positron; this latter factor is also present in the $\{\gamma^i, i = 1, 2, 3\}$ because γ^i is the product of γ^0 with the Dirac velocity operator α^i. The potential A^μ satisfies the Lorentz gauge condition

$$\partial_\mu A^\mu = 0 = \vec{\nabla} \cdot \vec{A}(\vec{r}, t) - \partial \phi(\vec{r}, t)/\partial t. \tag{2.8.3}$$

We consider here only cases where the electromagnetic field is time-independent; time-dependent fields, which occur (oscillating with the Bohr frequency) in inelastic scattering, require the more complicated treatment of Sect. 10.3 below.

The high energy of the electrons complicates matters by requiring a relativistic treatment, but it also permits an important simplification: since the electrons' energy is much bigger than the electromagnetic potential, the lowest Born approximation provides a good description of the scattering. Thus we can take the matrix elements of eq.(2.8.1) between free spinor plane waves

$$\langle \vec{r} | \Upsilon^s_{\vec{k}} \rangle = \Phi_{\vec{k}}(\vec{r}) = \mathcal{X}^s_{\vec{k}} e^{i\vec{k} \cdot \vec{r}} \tag{2.8.4}$$

where $\mathcal{X}^s_{\vec{k}}$ is a 4-component column vector obtained by Lorentz-boosting the electron's rest-frame Dirac spinor of spin projection s with the velocity corresponding to its momentum $\hbar\vec{k}$. The scattering amplitude is then given by eq.(2.4.9) with μc^2 replaced by $\varepsilon(\vec{k})$, the electron's energy

$$f^{s's}_{\vec{k}}(\Omega) = -\frac{\varepsilon(\vec{k})}{2\pi(\hbar c)^2} \left\langle \Upsilon^{s'}_{\vec{k}'} | \mathfrak{I}^e | \Upsilon^s_{\vec{k}} \right\rangle \tag{2.8.5}$$

$$= -\frac{e\varepsilon(\vec{k})}{2\pi(\hbar c)^2} \sum_\mu \text{tr} \left(\mathcal{X}^{s+}_{\vec{k}'} \gamma^0 \gamma_\mu \mathcal{X}^s_{\vec{k}} \right) \int d^3\vec{r} e^{i\vec{q} \cdot \vec{r}} A^\mu(\vec{r})$$

where the trace is over the Dirac-spinor indices and $\hbar\vec{q} \equiv \hbar(\vec{k}' - \vec{k})$ is the momentum transferred to the electron.

The matrix element of the electromagnetic potential may be expressed in terms of the distribution of its sources, the charge density $\varrho(\vec{r})$, current density $\vec{\mathcal{J}}(\vec{r})$, and magnetic-moment density $\vec{\mu}(\vec{r})$ which arises from the intrinsic magnetic moments of the source's constituents (in the case when the target is a nucleon, these are the quarks). Following Jackson, we introduce the effective current density

$$\mathcal{J}^\mu_{\text{eff}}(\vec{r}) = (c\varrho(\vec{r}), \vec{\mathcal{J}}(\vec{r}) + c\vec{\nabla} \times \vec{\mu}(\vec{r})), \tag{2.8.6}$$

which gives the electromagnetic potential

$$A^\mu(\vec{r}) = \frac{1}{c} \int d^3\vec{r}' \frac{\mathcal{J}^\mu_{\text{eff}}(\vec{r}')}{|\vec{r} - \vec{r}'|}. \tag{2.8.7}$$

The potential's Fourier transform is easily seen to be proportional to the Fourier transform of sources:

$$\tilde{A}^\mu(\vec{q}) = \int d^3\vec{r} e^{i\vec{q} \cdot \vec{r}} A^\mu(\vec{r}) = \frac{1}{c} \int d^3\vec{r}' d^3\vec{r}'' \mathcal{J}^\mu_{\text{eff}}(\vec{r}') \frac{e^{i\vec{q} \cdot (\vec{r}'' + \vec{r}')}}{r''}$$

$$= \frac{4\pi}{cq^2} \int d^3\vec{r} \mathcal{J}^\mu_{\text{eff}}(\vec{r}) e^{i\vec{q} \cdot \vec{r}}$$

$$= \frac{4\pi}{cq^2} \tilde{\mathcal{J}}^\mu_{\text{eff}}(\vec{q}). \tag{2.8.8}$$

The factor in front of $\tilde{\mathcal{J}}_{\text{eff}}^{\mu}(\vec{q})$ is recognized as the momentum-space Green function for the electromagnetic field.

Usually, the initial and final spins of the electron are not observed. In that case, we find the observed cross section by averaging $|f_{\vec{k}}^{s's}(\theta)|^2$ over the initial spin projection s and summing over the final projection s':

$$\frac{d\sigma}{d\Omega}(\theta) = \frac{1}{2}\sum_{ss'}|f_{\vec{k}}^{s's}(\theta)|^2 \tag{2.8.9}$$

$$= \frac{2e^2}{(\hbar c)^4}\frac{1}{q^4}\sum_{\mu\nu}\tilde{\mathcal{J}}_{\text{eff}}^{\mu}(\vec{q})\tilde{\mathcal{J}}_{\text{eff}}^{\nu}(-\vec{q})\eta_{\mu\nu}/c^2$$

where $\varepsilon(\vec{k})$ is the electron's energy, and the spinor factor $\eta_{\mu\nu}$ is defined as

$$\eta_{\mu\nu} = \varepsilon(\vec{k})^2\sum_{ss'}\text{tr}\,\chi_{\vec{k}'}^{s'+}\gamma_{\nu}\gamma^{\circ}\chi_{\vec{k}}^{s},\chi_{\vec{k}'}^{s'+}\gamma^{\circ}\gamma_{\mu}\chi_{\vec{k}}^{s}. \tag{2.8.10}$$

Using the cyclical properties of the trace over the Dirac indices, we can write

$$\eta_{\mu\nu} = \varepsilon(\vec{k})^2\text{tr}\gamma_{\nu}\gamma^{\circ}\left(\sum_{s'}\chi_{\vec{k}'}^{s'},\chi_{\vec{k}'}^{s'+}\right)\gamma^{\circ}\gamma_{\mu}\left(\sum_{s}\chi_{\vec{k}}^{s}\chi_{\vec{k}}^{s+}\right). \tag{2.8.10a}$$

The factors in parentheses are recognized as the projectors onto positive-energy states, boosted by the velocities of $\vec{k}'$ and $\vec{k}$ respectively. They may also be written (Bjorken and Drell vol. 1)

$$\Pi_{+}(\vec{k}) = \frac{1}{2m_e c}(m_e c + \sum_{\mu}\hbar k_{\mu}\gamma^{\mu}) \tag{2.8.11}$$

which makes the evaluation of $\eta_{\mu\nu}$ straightforward, giving after a little algebra

$$\eta_{\mu\nu} = \hbar^2 c^2[k_{\mu}k'_{\nu} + k'_{\mu}k_{\nu} - \delta_{\mu\nu}\sum_{\sigma}k_{\sigma}k'^{\sigma}]. \tag{2.8.12}$$

in the Dirac notation $k^{\mu} = (\varepsilon(\vec{k})/\hbar c, \vec{k})$. With this result, plus the momentum-space expression of current conservation

$$\vec{\nabla}\cdot\vec{\mathcal{J}}_{\text{eff}} = 0 = \vec{q}\cdot\tilde{\vec{\mathcal{J}}}_{\text{eff}}(\vec{q}) \tag{2.8.13}$$

we find

$$\sum_{\mu\nu}\eta_{\mu\nu}\tilde{\mathcal{J}}_{\text{eff}}^{\mu}(\vec{q})\tilde{\mathcal{J}}_{\text{eff}}^{\nu}(-\vec{q})/c^2 = |\tilde{\varrho}(\vec{q})|^2(\varepsilon(\vec{k})^2 + (\hbar k c)^2\cos\theta)$$

$$+ |\tilde{\vec{\mathcal{J}}}_{\text{eff}}(\vec{q})|^2(\varepsilon(\vec{k})^2 - (\hbar k c)^2\cos\theta)/c^2 \tag{2.8.14}$$

$$- 2\varepsilon(\vec{k})\hbar\vec{k}\cdot[\tilde{\vec{\mathcal{J}}}_{\text{eff}}(-\vec{q})\tilde{\varrho}(\vec{q}) + \tilde{\vec{\mathcal{J}}}_{\text{eff}}(\vec{q})\tilde{\varrho}(-\vec{q})].$$

The last quantity in square brackets will be zero because $\vec{\mathcal{J}}_{\text{eff}}$ and ϱ have opposite parity (this can also be shown in a fully covariant treatment without using parity). Since the electron is fully relativistic we can approximate $\varepsilon(\vec{k})$ by $\hbar c k$. Noting that $\vec{q}^2 = 4k^2 \sin^2 \theta/2$, we find

$$\frac{d\sigma}{d\Omega} = \frac{e^2}{(\hbar c)^2} \frac{1}{4k^2} \frac{\cos^2 \theta/2}{\sin^4 \theta/2} \left[|\tilde{\varrho}(\vec{q})|^2 + \frac{1}{c^2} |\tilde{\vec{\mathcal{J}}}_{\text{eff}}(\vec{q})|^2 \tan^2 \frac{\theta}{2} \right]. \tag{2.8.15}$$

for the spin-averaged cross-section from a static potential.

In the limit of small momentum transfer, $\tilde{\varrho}$ and $\tilde{\vec{\mathcal{J}}}_{\text{eff}}$ are given by the target's total charge Ze and magnetic moment $\vec{\mu}$:

$$\tilde{\varrho}(0) = Ze \tag{2.8.16a}$$

$$\tilde{\vec{\mathcal{J}}}_{\text{eff}}(\vec{q}) \approx ic\vec{q} \times \vec{\mu} \quad \text{as} \quad \vec{q} \to 0 \tag{2.8.16b}$$

We may use the Wigner-Eckart theorem to write the magnetic moment in terms of the spin angular momentum $\vec{s}$ of the target

$$\vec{\mu} = g \frac{e}{2m_{\text{N}}c} \vec{s} \tag{2.8.17}$$

where the conventional factors $e/2m_{\text{N}}c$ are introduced to make the factor g dimensionless. Applying the vector identity $(\vec{q} \times \vec{\mu}) \cdot (\vec{q} \times \vec{\mu}) = (\vec{q} \cdot \vec{q})(\vec{\mu} \cdot \vec{\mu}) - (\vec{\mu} \cdot \vec{q})^2$, we find that for $\vec{k} \to 0$

$$\frac{1}{c^2} |\tilde{\vec{\mathcal{J}}}_{\text{eff}}(\vec{q})|^2 \approx g^2 \left(\frac{e}{2m_{\text{N}}c} \right)^2 (\vec{q}^2 \vec{s}^2 - (\vec{s} \cdot \vec{q})^2). \tag{2.8.18}$$

When the target is a nucleon, $\vec{s}^2 = \frac{3}{4}\hbar^2$ and $\vec{s} \cdot \vec{q} = \pm \hbar q/2$ if we quantize along the axis $\vec{q}$. For an unpolarized target, the spin-averaged value of (2.8.18) is thus

$$\frac{1}{c^2} |\tilde{\vec{\mathcal{J}}}_{\text{eff}}(\vec{q})|^2 \approx \frac{1}{2} g^2 \left(\frac{e\hbar}{2m_{\text{N}}c} \right)^2 \vec{q}^2. \tag{2.8.19}$$

It is conventional to extract the low-momentum limit by expressing the cross section in terms of electric and magnetic form factors $G_E(\vec{q})$ and $G_M(\vec{q})$ defined by

$$\tilde{\varrho}(\vec{q}) = ZeG_E(\vec{q}) \tag{2.8.20a}$$

$$\frac{1}{c^2} |\tilde{\vec{\mathcal{J}}}_{\text{eff}}(\vec{q})|^2 = 2 \left(\frac{e\hbar}{2m_{\text{N}}c} \right)^2 \vec{k}^2 G_M^2(\vec{q}). \tag{2.8.20b}$$

These definitions imply the limits $G_E(\vec{q}) \to 1$ and $G_M(\vec{q}) \to g/2$ for $\vec{q} \to 0$. In this notation, we find the spin-averaged cross section

$$\frac{d\sigma}{d\Omega} = \left(\frac{e^2}{\hbar c} \right)^2 \frac{1}{4k^2} \frac{\cos^2 \frac{\theta}{2}}{\sin^4 \frac{\theta}{2}} \left[Z^2 G_E^2(\vec{q}) + 2 \left(\frac{\hbar}{2m_{\text{N}}c} \right)^2 \vec{q}^2 G_M^2(\vec{q}) \tan^2 \frac{\theta}{2} \right]. \tag{2.8.21}$$

The limiting case of scattering by a unit point charge, found by setting $ZG_E = 1$ and $G_M = 0$ in eq.(2.8.21), is known as the **Mott cross section**, $d\sigma_{Mott}/d\Omega$.

The formulas above were derived under the assumption that the target's charge and current distributions are independent of time. This is a reasonable approximation if the momentum transfer $\hbar\vec{q}$ is small compared to the mass of the target $m_N c$. The more general case requires the treatment of time-dependent electromagnetic fields which have to be quantized. We shall use such methods in Sect. 10.3; here, we are content to quote the result, known as the **Rosenbluth formula** for the spin-averaged cross section

$$\frac{d\sigma}{d\Omega} = \frac{d\sigma_{Mott}}{d\Omega} \left[\frac{G_E^2(\vec{q}) + (\hbar\vec{q}/2m_N c)^2 G_M^2(\vec{q})}{1 + (\hbar\vec{q}/2m_N c)^2} \right. \tag{2.8.22}$$

$$\left. + 2(\hbar\vec{q}/2m_N c)^2 G_M^2(\vec{q}) \tan^2 \frac{\theta}{2} \right].$$

Experiments can separate the electric and magnetic form factors by measuring the scattering at different angles θ while varying the beam energy to hold $q = 2k \sin\frac{\theta}{2}$ constant. The ratio of the measured cross section to the Mott cross section will be a linear function of $\tan^2\frac{\theta}{2}$; its slope will be proportional to $G_M^2(\vec{q})$ while its intercept gives $G_E^2(\vec{q})$. Alternately, $G_M^2(\vec{q})$ can be measured from the backward-scattering cross section at $\theta = \pi$; once it is known $G_E^2(\vec{q})$ can be inferred from a measurement at a forward angle with the same q.

Since the form factors are proportional to Fourier transforms of the charge and current distributions, they give information about the structure of the nucleon target. For example, the r.m.s. radius is related to the variation of G_E with q, because

$$\left. \frac{\partial^2 G_E}{\partial q^2} \right|_{q=0} = \frac{1}{Ze} \frac{\partial^2}{\partial q^2} \int d^3\vec{r} e^{i\vec{q}\cdot\vec{r}} \varrho(\vec{r}) \bigg|_{q=0}$$

$$= \frac{1}{Ze} \left(-\frac{1}{3} \right) \int d^3\vec{r}\, \vec{r}^2 \varrho(\vec{r}) = -\frac{1}{3} \langle r^2 \rangle. \tag{2.8.23}$$

The measured values of $G_E(\vec{q})$ for electron-proton scattering are well fit by the phenomenological form, known as the **dipole form**

$$G_E(\vec{q}) = (1 + \vec{q}^2 \langle r^2 \rangle / 12)^{-2} \tag{2.8.24}$$

with $\langle r^2 \rangle^{1/2} = 0.81 \text{fm}$ (see fig 2.8). For neutrons, $G_E(\vec{q}) \to 0$ for $\vec{q} \to 0$, but its electric form factor for larger momentum also indicates a distribution of charges over a volume similar to that of the proton. The magnetic form factors show currents distributed over a similar region; their limiting values as $q \to 0$, the magnetic moments, are given by g factors substantially larger than the value g=2 for a structureless Dirac particle like the electron. The magnetic moments are discussed further in Sect. 7.2.

We can try qualitatively to understand the nucleon's charge distribution by considering the consequences of the pion-nucleon coupling $V_{\pi N}$ for the static pion field around a nucleon. To do this we try treating the pion's field as if it were a classical field. The equation of motion of the free field ϕ_α would be just the Klein-Gordon equation,

$$\left(\hbar^2 \frac{\partial^2}{\partial t^2} - (\hbar c)^2 \vec{\nabla}^2 + m_\pi^2 c^4\right) \phi_\alpha(\vec{r}, t) = 0.$$

In the presence of $V_{\pi N}$, a source term appears on the right-hand side,

$$\frac{1}{m_\pi c^2}\left(\hbar^2 \frac{\partial^2}{\partial t^2} - (\hbar c)^2 \vec{\nabla}^2 + m_\pi^2 c^4\right) \phi_\alpha(\vec{r}, t) = \frac{\delta}{\delta \phi_\alpha(\vec{r}, t)} V_{\pi N} \qquad (2.8.25)$$

Applying eq.(2.5.1) for $V_{\pi N}$, we look for static solutions to the time-independent equation

$$\left(m_\pi c^2 - \frac{\hbar^2}{m_\pi} \vec{\nabla}^2\right) \phi_\alpha^{\text{Static}}(\vec{r}) = -f_\pi c\sqrt{4\pi} \cdot 4\left(\frac{\hbar}{m_\pi c}\right)^{\frac{3}{2}} \vec{\nabla} \cdot \left(\Phi_N^+(\vec{r}) \vec{s} t_\alpha \Phi_N(\vec{r})\right) \quad (2.8.26)$$

where $\phi_N(\vec{r})$ is the spinor-isospinor describing the nucleon. The solution to eq. (2.8.26) is easily found in terms of the static pion Green's function

$$G_\pi(\vec{r}) = \frac{\hbar^2}{4\pi m_\pi} \frac{e^{|\vec{r}|mc/\hbar}}{|\vec{r}|} \qquad (2.8.27a)$$

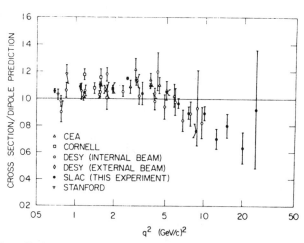

Figure 2.8   Compilation of electron-proton elastic-scattering cross sections. The ratio to the prediction of the dipole form factor (2.8.24) is plotted as a function of the squared momentum transfer. [*D. H. Coward, et al. Phys. Rev. Lett.* **20** *(1968)292.*]

which solves the equation

$$(m_\pi c^2 - \frac{\hbar^2}{m_\pi} \vec{\nabla}^2) G_\pi(\vec{r}) = \delta(\vec{r}).$$ (2.8.27b)

The static pion field is seen to be

$$\phi_\alpha^{\text{Static}}(\vec{r}) = \frac{2f_\pi}{\hbar\sqrt{\pi}} \left(\frac{\hbar}{m_\pi c}\right)^{\frac{3}{2}} \int d^3\vec{r}' \frac{\vec{r} - \vec{r}'}{|\vec{r} - \vec{r}'|^2} \left(1 + \frac{\hbar/m_\pi c}{|\vec{r} - \vec{r}'|}\right) e^{-|\vec{r} - \vec{r}'|m_\pi c/\hbar}$$
$$\cdot \Phi_N^+(\vec{r}') \vec{s} t_\alpha \Phi_N(\vec{r}).$$ (2.8.28)

Even if the nucleonic source $\Phi_N^+ \vec{s} t_\alpha \Phi_N$ were localized to an infinitesimal volume, there would be a distribution of charged pions around it within a distance $\hbar/m_\pi c \approx$ 1.4fm. Of course this picture cannot be used quantitatively, because the total number of pions in the field (2.8.28) is less than one, while the classical-field approximation is only valid when there are many field quanta. Nevertheless, we see that a significant part of the nucleon's charge distribution must be due to off-mass-shell "virtual" pions around it. The spatial distribution must be further augmented by the distribution of the quarks within the nucleon-source field Φ_N.

PROBLEMS

2.1 Calculate the scattering length for an infinitely repulsive potential of radius b:
$$V(r) = +\infty \qquad \text{for } r < b$$
$$V(r) = 0 \qquad \text{for } r > b.$$

2.2 Ignoring spins, use the information in fig. 2.1 to estimate the effective range $R(0)$ and scattering length a for neutron-proton scattering. Hint: find approximate expressions for the total cross sections at each of two energies, E_1 and E_2, in terms of $R(0)$ and a. For which values of E_1 and E_2 would it be reasonable to apply the approximations you used?

2.3 (a) Compute the effective range $R(k)$, and a, for a square well potential $V(r) = -V_0 \Theta(r<R)$.
(b) By inspecting your expression for $R(k)$, find a criterion for the validity of the approximation $R(k) \approx R(0)$.
(c) Find the values of V_0 and R if $R(0) = 1.70$ fm and a = 5.39 fm. For this numerical example, compute $R(k)$ and verify your answer to part (b).

2.4 A helium nucleus ordinarily has 2 neutrons and 2 protons. What is its isospin projection T_3?

2.5   The helicity H of a pair of particles is defined as the projection of the angular momentum along their relative momentum. For a system of one neutron and one proton:

(a) Which combinations of J, L, S, H are possible with $L \leq 2$? Give the parity π and isospin T of each configuration.

(b) Which of the configurations of part (a) are coupled by S_{12}? Give the matrix of S_{12} in the basis of your answers to part (a).

2.6   Which combinations of L, π, and T are possible for two pions? Which of these could be pions of the same charge?

2.7   (a) Show that the diagonal matrix element (mean-value) of S_{12} vanishes in an s-state (orbital angular momentum L=0).

(b) Show that

(i) $\langle SS_z | \vec{s}_1 \cdot \vec{s}_2 | SS_z \rangle = [2S(S+1) - 3]/4$

(ii) $\langle TT_3 | \vec{t}_1 \cdot \vec{t}_2 | TT_3 \rangle = [2T(T+1) - 3]/4$

(iii) for L=0 (orbital angular momentum) $\langle (\vec{s}_1 \cdot \vec{s}_2)(\vec{t}_1 \cdot \vec{t}_2) \rangle$ is the same for (S=1, T=0) and (S=0, T=1).

(c) Show that

(i) $\left[\vec{s}_1 \cdot \vec{s}_2, \vec{J} \right] = 0 = \left[\vec{s}_1 \cdot \vec{s}_2, \vec{L} \right]$

(ii) $\left[S_{12}, \vec{J} \right] = 0 \neq \left[S_{12}, \vec{L} \right]$

(d) Show that the "one-pion exchange potential",

$$\Im_{OPEP}(r) = 16 f_\pi^2 \frac{\hbar}{m_\pi^2 c} \vec{t}_1 \cdot \vec{t}_2 \left(\vec{s}_1 \cdot \vec{\nabla} \right) \left(\vec{s}_2 \cdot \vec{\nabla} \right) \frac{e^{-m_\pi rc/\hbar}}{r}$$

may be written

$$\Im_{OPEP} = 4 f_\pi^2 \frac{m_\pi c^2}{\hbar^2} \vec{t}_1 \cdot \vec{t}_2 \left\{ S_{12} \left[\left(\frac{\hbar}{m_\pi rc} \right)^3 + \left(\frac{\hbar}{m_\pi rc} \right)^2 + \frac{\hbar}{3 m_\pi rc} \right] e^{-m_\pi rc/\hbar} \right.$$
$$\left. + \frac{4}{3} \vec{s}_1 \cdot \vec{s}_2 \left[\frac{e^{-m_\pi rc/\hbar}}{m_\pi rc/\hbar} - \frac{4\pi \hbar^3}{m_\pi^3 c^3} \delta(\vec{r}) \right] \right\}.$$

2.8   Compute the differential cross section $d\sigma/d\Omega$ for scattering by one-pion exchange, neglecting shorter-range forces. To do this, use eq. (2.5.8) in eqs. (2.4.9) and (2.4.1). Plot your result as a function of the scattering angle θ for nucleon center-of-mass energies of 10 and 50 MeV per nucleon. What is the total cross section? Hint: don't use eq. (2.5.7) because the plane waves used in sect. 2.5 are normalized differently from the scattering states appearing in eq. (2.4.9).

2.9   In computing the one-pion exchange potential, the effective interaction was found to be real. Above a certain center-of-mass energy $E_{threshold}$, however, the $\Im$

matrix would become complex. What is $E_{threshold}$? What physical process causes $\mathfrak{S}$ to become complex?

2.10   Estimate the range (in fm) of forces arising from the following processes:
(a) Simultaneous exchange of two pions
(b) Excitation of a delta by pion exchange
(c) Simultaneous exchange of a pion and a rho meson

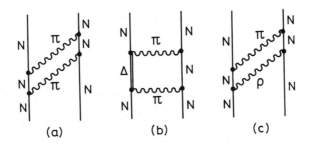

(a)   (b)   (c)

3

Scattering by Nuclei

Our study of the two-nucleon system has told us a lot about the force between nucleons: its range is short, about 1 fm; its attraction is strong, producing potential energies up to 100 MeV; it is repulsive at still shorter distances, less than 0.5 fm; and it depends on the nucleons' spins. We learned most of this information from studying the scattering (continuum) states of the two-body system, rather than its bound state. We now proceed to the study of complex nuclei composed of many neutrons and protons.

Instead of studying next the 3-nucleon system, followed by 4 nucleons, etc., we jump immediately to nuclei with many nucleons (at least a dozen or so). It turns out that these large nuclei are, in many ways, simpler to understand than those with only a few constituents. While each nucleus has its own individual details of structure and a unique spectrum of excited states, certain common features appear which give, overall, a fairly simple unified picture of the way nuclei are put together.

In this chapter, we study the scattering of a nucleon, electron, or pion by a complex nucleus. We will be led to introduce two key concepts: the average field acting on a nucleon, and the saturation property of nuclear dynamics. In the next chapter, we will study the motions of nucleons bound in nuclei (as opposed to scattered nucleons, which have enough energy to leave the nucleus). There, we will learn how the average field and the saturation property arise from the forces between nucleons which we studied in Chapter 2. For now, though, we proceed on an inductive basis, looking for a simple explanation of the observations.

3.1 THE BLACK SPHERE

The differential cross sections for the elastic scattering of a neutron from some typical nuclei are shown in fig. 3.1. We see a general trend falling off with increasing angle, modulated by an oscillating function with pronounced minima. The angular dependence of the cross section is reminiscent of a diffraction pattern observed in optics, e.g. Fresnel diffraction.

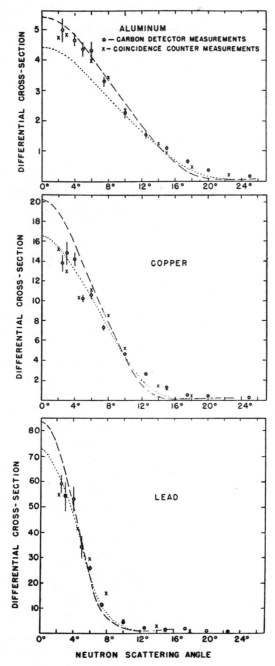

Figure 3.1   Angular distribution in barns per steradian, of elastically scattered 84 MeV neutrons on targets of Al, Cu, and Pb. [*from A. Bratenahl, S. Fernbach, R. Hildebrand, C. Leith, and B. Moyer, Phys. Rev.* **77** *(1950) 597*]

To describe the scattering in quantum mechanics, we recall that the incident beam of neutrons is described by a plane wave e^{ikz}, which may be expanded in the angular-momentum eigenstates $Y_\ell^m(\theta, \phi)$ (for the moment, we neglect spin):

for $r \to \infty$,

$$e^{ikz} \approx \frac{\sqrt{\pi}}{kr} \sum_{\ell=0}^{\infty} \sqrt{2\ell+1}\, i^{\ell+1} \left[e^{-i(kr-\ell\pi/2)} - e^{i(kr-\ell\pi/2)} \right] Y_\ell^0(\theta). \qquad (3.1.1)$$

Because of the cylindrical symmetry of the plane wave, only states with angular-momentum projection $m = 0$ appear. The radial dependence contains two terms, the incoming wave $e^{-i(kr-\ell\pi/2)}$ and the outgoing wave $e^{i(kr-\ell\pi/2)}$, with equal magnitudes and opposite phase. The scattering process modifies the wave function in the neighborhood of the nucleus, i.e. at small r; this in turn leads to a change in the outgoing wave far away from the nucleus, so that the total wave function becomes asympotically (for a spherically-symmetric nucleus)

$$\phi(r \to \infty) = \frac{\sqrt{\pi}}{kr} \sum_{\ell=0}^{\infty} \sqrt{2\ell+1}\, i^{\ell+1} \left[e^{-i(kr-\ell\pi/2)} - \eta_\ell\, e^{i(kr-\ell\pi/2)} \right] Y_\ell^0(\theta). \qquad (3.1.2)$$

The amplitude η_ℓ of the outgoing wave characterizes the scattering process. The scattered wave is the difference between the plane wave and the actual wave function; at large r it takes the form

$$\phi(r \to \infty) = e^{ikz} + f(\theta)\, \frac{e^{ikr}}{r} \qquad (3.1.3)$$

where

$$f(\theta) = i\frac{\sqrt{\pi}}{k} \sum_{\ell=0}^{\infty} \sqrt{2\ell+1}\,(1 - \eta_\ell) Y_\ell^0 \qquad (3.1.4)$$

is the scattering amplitude. Its square is the differential cross section for elastic scattering,

$$\left(\frac{d\sigma}{d\Omega} \right)_{el} = |f(\theta)|^2, \qquad (3.1.5)$$

which is measured by the scattering experiment.

The diffraction pattern of neutrons scattered from nuclei, like those familiar in optics, is mainly due to the absorption of the particles which hit the nucleus. The classical turning point for a neutron with momentum $p=\hbar k$ and angular momentum $\hbar\ell$ occurs when $k^2 = \ell(\ell+1)/R^2 \approx (\ell+\frac{1}{2})^2/R^2$. If we suppose that nucleons which pass within a distance R of the nucleus are absorbed, while those passing at greater distances are unaffected, then there will be no outgoing wave for small angular momentum,

$$\eta_\ell = 0 \quad \text{for } \ell < kR - \frac{1}{2}, \qquad (3.1.6a)$$

while for larger angular momentum, the outgoing wave will be the same as in the plane wave,

$$\eta_\ell = 1 \quad \text{for } \ell > kR - \frac{1}{2}. \tag{3.1.6b}$$

Thus we would expect the cross section to be

$$\frac{d\sigma}{d\Omega} = \frac{\pi}{k^2} \left| \sum_{\ell=0}^{kR-\frac{1}{2}} \sqrt{2\ell+1}\, Y_\ell^0(\theta) \right|^2. \tag{3.1.7}$$

If kR is small, then the sum has only a few terms; if kR is large, we may use the analytical approximation for the spherical harmonics for large ℓ,

$$\text{for } \ell\theta \gg 1, \ Y_\ell^0 \approx \frac{\sin\left[(\ell+\frac{1}{2})\theta + \frac{\pi}{4}\right]}{\pi\sqrt{\sin\theta}} \tag{3.1.8a}$$

$$\text{for } \ell\theta \ll 1, \ Y_\ell^0 \approx \sqrt{\frac{2\ell+1}{4\pi}}\left(1 - \ell(\ell+1)\theta^2/2\right) \tag{3.1.8b}$$

and approximate the sum by an integral:

$$\frac{d\sigma}{d\Omega} \approx \frac{\pi}{k^2} \left| \int_0^{kR-\frac{1}{2}} \frac{d\ell}{\pi}\sqrt{\frac{2\ell+1}{\sin\theta}}\sin\left[(\ell+\frac{1}{2})\theta+\frac{\pi}{4}\right] \right|^2$$

$$\approx \frac{2}{\pi k^2 \sin\theta} \left| \int_0^{kR}\sqrt{x}\,dx\sin\left(x\theta+\frac{\pi}{4}\right) \right|^2$$

$$\approx \frac{2R}{\pi k\theta^2 \sin\theta}\cos^2\left(kR\theta+\frac{\pi}{4}\right) \quad \text{for } kR\theta \gg 1 \tag{3.1.9a}$$

$$\frac{d\sigma}{d\Omega} \approx \frac{k^2R^4}{4}\left(1-(kR\theta/2)^2\right)^2 \quad \text{for } kR\theta \ll 1 \tag{3.1.9b}$$

These are like the formulas obtained in optics for diffraction of light of wavelength $\lambda = 2\pi/k$ from a black disc of radius R. Their dependence on θ is shown in fig. 3.2. We see that $\frac{d\sigma}{d\Omega}$ has a maximum in the forward direction, and a minimum at an angle

$$\theta_{\min} \approx \frac{5\pi}{4kR}. \tag{3.1.10}$$

Thus, by observing the angle at which the diffraction pattern falls off, we can estimate the radius of the nucleus. For example, in the scattering of neutrons of energy $\varepsilon = 84$ MeV by Pb (fig. 3.1), $\theta_{\min} \approx 15° \approx 0.26$ radians, while $k = \sqrt{2m_n\varepsilon/\hbar^2} \approx 2.0$ fm^{-1}, leading to an estimate of about 7.5 fm for the radius of Pb.

3.2   NUCLEAR SIZES AND THE SATURATION OF NUCLEAR FORCES

Comparing the scattering of neutrons from various nuclei in fig. 3.1, we see that neutrons of a given energy are scattered over a larger angular region $\Delta\theta$ for lighter nuclei than they are for heavier targets. This means that lighter nuclei are smaller than heavier nuclei. In fact, if one estimates the nuclear volume Ω_r from the angular width of the diffraction pattern (3.1.10),

$$\Omega_r = \frac{4}{3}\pi R^3 = \frac{4}{3}\pi\left(\frac{4k\theta_{min}}{5\pi}\right)^{-3}, \tag{3.2.1}$$

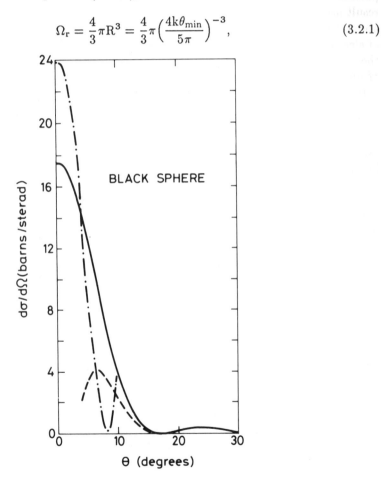

Figure 3.2   Diffraction from a totally absorbing sphere of radius R=7 fm for incident wave number k=2 fm^{-1}. Full curve: eq. (3.1.7); dashed curve: approximation (3.1.9a); dot-dashed curve: approximation (3.1.9b). [*Courtesy of S. Pratt*]

it turns out that Ω_r is roughly proportional to the number of nucleons in the nucleus

$$\Omega_r = \Omega_0 A \qquad (3.2.2a)$$

or

$$R = r_0 A^{1/3}, \qquad (3.2.2b)$$

where $r_0 \approx 1.3$ fm or $\Omega_0 = \frac{4}{3}\pi r_0^3 \approx 9$ fm^3.

This is not too surprising; for example, the volume of a drop of water is proportional to the number of water molecules it contains. On the other hand, the result need not have been thus: for example, all atoms have about the same radius, about $1\text{Å} = 10^{-8}$cm; an aluminum atom is just as big as a lead atom. The density of electrons in an atom of a heavy element is much greater than in a light element: the electrons can be packed together as tightly as you like. In contrast, the density of nucleons in a heavy nucleus is about the same as in a lighter nucleus (at least, when the nucleus has a couple of dozen nucleons or more). We say that the density of nucleons in a nucleus is saturated: it can hold no more in the same volume.

The reason a nucleus is more like a drop of liquid than like an atom is not hard to find: the forces are qualitatively similar to those between molecules. The attractive force between two water molecules falls off rapidly when they are far from each other; the same is true of the force between two nucleons. And just as two nucleons repel each other when they approach too closely, so do two water molecules. The saturation of the density of a large assembly of particles is a result of the interplay between the attractive and repulsive forces: the particles prefer to be a certain distance apart on the average. That distance is roughly equal to the distance at which the attraction is greatest. Thus it is not mere chance that the mean spacing between nucleons r_0 (eq. (3.2.2b)) is similar to the range of the effective interactions $\hbar/m_\pi c$ (eq. (2.5.9)).

3.3   THE OPTICAL MODEL

While the black sphere provides a qualitative interpretation of the elastic scattering of nucleons from nuclei, it does not predict precisely the observed angular distributions. In addition, the radius required to give a best fit to the scattering from a given nucleus depends somewhat on the energy of the scattered nucleon. These shortcomings are not grounds for discouragement, however: they suggest that the scattering is sensitive to other properties of the nucleus, as well as its radius. Thus a somewhat more detailed model is needed to explain elastic scattering in a quantitative way.

The most obvious candidate for a model of elastic scattering is the **potential model**, in which the force acting on the scattered nucleon is assumed to be a function of its position $\vec{r}$. The motion of the nucleon is governed by the Hamiltonian

$$H = \frac{\vec{p}^2}{2m_N} + U(\vec{r}). \qquad (3.3.1)$$

The potential model is qualitatively different from the black sphere, because it does not allow for the possibility of absorption. Since the Hamiltonian (3.3.1) is Hermitean, every incident particle emerges, and with the same energy as it went in. By contrast, in the black sphere model, the main effect is the absorption of the nucleons: the elastic scattering is a quantal interference phenomenon resulting from the absence of outgoing waves.

One way to see whether the potential model is better than the black sphere would be to compare the angular distributions that are obtained for elastic scattering. A better, more direct test is to look at different experimental information and see whether there really is absorption. The best way to do this is to compare the measured elastic cross section

$$\sigma_{el} = \int d\Omega \left(\frac{d\sigma}{d\Omega}\right)_{el} \qquad (3.3.2)$$

with the measured total cross section σ_{tot}, which also includes the cross section σ_{abs} for absorption

$$\sigma_{tot} = \sigma_{el} + \sigma_{abs}. \qquad (3.3.3)$$

The total cross section may be measured by an **attenuation experiment**: instead of looking at the scattered nucleons, we look at the transmitted beam. The intensity of a beam of initial intensity I_0 after passing through a target of thickness Δz containing n scattering centers per unit volume is given by

$$I = I_0 \exp(-\sigma_{tot} n \Delta z). \qquad (3.3.4)$$

Comparing the measured σ_{tot} with the measured σ_{el} gives the absorptive cross section according to eq. (3.3.3).

The predictions of the two models for σ_{abs} are very different. Since the amplitude η_ℓ of the outgoing partial wave of angular momentum ℓ is diminished from unity by the absorption, the probability of absorption in each partial wave is given by $1-|\eta_\ell|^2$. The absorptive cross section is thus

$$\sigma_{abs} = \frac{\pi}{k^2} \sum_{\ell=0}^{\infty} (2\ell + 1)(1 - |\eta_\ell|^2), \qquad (3.3.5)$$

while the total elastic cross section from eqs. (3.3.2), (3.1.4), and (3.1.5) is

$$\sigma_{el} = \frac{\pi}{k^2} \sum_{\ell=0}^{\infty} (2\ell + 1)|1 - \eta_\ell|^2. \qquad (3.3.6)$$

In the black sphere model, where $\eta_\ell = 1$ or 0, it is easy to see that

$$\sigma_{abs}(\text{black sphere}) = \sigma_{el}(\text{black sphere}) = \pi R^2, \qquad (3.3.7)$$

while in the potential model

$$\sigma_{abs}(\text{potential}) = 0 \tag{3.3.8}$$

since the scattering amplitude is unitary,

$$\eta_\ell = \exp(2i\delta_\ell). \tag{3.3.9}$$

The results of attenuation experiments, together with measurements of the elastic cross sections, show that the absorptive cross sections are comparable to the elastic cross sections. Thus the potential model is unacceptable. The qualitative reason for the importance of absorptive processes is not hard to understand: like the saturation phenomenon, it can be traced to the short range of the force between nucleons. When the neutron from the accelerator hits the nucleus, it mainly scatters from a single nucleon at a time — the nearest one. It shares its momentum and kinetic energy with the struck nucleon, not with the entire nucleus. The result is that, in the center of mass, neither nucleon has the energy of the beam, and so neither will count as an elastically-scattered nucleon. The number of nucleons is not reduced; in fact, more nucleons may come out than went in! But the number of nucleons **with the full c.m. energy** is reduced, and they are the ones that count as elastic scattering. Ergo, there is an "inelastic" or "absorptive" cross section.

Obviously, a complete description of nucleon scattering would require a complicated wave function depending on the coordinates of more than one nucleon. To describe the motion of the nucleons which don't lose energy, however, it is not usually necessary to introduce such a many-nucleon description. The reason is that, once a nucleon loses energy, it practically never gets it all back. (Exceptions to this rule happen when the nucleon had very little energy to start with; this possibility is discussed in Chapter 10). Thus we only need to modify the Schrödinger equation in such a way as to make nucleons disappear. The resulting wave function will not give the entire probability distribution of the nucleon; it will only give the probability of finding a nucleon with the original center-of-mass energy. The latter probability will not be conserved.

With this interpretation in mind, we can easily find a way to introduce a Hamilton operator that accounts for absorption. Since the absorptive process doesn't conserve probability (in the sense we just have discussed) it will appear as a non-Hermitean term in the Hamiltonian. We still expect the rate of absorption (i.e. inelastic collisions) to depend on where the nucleon is. Thus we are prompted to imitate the potential model and write (using hats to remind ourselves that we are merely introducing parametrized forms)

$$H = \frac{\vec{p}^2}{2m_N} + \hat{U}(\vec{r}) - i\hat{W}(\vec{r}), \tag{3.3.10}$$

where the term $i\hat{W}(\vec{r})$ accounts for the absorption, $\hat{W}$ being a real function of $\vec{r}$. The term $i\hat{W}$ is anti-Hermitean, and thus doesn't conserve probability. It may represent absorption or creation of particles, depending on the sign of $\hat{W}$. A simple example serves to determine the appropriate sign of $\hat{W}$: suppose $\hat{W}$ and $\hat{U}$ were independent of R, and the wave function at time t=0 was $e^{i\vec{k}\cdot\vec{r}}$. Then the time evolution is determined by the Schrödinger equation,

$$i\hbar\frac{\partial\phi(\mathbf{r},t)}{\partial t} = H\phi(\mathbf{r},t) \tag{3.3.11}$$

which is solved by

$$\phi(\vec{r},t) = e^{i\vec{k}\cdot\vec{r}} e^{-i(\hbar^2 k^2/2m_N+\hat{U})t/\hbar} e^{-\hat{W}t/\hbar}. \tag{3.3.12}$$

To describe absorption, the sign of $\hat{W}$ must be positive.

Another way to see the effect of the imaginary potential $-i\hat{W}$ is to consider a steady-state situation in which a stream of nucleons of energy ε moves toward the positive x direction. Again taking $\hat{U}$ and $\hat{W}$ independent of $\vec{r}$, we look for a stationary solution to (3.3.11).

$$\phi(\vec{r},t) = \phi(\vec{r})e^{-i\varepsilon t/\hbar} \tag{3.3.13a}$$

where we guess the functional form

$$\phi(\vec{r}) = \text{constant} \times e^{(ik-\kappa)x}. \tag{3.3.13b}$$

If this wave function is to solve eq. (3.3.11), k and κ must obey the conditions

$$\frac{\hbar^2}{2m_N}(k^2 - \kappa^2) = \varepsilon - \hat{U} \tag{3.3.14a}$$

and

$$\frac{\hbar^2}{m_N}\kappa k = \hat{W}. \tag{3.3.14b}$$

The probability density is attenuated as the wave progresses along $\hat{x}$,

$$|\phi(\vec{r})|^2 \sim e^{-2\kappa x}, \tag{3.3.15}$$

so that the nucleons have a probability 1/e of surviving for a distance

$$\lambda = \frac{1}{2\kappa}. \tag{3.3.16}$$

We can think of λ as the mean free path for absorption of a nucleon.

Not surprisingly, a good description of the scattering of nucleons from nuclei requires spin-dependent forces, just like nucleon-nucleon scattering. The main spin-dependent effect can be obtained by adding to the Hamiltonian (3.3.10) a spin-orbit potential analogous to the one found in atomic physics,

$$H^{POM} = \frac{\vec{p}^2}{2m_N} + \hat{U}(r) + \vec{\ell}\cdot\vec{s}\,\hat{U}^{\ell s}(r) - i\hat{W}(r) \tag{3.3.17}$$

In atomic physics, the function $\hat{U}^{\ell s}$ is fixed by quantum electrodynamics. Since there is no fundamental theory of nuclear forces, $\hat{U}^{\ell s}$ has to be chosen to fit data on the polarization of elastic scattering. The functions $\hat{U}$, $\hat{U}^{\ell s}$, and $\hat{W}$ are chosen to be central potentials (i.e. spherically symmetric). H^{POM} is called the Phenomenological Optical Model.

The predictions of the Phenomenological Optical Model for elastic and inelastic cross sections are computed in a way that is very similar to the procedure used for scattering from a central potential. The stationary Schrödinger equation,

$$\varepsilon\phi = H^{POM}\phi, \tag{3.3.18}$$

is separated into partial waves; the radial equation for each partial wave contains a complex potential. These differential equations are integrated numerically, and their asymptotic solutions for large r are used to identify the amplitudes η_ℓ as in eq. (3.1.2). The elastic and absorptive cross sections follow from equations analogous to (3.1.4), (3.1.5) and (3.3.5). (They are a little more complicated because the spin-dependent forces require the use of eigenstates of the total angular momentum $\vec{j} = \vec{\ell} + \vec{s}$). Various forms of $\hat{U}(r)$, $\hat{U}^{\ell s}(r)$, and $\hat{W}(r)$ are tried until agreement with data is obtained.

The scattering of nucleons from a wide variety of nuclei, with different numbers of nucleons A, can be fitted by optical potentials of the form [Perey and Perey]

$$\hat{U}(r) = U_0\, f\left((r - R(A))/a_u\right) + U_C(r) \tag{3.3.19a}$$

$$\hat{W}(r) = \left(W_0 - 4W_1 a_W \frac{\partial}{\partial r}\right) f\left((r - R(A))/a_W\right) \tag{3.3.19b}$$

$$\hat{U}^{\ell s}(r) = U_0^{\ell s} \frac{1}{r}\frac{\partial}{\partial r} f\left((r - R(A))/a_{\ell s}\right) \tag{3.3.19c}$$

where $U_C(r)$ is the Coulomb potential acting only on protons (see sect. 3.5), and

$$f(x) = (1 + \exp(x))^{-1} \tag{3.3.20a}$$

$$R(A) = r_0 A^{1/3}, r_0 = 1.25\text{fm} \tag{3.3.20b}$$

$$a_u = 0.65\text{fm}, a_w = a_{\ell s} = 0.47\text{fm} \tag{3.3.20c}$$

The function f (called a Woods-Saxon shape) approaches a constant value of 1 inside the nucleus, then falls from 0.9 to 0.1 as r varies from $R - 2.2a$ to $R + 2.2a$. We say that the nucleus has a **uniform interior**, and a **surface thickness** of 4.4a, or about 2.9 fm (see fig. 3.3).

It is not surprising that the surface thickness turns out to be comparable to the range of the force, or the distance between nucleons. The radius R(A) found in the optical-model fits turns out to be a little smaller than the black-sphere value (3.2.2). The reason is that the absorption is so strong in the surface that the incident nucleons are likely to be absorbed even before they reach the radius R. The term proportional to W_1 is introduced to account for this strong absorption: it is peaked in the surface (see fig. 3.3), and adds absorption there, since W_1 turns out to be positive (as is W_0). The strengths of the real and imaginary potentials, U_0, $U_0^{\ell s}$, W_0, and W_1, are about the same for all nuclei with A $\geq$16. This is a satisfying confirmation of the idea of the saturation of nuclear forces.

The biggest surprise arising from the analysis of experiments in the phenomenological optical model is that W_0, W_1 and (to a lesser extent) U_0 depend on the energy E of the incident nucleon. While no single parametrization of the phenomenological optical model can fit all the data, the potentials derived from various data for $\varepsilon < 100$ MeV have parameters which are usually within about 10-15% of those in (3.3.20) with strengths similar to the following [compare Perey and Perey]:

$$W_0(\varepsilon) \approx \max(0.22\varepsilon - 2\text{MeV}, 0) \tag{3.3.21a}$$

$$W_1(\varepsilon) \approx \max\left[12\text{MeV} - 0.25\varepsilon + 24\text{MeV} \cdot t_3 \frac{N-Z}{A}, 0\right] \tag{3.3.21b}$$

$$U_0(\varepsilon) \approx -50\text{MeV} - 48\text{MeV} \cdot t_3 \frac{N-Z}{A} + 0.3\,(\varepsilon - U_C(R)) \tag{3.3.21c}$$

$$U_0^{\ell s} \approx 30\text{MeV fm}^2/\hbar^2. \tag{3.3.21d}$$

On reflection, the energy dependence seems less surprising. Since W_0 represents the probability of exciting the nucleus by sharing its energy with another nucleon, it

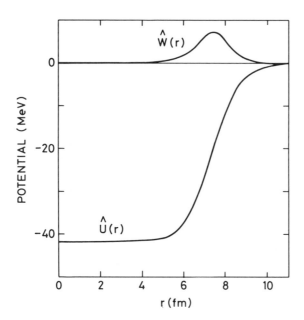

Figure 3.3   Phenomenological optical potential of a 10 MeV neutron in ^{208}Pb (see eqs. (3.3.19 - 21)).

increases with the nucleon's energy, because a nucleon with more energy can more easily excite the nucleus. We are reminded once more that the phenomenological optical model is not really a Hamiltonian. While it tells how elastic-scattering waves propagate, it is very much more than a Hamiltonian: it does not conserve probability, and it depends on energy. In this latter respect, it is reminiscent of the $\mathfrak{S}$ matrix introduced in Chapter 2. We shall see in the next chapter that there is, indeed, an intimate connection between the $\mathfrak{S}$ matrix and the optical potential.

The magnitude of U_0 is also not too surprising, being roughly comparable with the 2-body potentials V introduced in Chapter 2. Thus the potential energy of a nucleon in a nucleus is of the same order of magnitude as its potential energy near a nucleon in free space. The term proportional to N−Z reflects the fact that neutrons scatter more weakly from other neutrons than from protons, due to the absence of the $^3S_1$ state of relative motion.

The magnitude of W_0, on the other hand, is much harder to understand. Consider for example the scattering of a 40 MeV neutron. Using the values from eq. (3.3.21a) in eqs. (3.3.14), one finds a mean free path λ (eq. (3.3.16)) of about 5 fm in the nuclear interior. By comparison, the mean free path computed from the 2-body cross section $\sigma \approx 4\pi d\sigma/d\Omega \approx 17\,\text{fm}^2$ (compare fig. 2.1) is

$$\lambda = (n\sigma)^{-1} \approx 0.4\,\text{fm}. \qquad (3.3.22)$$

using the density of nuclear matter

$$n = \frac{1}{\Omega_0} = \left(\frac{4}{3}\pi r_0^3\right)^{-1}. \qquad (3.3.23)$$

Inside the nucleus, a 40 MeV nucleon travels ten times farther than expected! In the surface, the measured mean free path is shorter ($\hat{W}$ is bigger) which reduces the discrepancy with eq. (3.3.22); but the trend is wrong: according to eq. (3.3.22), the mean free path ought to be longer in the surface, which makes sense since there are fewer nucleons to collide with. The riddle of the long mean free path is one of the most striking features of nuclear physics. We shall return to the resolution of this riddle in the next chapter.

3.4   NUCLEAR MATTER

The success of the optical model reinforces our picture of a nucleus as a tiny drop of nearly incompressible matter. Roughly speaking, all nuclei look the same. Inside, there is a uniform region; surrounding it is a surface about one particle thick. We say that the nucleus is a **droplet of nuclear matter**.

Often, it simplifies our thinking (and computations!) to consider a hypothetical system of uniform nuclear matter of infinite extent. Nuclei cannot actually be very large because the repulsive Coulomb interactions among the protons cause large nuclei to fall apart (Chapters 5, 7). In the hypothetical nuclear matter, then, the Coulomb interactions are neglected. We say that the potential energy of a nucleon in nuclear matter is about 50 MeV; we agree to neglect the electrostatic energy.

The hypothetical ideal of nuclear matter has been especially useful for theorists trying to find approximate solutions to the Schrödinger equation for nuclei. Such computations are often much easier for nuclear matter. For example, correlations between two nucleons, which depend on both their coordinates $\vec{r}_1$ and $\vec{r}_2$ in real nuclei, become a function of a single vector $\vec{r}_1 - \vec{r}_2$ in nuclear matter because of its translational invariance. In the following chapters we will often gain simple insights by applying the concept of infinite nuclear matter to help understand observations of finite nuclei.

3.5   ELECTROMAGNETIC FORCES; SCATTERING OF ELECTRONS AND PROTONS BY NUCLEI

In a real nucleus, of course, the protons feel the electrostatic forces. The electrostatic potential can be measured independently by scattering electrons from nuclei. Since the electrons do not feel the strong forces among the hadrons, they provide a weakly-interacting probe of the nucleus. If the electrons' kinetic energy is large compared to their electrostatic potential energy, $U_C(r)$, their scattering may be described by the Born approximation. The maximum value of the electrostatic potential at the center of the nucleus may be estimated from a uniform spherical charge distribution of radius $R(A)$ and total charge Ze,

$$U_C(r = 0) = -\frac{3}{2}\frac{Ze^2}{R(A)}. \tag{3.5.1}$$

For the heaviest nuclei, $U_C(r=0)$ is about -25MeV; thus small corrections to the Born approximation are necessary for electrons of energy up to a few hundred MeV. When these corrections are taken into account, the charge form factors $G_E(\vec{q})$, eq. (2.8.20a), can be measured very precisely, leading to a determination of the distribution of the protons (with small corrections for the virtual mesons exchanged among the nucleons). After corrections are made for the distribution of charge within an individual nucleon, it is found that the protons' distribution is very similar to the optical potential (fig. 3.4). If the distribution of protons is fit by a Woods-Saxon shape (3.3.20), the thickness a is a little smaller (about 0.55 fm) and the radius $R(A)$ is a little smaller too ($R(A){\approx}1.15\text{fm}{\cdot}A^{1/3}$). It is not surprising that the potential reaches a little farther than the density: the difference is of the order of the range of the force between the nucleons.

The scattering of protons by nuclei is well accounted for by adding the electro-static potential to the phenomenological optical potential felt by the neutrons. The theoretical description of proton scattering is tedious, because the long range of the electrostatic potential changes the asymptotic form of the wave function. The theory is straightforward, however, as are the numerical computations. The only difference compared to neutrons, in the end, turns out to be that U_0 is deeper for protons on heavy nuclei, and less attractive for neutrons on these nuclei. (See the phenomenological expression (3.3.21c)). Since the heavy nuclei contain more neutrons than protons, we can understand this effect qualitatively: the strong attraction of the spin triplet, isospin singlet 2-body (deuteron) state is not available to like particles scattering from each other, so the protons interact more strongly with the neutrons

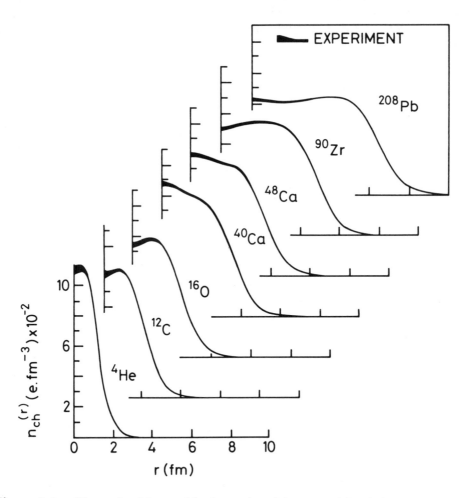

Figure 3.4   Charge densities $n_{ch}(r)$ of several nuclei measured by electron scattering. The width of the line indicates the experimental uncertainty. [*from B. Frois, Proceedings of the Niels Bohr Centennial Symposium, North-Holland, Amsterdam, 1986, p. 125.*]

and vice versa. Thus the last term in eq. (3.3.21b), proportional to $t_3(N-Z)$, changes sign for protons. We shall return to a more quantitative understanding of this term in Chapter 7.

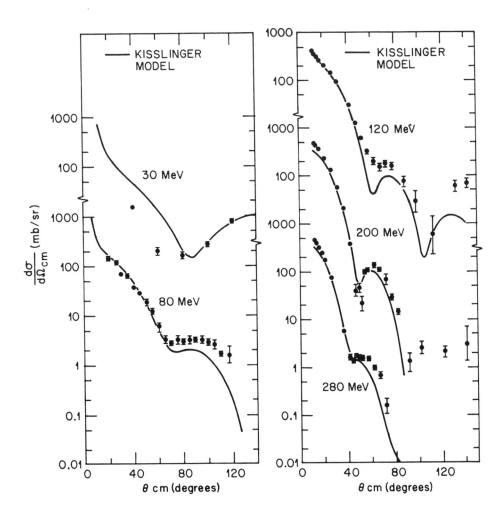

Figure 3.5   Elastic scattering cross sections of negative pions from ^{12}C at various energies, as a function of center-of-mass scattering angle. The curves are fits with the Kisslinger type of optical model, see sect. 4.2.F [*after M. M. Sternheim and R. R. Silbar, in Annual Reviews of Nuclear Science* **24** *(1974)249*]

3.6   SCATTERING OF PIONS BY NUCLEI

The differential cross sections for elastic scattering of a negative pion from ^{12}C are shown in fig. 3.5 for several pion energies. They show a diffractive structure similar to that observed for elastic nucleon scattering. As with nucleons, the pions' elastic scattering is accompanied by large amounts of inelastic scattering, including charge-exchange reactions. A substantial portion of the pion-nucleus absorption, however; is **true absorption** in which no pion appears in the final state.

Fig. 3.6 shows the true absorption cross sections for several nuclei as a function of the pion's kinetic energy. For the lighter nuclei the absorption shows a pronounced maximum (note the figure's logarithmic scale) near the Δ resonance. The value of the resonant cross section is about $\pi R(A)^2$ with $R(A)$ similar to the radius of the nucleon distributions for both light and heavy nuclei. For heavy nuclei, the absorption is also very strong for lower energies. Evidently low-energy pions can penetrate a light nucleus but not a heavy one: the mean free path for true absorption of low-energy pions must be about the size of a medium-mass nucleus.

The true absorption of a pion requires the involvement of at least two nucleons, since one nucleon cannot absorb a pion while conserving both momentum and en-

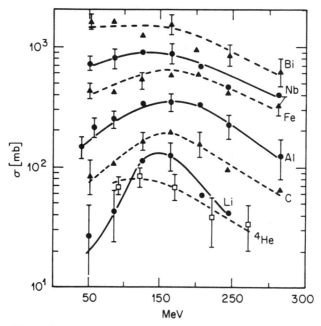

Figure 3.6   True absorption cross sections for positive pions on various nuclei, as a function of pion kinetic energy. [*after D. Ashery and J. P. Schiffer, Annual Reviews of Nuclear and Particle Science* **36** *(1986)207.*]

ergy. In the simplest picture of resonant absorption, a pion joins with a nucleon to create a delta; the delta and a second nucleon then scatter inelastically to become two nucleons again. This process may be verified by observing the final state, in which the nucleons will have so much kinetic energy they are likely to escape the nucleus. In fact, observations show that an average of 3 to 5 nucleons are ejected by true pion absorption. Evidently several nucleons help to absorb the pion.

Pion-nucleus elastic scattering may be fit by a variety of phenomenological optical models. The obvious role of the delta resonance dictates a strong energy dependence in the optical model; the derivative coupling of the pion to the nucleon suggests a strong momentum dependence, and also leads one to expect a special role for the nuclear surface, where gradients are largest. As a result, so many parameters have to be introduced into the optical model that none of them are determined uniquely by the elastic cross sections. Fits using one theoretical model, the Kisslinger potential, are shown in fig. 3.5 (see Sect. 4.2.F).

PROBLEMS

3.1   Estimate the angle at which a nucleus of zirconium, with 40 protons and 50 neutrons, shows the first minimum in the scattering of 60 MeV neutrons.

3.2   A beam of 50 MeV neutrons scatters elastically from a target of unknown material with the differential cross section shown in the sketch.
(a)   Estimate the mass A of the target.
(b)   Estimate σ_0.

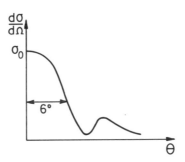

3.3   The spin-averaged scattering of neutrons from nuclei can be roughly fitted by an optical potential

$$\hat{U} = \frac{U_0}{1 + e^{[r-R(A)]/a}} - i\hat{W}$$

where

$$U_0 = -40\text{MeV}$$

$$a = 0.6\text{fm}$$

$$R(A) = 1.2A^{1/3}\text{fm}$$

$$\hat{W} = 0.2\varepsilon \left(1 + \alpha\frac{\partial}{\partial r}\right)\left[1 + e^{[r-R(A)]/a}\right]^{-1}.$$

(a)   With the above parameters, calculate the mean free path λ of a nucleon of total energy ε deep inside the nucleus. Give numerical answers for $\varepsilon = 10\text{MeV}$ and $\varepsilon = 50\text{MeV}$.

(b) λ gets to be about half as big in the surface as in the center. Find α.

3.4   Show that eq. (3.3.11) implies the generalized continuity equation

$$\vec{\nabla} \cdot \vec{j}(\vec{r}, t) + \frac{\partial}{\partial t}n(\vec{r}, t) = -\frac{2}{\hbar}\hat{W}(\vec{r})n(\vec{r}, t)$$

where $\vec{j}$ and n are the current and the particle density, respectively. Compute the absorption width Γ defined by

$$\Gamma \equiv -\hbar \int \frac{\partial n(\vec{r}, t)}{\partial t}d^3\vec{r}$$

3.5   Calculate the lifetime of a particle in a bound-state solution of eq. (3.3.11). The time dependence is then $e^{-i\varepsilon t/\hbar}$, where the real part of ε is negative.

3.6   What is the radial dependence of nucleon wave functions in nuclear matter?

4

Bound Nucleons

We have introduced the optical potential to explain the scattering of unbound nucleons by nuclei. We picture the real part of the optical potential, $\hat{U}(r)$, as a parametrization of the average force on the scattered particle, due to the nucleons in the nucleus. It seems reasonable to suppose that the nucleons already in the nucleus also feel such forces, due to each other. In this chapter we explore the two-way relationship between the motion of the nucleons in the nucleus, and the average forces which both determine, and are determined by, their motion.

4.1 BOUND STATES IN THE NUCLEAR POTENTIAL

Can the same potential energy that is felt by a scattered nucleon be responsible for holding the bound nucleons in the nucleus? To answer this, we need to know how much energy is involved in holding a nucleus together. The binding energy $B(N,Z)$ of a nucleus with N neutrons and Z protons is the difference of its rest energy from that of the neutrons and protons; in terms of the nucleus' mass $M(N,Z)$ we have

$$B(N, Z) = [N m_n + Z m_p - M(N, Z)] c^2. \tag{4.1.1}$$

The masses can be measured with a mass spectrograph. The binding energies of a number of nuclei are shown in fig. 4.1.

We see that, except for the lighter nuclei, the binding energy is approximately proportional to the number of nucleons $A = N + Z$,

$$B(N, Z) \approx bA \tag{4.1.2}$$

with b≈8 MeV. Can we relate this observation to the optical potential observed in scattering experiments?

We are led to look for bound states in the optical potential. Since the energies of the nucleons we are interested in are less than the energies of the scattered nucleons, and since the imaginary potential $\hat{W}$ decreases for lower energy (see eq. (3.3.21a)) we might expect $\hat{W}$ to be very small for bound nucleons (this turns out to be somewhat misleading, see sect. 4.4, but will do for now as a working hypothesis).

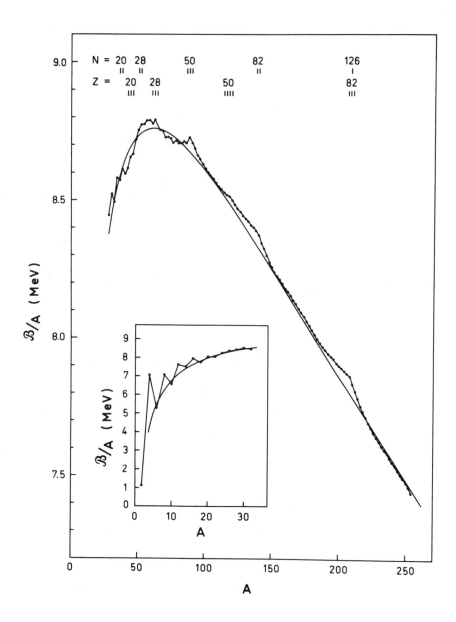

Figure 4.1   The experimental binding energies are taken from the compilation by J.H.E.Mattauch, W.Thiele, and A.H.Wapstra, *Nucl. Phys.* **67** (1965) 1. The smooth curve represents the semi-emperical mass formula, eq. (4.3.2), with the constants given by A.E.S.Green and N.A. Engler, *Phys. Rev.* **91** (1953) 40. [*from Bohr and Mottelson, vol. 1*]

Thus we neglect $\hat{W}$ and concentrate on the real potential $\hat{U}(r)$. We will also, for the moment, ignore its energy dependence.

4.1 A. The Fermi-Gas Model

Certainly $\hat{U}(r)$ has a depth (-50 MeV, eq. (3.3.21b)) sufficient to have bound states with 8 MeV of binding energy. But since nucleons obey Fermi-Dirac statistics, they also obey the Pauli principle, which prevents more than one nucleon from occupying the same quantum state. Thus we must ask whether there are enough bound states to accomodate all the nucleons.

One way to answer this question would be to use a computer to find the energies of the bound states by solving the stationary Schrödinger equation. A simpler estimate, however, is more illuminating. Since we are dealing with a system of many particles, we can borrow arguments from the statistical mechanics of large systems. In the Fermi-gas model, we think of the potential as a large box of volume Ω_r. Then we know its eigenstates are approximately plane waves, characterized by their momentum $\vec{p}$. Roughly, there is (neglecting spin and isospin) one quantum state per h^3 volume of six-dimensional phase space. The many-particle state of lowest energy will be obtained by filling these single-particle states up to some energy μ (the Fermi energy or chemical potential), starting at the bottom of the well with energy U_0. The momentum of the last state filled, the Fermi momentum, is

$$p_F = \sqrt{2m_N(\mu - U_0)}, \tag{4.1.3}$$

so that the volume of momentum space with states of energy less than μ is

$$\Omega_p = \frac{4}{3}\pi p_F^3. \tag{4.1.4}$$

The total number of states in the nucleus, with energy less than μ, is the phase-space volume, $\Omega_r \cdot \Omega_p$, divided by the volume per state, h^3. Thus the number of states with a given spin and isospin is

$$
\begin{aligned}
N(\mu) &= \frac{\Omega_r \cdot \Omega_p}{h^3} \\
&= \frac{\frac{4}{3}\pi R(A)^3 \cdot \frac{4}{3}\pi p_F^3}{(2\pi\hbar)^3} \\
&= A\frac{2}{9\pi}\left[\frac{r_0^2(\mu - U_0)2m_N}{\hbar^2}\right]^{3/2}.
\end{aligned}
\tag{4.1.5}
$$

Multiplying by 4 for the spin and isospin degeneracy, and inserting r_0 from eq. (3.3.20), we see

$$4N(\mu) = A\left(\frac{\mu - U_0}{33\mathrm{MeV}}\right)^{3/2}. \tag{4.1.6}$$

To have enough states to accomodate A particles, we must fill up the potential well to a maximum energy level

$$\mu = U_0 + 33\text{MeV} \approx -17\text{MeV}. \tag{4.1.7}$$

We see that the separation energy μ is the same for each nucleus, if there are to be enough bound states in $\hat{U}(r)$ to accommodate all the nucleons.

Since the separation energy μ is the same for each nucleus, it follows that the average binding energy per nucleon b also has to be equal to μ. This somewhat surprising result can most simply be understood in nuclear matter, where the total energy per nucleon b_V is a function of the density. Then adding one nucleon, without changing the density, leads to the same total energy per nucleon, i.e. the added nucleon has a separation energy μ equal to $-b_V$. This result is in qualitative agreement with the trend of the experimental data, fig. 4.1; the experimental value of b, eq. (4.1.2), includes the Coulomb energy as well as the effects of the finite nuclear size. When these are accounted for (see Sect. 4.3), the corrected value b_v for nuclear matter, extracted from observed binding energies is -15.6 MeV, in quite good agreement with the estimate of eq. (4.1.7).

4.1 B. Shells and Magic Numbers

Further evidence for the role of the mean potential U(r) in binding nuclei together comes from a more detailed study of the energy eigenvalues of the bound states. The energy levels in a spherical potential are eigenstates of the angular momentum $\vec{j}^2$, with eigenvalue $j(j+1)\hbar^2$; each value of j corresponds to $2j+1$ degenerate states with different angular-momentum projections $m\hbar$. Thus the energies of the single-particle states come in bunches of $2j+1$; as in atoms, an additional bunching of the levels occurs due to accidental, approximate degeneracies. A group of $2j+1$ states with the same j are called a **subshell**, and the larger groupings are called **major shells**. As in atoms, the binding energy gained by adding an additional particle does not vary smoothly with the number of particles present, but changes abruptly when a "magic number" of nucleons are present. The magic number corresponds to the number of fermions necessary to fill all the levels in and below a major shell. When a nucleon is added beyond the magic number, it has to go into a new major shell with higher energy levels, and thus is not as tightly bound. Fig. 4.2 shows the energy it takes to remove a nucleon from a nucleus, called the **separation energy** S_n or S_p for neutrons and protons, respectively:

$$S_n(N, Z) = B(N, Z) - B(N - 1, Z) \tag{4.1.8a}$$
$$S_p(N, Z) = B(N, Z) - B(N, Z - 1). \tag{4.1.8b}$$

The "magic numbers" are apparent as abrupt changes in the values of S_n or S_p.

The "magic numbers" for nuclei are 2, 8, 20, 28, 50, 82, and 126. They can be explained by counting the energy levels in a potential well $\hat{U}(r)$ similar to that in

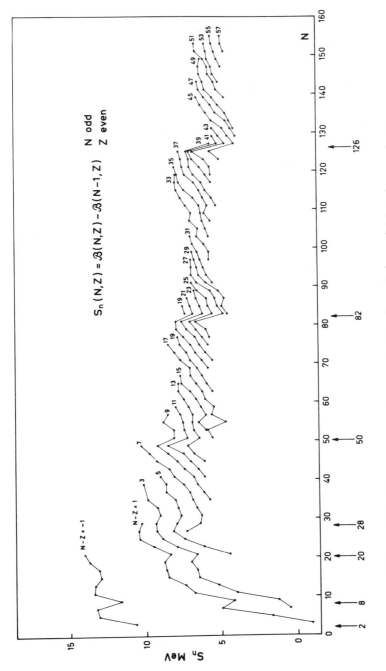

Figure 4.2a   Neutron separation energies, S_n [*from Bohr and Mottelson, vol. 1*]

the optical model, provided the spin-orbit potential is sufficiently strong (see fig. 4.3). For this observation (in 1948, before the optical model was known), Mayer and Jensen received the Nobel Prize (their book is quite readable). The picture of the nucleus as a collection of nucleons moving in a potential is called the **independent-particle model**.

4.1 C.   Transfer Reactions and the DWBA

The case for the independent-particle model is further strengthened by evidence from nuclear reactions in which a nucleon is transferred from a projectile nucleus to a target nucleus. Such a reaction is called a **stripping reaction**. For example, deuterons incident on a nuclear target may emerge as protons, "stripped" of their neutrons by the target nucleus. Such a reaction is denoted

$$^{A}Z\,(d,p)\,^{A+1}Z \tag{4.1.9}$$

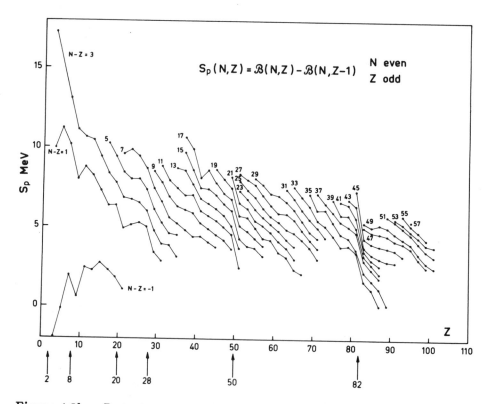

Figure 4.2b   Proton separation energies, S_p [*from Bohr and Mottelson, vol. 1*]

where the parentheses indicate what happens to the incident beam particle.

The proton energies produced by the (d,p) reaction can be described by the spectrum $\frac{d\sigma}{d\varepsilon_p}(\varepsilon_p)$: the probability of the proton emerging with energy between ε_p

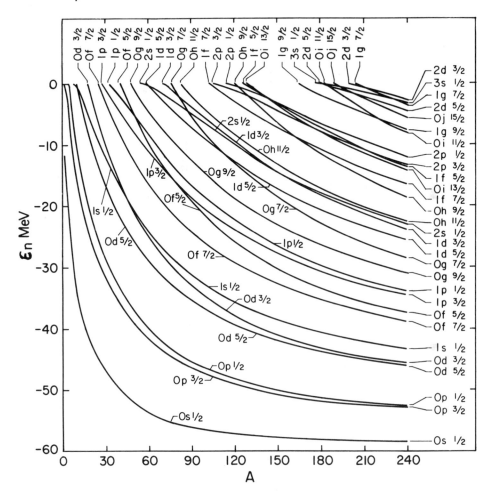

Figure 4.3   Energies of neutron orbits in the real part of the optical potential, eqs. (3.3.19-21). The values of n_r, ℓ, and j are given for each state, where n_r is the number of nodes in the radial wavefunction, counting neither the node at $r = \infty$ nor at $r = 0$. [*Courtesy of S. Shlomo*]

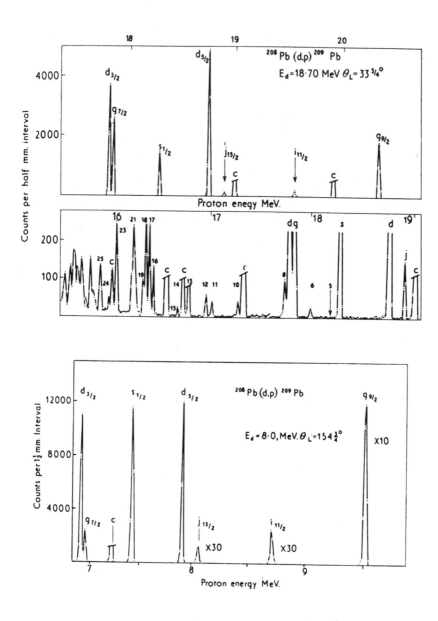

Figure 4.4   Spectra of proton energies from the reaction ^{208}Pb $(d,p)^{209}$Pb, for two different energies of deuterons. The angle given in each figure is the angle of the outgoing proton in the laboratory. Angular-momentum assignments of the single-particle levels are indicated. [*from A. F. Jeans, W. Darcey, W. G. Davies, K. N. Jones and P. K. Smith, Nucl. Phys.* **A128** *(1969) 224*]

and $\varepsilon_p + d\varepsilon_p$ is given by

$$dP(\varepsilon_p) = nt \frac{d\sigma}{d\varepsilon_p}(\varepsilon_p)\, d\varepsilon_p \qquad (4.1.10)$$

where n is the target density and t is the target thickness (assumed thin). Such cross sections have been measured, and have pronounced maxima as a function of the proton energy (fig. 4.4). The energies ε_p at which these maxima occur vary with the beam energy E_d. To interpret these energies, we invoke the conservation of energy to write

$$\varepsilon_p = E_d - B_d - \varepsilon_i - E_R \qquad (4.1.11)$$

where B_d is the binding energy of the deuteron, E_R is the kinetic energy of the recoiling target nucleus, and ε_i the energy of the neutron moving in the target nucleus. The energies ε_i do not depend on the bombarding energy E_d, but are characteristic of the target nucleus, varying smoothly as a function of the target's mass and charge. They can be understood as the energies of eigenstates of the independent-particle motion in a potential $U(r)$.

When the neutron goes into the lowest unoccupied level ε_i in the target nucleus's potential, then the proton has its maximum energy ε_p; the level ε_i correspond to the ground state of the nucleus ^{A+1}Z. Other peaks correspond to the energies ε_i of other eigenstates of the independent-particle motion. The situation is sketched schematically in fig. 4.5.

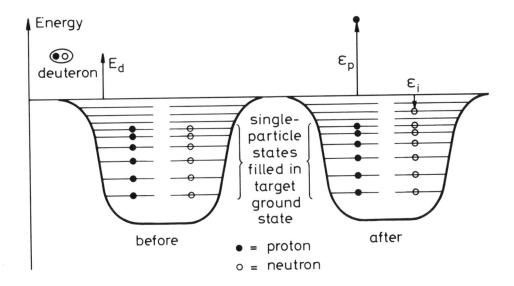

Figure 4.5   Deuteron stripping reaction $^{A}Z(d,p)^{A+1}Z$.

Fig. 4.4 shows examples of spectra measured in a (d,p) reaction. (Not all the peaks can be identified with neutron orbitals; some correspond to more complicated processes, for example simultaneously exciting a nucleon in the target from its original level to an empty level.) Additional measurements of the angular distributions of the protons, and their spins, often make it possible to determine the orbital and total angular momentum of the neutron independent-particle state.

The quantitative interpretation of non-elastic reaction cross sections requires an extension of scattering theory to inelastic processes. We have already applied the $\mathfrak{S}$-matrix theory to the case of more than two particles in the intermediate states of the pion-exchange force, Sect. 2.5. The generalization to the case where the initial and final states each contain no more than two simple or complex nuclei is straightforward, as long as the nuclear states are bound (or sufficiently long-lived). A readable explanation is given in the first chapters of Satchler's book. Here we will briefly summarize the resulting method.

To characterize the initial and final states, we need to specify not only the motion of the projectile, but also the composition and internal state of the target and projectile. Before the reaction, the target has (say) N_A neutrons and Z_A protons, an internal state (undoubtedly its ground state) $|A\rangle$, and a momentum $\vec{p}_A$ (zero in the laboratory frame). We denote the corresponding quantities for the projectile as N_a, Z_a, $|a\rangle$ and $\vec{p}_a$ before the reaction. After the reaction, the target becomes N_B, Z_B in state $|B\rangle$ with momentum $\vec{p}_B$, and the projectile is correspondingly characterized by N_b, Z_b, $|b\rangle$ and $\vec{p}_b$. The characterization of target and projectile are said to define a **reaction channel** α or β. The laboratory scattering angle is given by the direction of $\vec{p}_b$ relative to $\vec{p}_a$. In the center of mass, the relative momentum of the initial state is $\vec{p}_\alpha$, of the final state $\vec{p}_\beta$. The asymptotic states are eigenstates of the Hamiltonian

$$H_\beta = H_B + H_b + \vec{p}_\beta{}^2/2\mu_\beta \qquad (4.1.12)$$

where H_B and H_b govern the inner motion of the two nuclei, and the remaining term determines their relative motion with the reduced mass $\mu_\beta = m_{B}m_b/(m_B + m_b)$. The eigenvalues of H_β are $E_B + E_b + E_\beta$ where $E_\beta = \vec{p}_\beta{}^2/2\mu_\beta$. We observe one of the principle formal difficulties of reaction theory with complex projectiles: the eigenstates of different target- projectile partitions H_β are not orthogonal, for example a deuteron inside a nucleus with A nucleons is not distinct from a proton inside a nucleus with A+1 nucleons. This is not important as long as we only use H_β to construct wave packets representing well-separated nuclei; we will ignore the difficulties arising when the nuclei approach each other. We will denote the plane-wave eigenstates of relative momentum $\vec{p}_\beta$ as $|\Upsilon^\beta_{\vec{p}_\beta}\rangle$; then the eigenstates of H_β are product states

$$H_\beta|\Upsilon^\beta_{\vec{p}_\beta}\rangle|B\rangle|b\rangle = (E_\beta + E_B + E_b)|\Upsilon^\beta_{\vec{p}_\beta}\rangle|B\rangle|b\rangle. \qquad (4.1.13)$$

The wave function $|\Psi_\alpha\rangle$ describing the scattering and reactions has the form

$$|\Psi_\alpha\rangle = \sum_\beta |B\rangle|b\rangle|\xi^{(+)}_\beta\rangle \qquad (4.1.13a)$$

where the relative-motion wave functions $|\xi_\beta^{(+)}\rangle$ have the asymptotic form

$$\langle\vec{r}_\beta|\xi_\beta^{(+)}\rangle \rightarrow e^{i\vec{p}_\alpha\cdot\vec{r}_\alpha/\hbar}\delta_{\alpha\beta} + f_{\beta\alpha}(\theta_\beta)\frac{e^{ip_\beta r_\beta/\hbar}}{r_\beta}. \qquad (4.1.13b)$$

The cross section is determined from the asymptotic wave function by the ratio of the beam current to the current in the scattered wave, and thus is given in terms of the reaction amplitude $f_{\beta\alpha}$ by

$$\frac{d\sigma_{\beta\alpha}}{d\Omega_\beta} = \frac{v_\beta}{v_\alpha}|f_{\beta\alpha}(\theta_\beta)|^2. \qquad (4.1.14)$$

The velocities of the states β and α enter in this expression because the cross section is related to current densities while the wave function gives probability densities. The relation to the $\mathfrak{S}$ matrix is analogous to eq. (2.4.9),

$$\langle B|\langle b|\langle \Upsilon_{\vec{p}_\beta}^\beta|\mathfrak{S}_E|\Upsilon_{\vec{p}_\alpha}^\alpha\rangle|A\rangle|a\rangle = -\frac{2\pi\hbar^2}{\mu_\beta}f_{\beta\alpha} \qquad (4.1.15)$$

where $E = E_A + E_a + E_\alpha = E_B + E_b + E_\beta$ is the total energy in the center of mass.

To obtain an expression for the reaction amplitude we begin by writing the Schrödinger equation for $|\Psi_\alpha\rangle$ in the form

$$(E_\beta - \vec{p}_\beta{}^2/2\mu_\beta)|\Psi_\alpha\rangle = V_\beta|\Psi_\alpha\rangle \qquad (4.1.16)$$

where the residual interaction V_β is the part of the total Hamiltonian not included on the left-hand side of the equation. V_β contains the interactions of the target nucleons with the projectile nucleons. Projecting eq. (4.1.16) from the left with the channel eigenstates $|B\rangle|b\rangle$, and using eq. (4.1.13a) for $|\Psi_\alpha\rangle$, we find an equation for $|\xi_\beta^{(+)}\rangle$:

$$(E_\beta - \vec{p}_\beta{}^2/2\mu_\beta)|\xi_\beta^{(+)}\rangle = \langle B|\langle b|V_\beta|\Psi_\alpha\rangle. \qquad (4.1.17)$$

In coordinate representation, this equation is just like eq. (2.4.11) and thus may be solved using the Green's function of eq. (2.4.12) with E replaced by E_β and μ by μ_β.

The solution with the boundary conditions implied by eq. (4.1.13b) is, analogous to eq. (2.4.10),

$$\langle\vec{r}_\beta|\xi_\beta^{(+)}\rangle = e^{i\vec{p}_\alpha\cdot\vec{r}_\alpha/\hbar}\delta_{\alpha\beta} - \frac{\mu_\beta}{2\pi\hbar^2}\int d^3\vec{r}_\beta{}'\frac{e^{ip_\beta|\vec{r}_\beta-\vec{r}_\beta{}'|/\hbar}}{|\vec{r}_\beta - \vec{r}_\beta{}'|}\langle\vec{r}_\beta{}'|\langle B|\langle b|V_\beta|\Psi_\alpha\rangle. \quad (4.1.18)$$

Inspecting the limiting case $r_\beta \gg r_\beta'$, which implies $p_\beta|\vec{r}_\beta - \vec{r}_\beta{}'| \approx p_\beta r_\beta - \vec{p}_\beta\cdot\vec{r}_\beta{}'$, we see by comparison with eq. (4.1.13b) that the reaction amplitude is

$$f_{\beta\alpha}(\theta_\beta) = -\frac{\mu_\beta}{2\pi\hbar^2}\langle B|\langle b|\langle\Upsilon_{\vec{p}_\beta}^\beta|V_\beta|\Psi_\alpha\rangle. \qquad (4.1.19)$$

In the Born approximation, we would replace $|\Psi_\alpha\rangle$ by $|\Upsilon^\alpha_{\vec{p}_\alpha}\rangle$. For strongly interacting systems, however, that would be a poor approximation.

To find a better approximation, we take account of the main part of the interaction between target and projectile by introducing a phenomenological optical model for their relative motion. Then eq. (4.1.12) becomes

$$\tilde{H}_\beta = H_B + H_b + \vec{p}_\beta{}^2/2\mu_\beta + \hat{U}^{POM}_\beta(\vec{r}_\beta) \tag{4.1.20}$$

where $\hat{U}^{POM}_\beta$ describes the elastic scattering of projectile b on target B. The corresponding **distorted wave** functions $|\tilde{\Upsilon}^{\beta(+)}_{\vec{p}_\beta}\rangle$ satisfy

$$\left(\vec{p}_\beta{}^2/2\mu_\beta + \hat{U}^{POM}_\beta(\vec{r}_\beta)\right)|\tilde{\Upsilon}^{\beta(+)}_{\vec{p}_\beta}\rangle = E_\beta|\tilde{\Upsilon}^{\beta(+)}_{\vec{p}_\beta}\rangle \tag{4.1.21a}$$

with the boundary condition that their asymptotic form is

$$\langle\vec{r}_\beta|\tilde{\Upsilon}^{\beta(+)}_{\vec{p}_\beta}\rangle \rightarrow e^{i\vec{p}_\beta\cdot\vec{r}_\beta/\hbar} + f^U_\beta(\theta_\beta)\frac{e^{ip_\beta r_\beta/\hbar}}{r_\beta} \tag{4.1.21b}$$

with $f^U_\beta(\theta_\beta)$ giving the elastic scattering of b on B by $\hat{U}^{POM}_\beta$. The residual interaction is now

$$\tilde{V}_\beta = V_\beta - \hat{U}^{POM}_\beta(\vec{r}_\beta) \tag{4.1.22}$$

which we may hope to be small enough to treat perturbatively. The reaction amplitude becomes

$$f_{\beta\alpha}(\theta_\beta) = f^U_\alpha(\theta_\alpha)\delta_{\alpha\beta} - \frac{\mu_\beta}{2\pi\hbar^2}\langle B|\langle b|\langle\tilde{\Upsilon}^{\beta(-)}_{\vec{p}_\beta}|\tilde{V}_\beta|\Psi_\alpha\rangle \tag{4.1.23}$$

where $|\tilde{\Upsilon}^{\beta(-)}_{\vec{p}_\beta}\rangle$ satisfies incoming-wave boundary conditions instead of the outgoing-wave condition (4.1.21b). It may be obtained from $|\Upsilon^{\beta(+)}_{\vec{p}_\beta}\rangle$ by the time-reversal relation

$$\langle\vec{r}_\beta|\tilde{\Upsilon}^{\beta(-)}_{\vec{p}_\beta}\rangle = \langle\vec{r}_\beta|\tilde{\Upsilon}^{\beta(+)}_{-\vec{p}_\beta}\rangle^*; \tag{4.1.24}$$

note that, for plane waves, $|\Upsilon^{(+)}_{\vec{p}}\rangle = |\Upsilon^{(-)}_{\vec{p}}\rangle$. The derivation of eq. (4.1.23) proceeds much like that for eq. (4.1.19), but is clouded by the lack of hermiticity of $\tilde{H}_\beta$, which may therefore not necessarily possess a Green's function. A discussion of its validity is found in [Satchler]. The great utility of eq. (4.1.23) lies in the possibility of considering $\tilde{V}_\beta$ as a perturbation, which allows us plausibly to approximate $|\Psi_\alpha\rangle \approx |A\rangle|a\rangle|\tilde{\Upsilon}^{\alpha(+)}_{\vec{p}_\alpha}\rangle$ leading to the **distorted-wave Born approximation** or DWBA:

$$\langle B|\langle b|\langle\Upsilon^\beta_{\vec{p}_\beta}|\mathfrak{I}^{DWBA}_E|\Upsilon^\alpha_{\vec{p}_\alpha}\rangle|A\rangle|a\rangle$$
$$= -\delta_{\alpha\beta}\frac{2\pi\hbar^2}{\mu_\alpha}f^U_\alpha(\theta_\alpha) + \langle B|\langle b|\langle\tilde{\Upsilon}^{\beta(-)}_{\vec{p}_\beta}|\tilde{V}_\beta|\tilde{\Upsilon}^{\alpha(+)}_{\vec{p}_\alpha}\rangle|A\rangle|a\rangle. \tag{4.1.25}$$

This equation appears strangely asymmetric, because $\tilde{V}_\beta$ appears to give a special role to the exit channel β. However, a similar equation could be derived, with the same approximations, where $\tilde{V}_\alpha$ appears instead of $\tilde{V}_\beta$. These alternative equations, known as the **post** and **prior** representations, give the same answers if the same

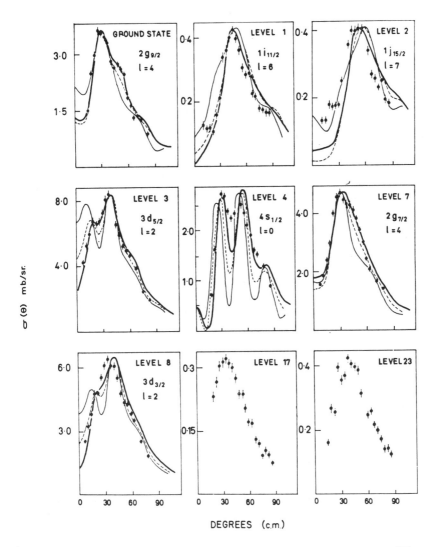

Figure 4.6   Angular distributions of protons leading to states in ^{209}Pb at $E_d =$ 18.7 MeV. The indicated curves are the predictions of DWBA calculations with various parametrizations of the optical potentials. [*from A. F. Jeans et al, loc. cit*]

approximations are used in estimating $\tilde{V}$ in both cases.

As an example of the DWBA we consider the deuteron stripping reaction (d,p). In the independent-particle model for the target nucleus, the wave function of the target after the transfer is given by

$$|B\rangle = |A\rangle|\phi_i(n)\rangle \tag{4.1.26}$$

where $|\phi_i\rangle$ is the neutron's wave function. The internal wave function of the deuteron is

$$\langle \vec{r}_n \vec{r}_p | a \rangle = \psi_d(\vec{r}_n - \vec{r}_p). \tag{4.1.27}$$

The residual interaction $\tilde{V}_\beta$ in the post representation is the interaction of the neutron with the outgoing proton, minus that part which is included in the proton's optical potential. In the prior representation, $\tilde{V}_\alpha$ is the interaction of the neutron with the target, minus that part which is included in the deuteron's optical potential. In either case, the residual interaction depends only on the coordinates $\vec{r}_n$ and $\vec{r}_p$ of the neutron and proton from the deuteron. Thus the matrix element in 4.1.25 becomes

$$\langle B|\langle b| \ \langle \tilde{\Upsilon}_{\vec{p}_\beta}^{\beta(-)}|\tilde{V}|\tilde{\Upsilon}_{\vec{p}_\alpha}^{\alpha(+)}\rangle|A\rangle|a\rangle =$$

$$\int d^3\vec{r}_n d^3\vec{r}_p \, \tilde{\Upsilon}_{\vec{p}_p}^{\beta(-)*}(\vec{r}_p)\phi_i^*(\vec{r}_n)\tilde{V}(\vec{r}_n,\vec{r}_p)\psi_d(\vec{r}_n - \vec{r}_p)\tilde{\Upsilon}_{\vec{p}_d}^{d(+)}\left(\frac{\vec{r}_n + \vec{r}_p}{2}\right). \tag{4.1.28}$$

where $\tilde{V}$ is either the post or prior interaction. The post form of the interaction is especially convenient, since then $\tilde{V}_\beta$ has the short range of the nucleon-nucleon interaction. This makes the integrals easier to evaluate; in practice, $\tilde{V}_\beta$ is often approximated by a zero-range force. The resulting angular distributions depend characteristically on the angular momentum of the ipm wave function ϕ_i. Fig. 4.6 illustrates how this angular dependence is used to identify the angular momentum of various states appearing as peaks in fig. 4.4. Similar techniques are applied to other reactions to give detailed information about the nuclear wave function.

4.2   MEAN FIELD THEORY

Although the independent-particle model has many successes, it also suffers some glaring failures. The most fundamental of these is in its description of the binding energy of the nucleus. Since the nuclear Hamiltonian is approximated by the sum of terms for each particle individually,

$$H_{ipm} = \sum_{i=1}^{A}\left[\frac{p_i^2}{2m_N} + \hat{U}(\vec{r}_i,\vec{s}_i)\right], \tag{4.2.1}$$

its mean value (which is minus the binding energy), will just be the average of the single-particle energies $\hat{\varepsilon}_i$

$$\langle H_{ipm} \rangle = \frac{1}{A} \sum_{i=1}^{A} \hat{\varepsilon}_i . \tag{4.2.2}$$

We can quickly see the nature of the difficulty by considering nuclear matter. In the Fermi-gas model, the energies are just

$$\varepsilon_i^{(FG)} = \frac{p_i^2}{2m_N} + U_0 . \tag{4.2.3}$$

The average kinetic energy for a set of states distributed uniformly throughout a sphere of radius p_F in momentum space is $\frac{3}{5} p_F^2 / 2m_N$. Thus the average energy is just the mean kinetic plus potential energy; using (4.1.3) we get

$$\langle H_{ipm} \rangle^{(FG)} = \frac{3}{5} \frac{p_F^2}{2m_N} + U_0 = \mu + \frac{2}{5}(U_0 - \mu) . \tag{4.2.4}$$

The experimental result, however, is that the average binding energy equals the binding energy of the last nucleon, eq. (4.1.2). The independent-particle model is therefore unacceptable in this respect. We need to find a model which shares the successful aspects of the independent-particle model, but gives a better account of the average binding energy. To find such a model, we need to understand the origin of the single-particle potential $\hat{U}(r)$.

A simple model has been very successful in explaining the single-particle aspects of atomic structure: the Hartree-Fock theory (a clear exposition is found in Bethe and Jackiw). We can try to apply that theory to nuclei. We will find that we obtain a qualitative explanation of the independent-particle model's problems with the total energy, but unfortunately other, related problems arise. Understanding these problems will lead, eventually, to their resolution in the following section.

4.2 A.   Hartree-Fock Theory

The independent-particle model is quite successful at describing the motion of a nucleon in a nucleus, whether the nucleon is in a bound state or a continuum state. But when we try to interpret it as a Hamiltonian for the nucleus, we run into two difficulties: it is energy dependent, and it gives the wrong total energy. Therefore we try to back up a step to an actual Hamiltonian, the potential between pairs of nucleons, introduced in Chapter 2. We preserve the insight of the independent-particle model, however, by approximating the nucleus' wave function as the independent motion of the constituent nucleons. In the development we give here, we will neglect the spin and isospin for simplicity.

The simplest wave function describing independent-particle motion of A particles would be a product

$$\psi(\vec{r}_1, \vec{r}_2, \ldots, \vec{r}_A) = \phi_1(\vec{r}_1)\phi_2(\vec{r}_2)\phi_3(\vec{r}_3) \ldots \phi_A(\vec{r}_A) .$$

Then the probability density $\psi^*\psi$ is just a product of factors $\phi_i^*(\vec{r}_i)\phi_i(\vec{r}_i)$ for the individual nucleons. Such a wave function, however, is impossible for a system of fermions, because it is not antisymmetric.

The simplest antisymmetric wave function is the antisymmetrized product

$$\psi_{HF}(\vec{r}_1, \vec{r}_2, \ldots, \vec{r}_A) = (A!)^{-1/2} \sum_P (-1)^P \prod_{i=1}^{A} \phi_i(\vec{r}_{p(i)})$$

(4.2.5)

$$= (A!)^{-1/2} \left\{ \phi_1(\vec{r}_1)\phi_2(\vec{r}_2) \ldots \phi_A(\vec{r}_A) - \phi_1(\vec{r}_2)\phi_2(\vec{r}_1) \ldots \phi_A(\vec{r}_A) + \ldots \right\}$$

where p(i) is a permutation of the integers from 1 to A, and $(-1)^P$ is ± 1 according to whether the permutation is even or odd. One often sees eq. (4.2.5) written in a short notation

$$\psi_{HF} = \mathcal{A} \prod_{i=1}^{A} \phi_i$$

(4.2.5a)

where the $\mathcal{A}$ means "antisymmetrized". ψ_{HF} is often called a Slater determinant, because if one thinks of the matrix $M_{ij} = \phi_i(\vec{r}_j)$, then ψ_{HF} is essentially its determinant:

$$\psi_{HF}(\vec{r}_1, \vec{r}_2, \ldots, \vec{r}_A) = (A!)^{-\frac{1}{2}} \det \phi_i(\vec{r}_j).$$

(4.2.5b)

The factor $(A!)^{-1/2}$ is a normalization necessary because there are A! terms in the summation. It is conventional to normalize the single-particle wave functions such that

$$\int d^3\vec{r}\,\phi_i^*\phi_j = \delta_{ij}.$$

(4.2.6)

The Hartree-Fock method consists of determining the single-particle wave functions by the requirement that the expectation value of the Hamiltonian be a minimum. The Hamiltonian is taken to contain a local 2-body interaction:

$$H = \sum_i \frac{p_i^2}{2m_N} + \frac{1}{2} \sum_{i \neq j} V(\vec{r}_{ij}).$$

(4.2.7)

The factor $\frac{1}{2}$ in the sum over two-body interactions is necessary to avoid counting a given pair of nucleons (i,j) twice. The expectation value of the energy in the state ψ_{HF} may be written in terms of its one-body density matrix

$$\rho(\vec{r}, \vec{r}') = \sum_j \phi_j^*(\vec{r})\phi_j(\vec{r}'),$$

(4.2.8)

$$\langle \psi_{HF}|H|\psi_{HF}\rangle = \int d^3\vec{r}\,d^3\vec{r}' \left(\frac{\hbar^2}{2m_N} \delta(\vec{r} - \vec{r}')\vec{\nabla}_r \cdot \vec{\nabla}_{r'}\rho(\vec{r}, \vec{r}') \right.$$

$$\left. + \frac{1}{2}V(\vec{r} - \vec{r}') \left[\rho(\vec{r}, \vec{r})\rho(\vec{r}', \vec{r}') - \rho(\vec{r}, \vec{r}')^2\right] \right).$$

(4.2.9)

The energy (4.2.9) has to be minimized with respect to independent variations of the wave functions ϕ_i , and their complex conjugates ϕ_i^*, subject to the normalization constraints (4.2.6). The variational equation is thus

$$\frac{\delta}{\delta\phi_i^*}\left(\langle\psi_{HF}|H|\psi_{HF}\rangle - \sum_j \varepsilon_j \int d^3\vec{r}\,\phi_j^*(\vec{r})\phi_j(\vec{r})\right) = 0. \qquad (4.2.10)$$

After some manipulations, one obtains the Hartree-Fock equations

$$\varepsilon_i\phi_i(\vec{r}) = -\frac{\hbar^2}{2m_N}\vec{\nabla}^2\phi_i(\vec{r}) + U^D(\vec{r})\phi_i(\vec{r}) - \int d^3\vec{r}'\,U^X(\vec{r},\vec{r}')\phi_i(\vec{r}') \qquad (4.2.11)$$

where the U^D and U^X are the direct and exchange potentials given by

$$U^D(\vec{r}) = \int d^3\vec{r}'\,V(\vec{r}-\vec{r}')\rho(\vec{r}',\vec{r}') \qquad (4.2.11a)$$

$$U^X(\vec{r},\vec{r}') = V(\vec{r}-\vec{r}')\rho(\vec{r}',\vec{r}). \qquad (4.2.11b)$$

The Lagrange multiplier ε_i appears in eq. (4.2.10) as a result of the normalization constraints (4.2.6). It may be thought of as an eigenvalue to be adjusted to obtain solutions of the set of A equations (4.2.11).

The Hartree-Fock equations are readily interpreted. In essence, they say that the motion of each nucleon is determined by a Schrödinger-like wave equation in which the potential, U^D, is just the average potential due to all the particles in the nucleus. There is a correction term, U^X, due to the antisymmetrization of the wave functions, which is not surprising. One might have expected to see only the field due to all the other particles. But if the ith orbital is left out of the sum in eq. (4.2.8), the resulting eq. (4.2.11) is exactly the same, because the direct and exchange terms from the ith orbital cancel.

We seem to have found a reasonable, approximate expression for the real potential which governs the motion of the nucleons. Not only that, but we can now see one reason why the independent-particle model gave the wrong energy: multiply eq. (4.2.11) by $\phi_i^*(\vec{r})$, integrate over r using (4.2.6), sum over orbitals i, and compare with eq. (4.2.9) to obtain

$$\langle\psi_{HF}|H|\psi_{HF}\rangle = \sum_{i=1}^{A}\varepsilon_i - \frac{1}{2}\int d^3\vec{r}\,d^3\vec{r}'V(\vec{r}-\vec{r}')\left[\rho(\vec{r},\vec{r})\rho(\vec{r}',\vec{r}') - \rho(\vec{r},\vec{r}')^2\right]. \quad (4.2.12)$$

The sum of the eigenvalues overestimates the binding effect of the potential energy because it counts each pair of nucleons' interactions twice!

Unfortunately, when realistic nucleonic forces are used, the Hartree-Fock equations have no solutions. The reason for this was understood by Wigner in 1933. It can be most readily understood in terms of the Fermi-gas model, the limiting case of an infinitely large nucleus of volume Ω_r (Sect. 4.1.A). The wave functions are plane waves $e^{i(\vec{k}\cdot\vec{r})}/\Omega_r^{1/2}$. Converting the sum in eq. (4.2.8) into an integral over $\vec{k}$, it is easy to show that, as $\Omega_r \to \infty$, the density $\rho(\vec{r}, \vec{r})$ is just a constant n, and the density matrix depends only on the difference $\vec{r} - \vec{r}'$

$$\rho(\vec{r}, \vec{r}') = \frac{3nj_1(k_F|\vec{r} - \vec{r}'|)}{k_F|\vec{r} - \vec{r}'|}, \tag{4.2.13}$$

where the Fermi wave-number $k_F = p_F/\hbar$ is given by

$$n(\vec{r}) = \rho(\vec{r}, \vec{r}) = \frac{g}{6\pi^2}k_F^3 \tag{4.2.14}$$

in terms of the spin-isospin degeneracy g. The kinetic energy per particle,

$$\langle KE \rangle_{\text{nuclear matter}} = \frac{3}{5}\frac{\hbar^2}{2m_N}k_F^2 = \frac{3}{10}\frac{\hbar^2}{m_N}\left(\frac{6\pi^2}{g}\right)^{2/3}n^{2/3}, \tag{4.2.15}$$

is proportional to $n^{2/3}$, while the potential energy per particle, U^D, is proportional to n:

$$U^D = n\int d^3\vec{r}\, V(\vec{r}). \tag{4.2.16}$$

The exchange energy per particle is proportional to n and higher powers of n.

If the nucleus is to be bound, the potential energy per particle must be negative. In that case, the potential energy per particle can be made arbitrarily large by increasing the density: the Hartree-Fock theory collapses, and there is no minimum energy. Only in case of extreme spin-dependent forces can the collapse be prevented. With realistic nuclear forces, there is no collapse; instead, the strong repulsive potentials cause U^D to be positive, so that nuclear matter isn't bound at all.

We see that Hartree-Fock theory, while very appealing, fails to explain nuclear binding. The difficulties are twofold: first, Hartree-Fock theory cannot lead to a stable saturation density of nuclear matter. Second, the strong repulsion of the short-range nuclear force makes nuclear matter unbound.

The failure of Hartree-Fock theory appears to require an abandonment of the postulate of independent-particle motion. The reason for the failure is not hard to see. When the nucleons move completely independently, they cannot avoid feeling the repulsive core of the other nucleons. The determinantal wave function is too inflexible. We can see the trouble more clearly by recalling the problem of two-nucleon scattering. Independent motion of the scattering particles would be free motion, i.e. plane waves; the scattering matrix would then be given in the Born approximation

$$\langle \psi_f | \mathfrak{I} | \psi_i \rangle \approx \langle \psi_f | V | \psi_i \rangle$$

which always gives repulsive matrix elements. In the Hartree-Fock theory for nuclear structure, as in the Born approximation for two-nucleon scattering, independent-particle motion is not enough. The nucleons need to be able to avoid the repulsive forces that keep them apart at short distances.

4.2 B.   Correlated Variational Wave Functions

A natural way to improve on Hartree-Fock theory is to improve on the trial wave function (4.2.5). A simple approximation, due to A.Bijl, is known as the Jastrow ansatz:

$$\psi_J(\vec{r}_1, \ldots, \vec{r}_n) = \psi_{HF}(\vec{r}_1, \vec{r}_2, \ldots, \vec{r}_A) \prod_{i<j} f(|\vec{r}_i - \vec{r}_j|). \qquad (4.2.17)$$

The factor f(r) should go to zero as r→0, to keep the pairs of nucleons apart, and must go to 1 as r→ ∞, where the effects of the 2-body force disappear. Then the 2-body correlation function f, as well as the 1-body wave functions ϕ_i, may be determined by minimizing the energy. When suitable account is taken of the dependence of f on the spins and isospins of the pairs of particles, as well as on their momenta, considerable success has been attained in describing nuclear properties, such as binding energy, saturation density and optical potential.

The successes have been won only at the expense of very elaborate computational schemes, because it is exceedingly difficult to calculate the mean value of the energy,

$$\langle H \rangle = \frac{\langle \psi_J | H | \psi_J \rangle}{\langle \psi_J | \psi_J \rangle}, \qquad (4.2.18)$$

even when ϕ_i and f are given. The reason is that the mean value (4.2.18) involves integrals over the coordinates of up to A nucleons; unlike the Hartree-Fock case, the integrals cannot be reduced precisely to lower-dimensional integrals (the worst integral in Hartree-Fock theory is over 2 coordinates). Recently, several workers have shown how to approximate these integrals [see Blaizot and Ripka]. It turns out to be necessary to solve non-linear integral equations even to get an approximate answer with a known wave function. The main technical difficulty that arises with simpler approximation schemes is that the approximate values of ⟨H⟩ can be large and negative if an unfortunate choice of the wave function is made, even though the approximate ⟨H⟩ may be satisfactory for more realistic choices of f. It is very hard to do a variational computation if one's approximation methods do not admit lower bounds to the energy!

While the successes of the correlated variational theory are impressive, the clumsiness of its computational procedures, and the lack of a simple but reasonable approximation, have restricted its applications to a few specific nuclear properties. As a foundation for nuclear structure, it is quite firm, but it is difficult to fasten a superstructure to it. Another approach, the Brueckner theory, has been more illuminating.

4.2 C.   Brueckner Theory, the Independent Pair Approximation

The most widely used method for taking account of the short-range correlations between nucleons in nuclei, the Brueckner theory, takes as its starting point the theory for scattering of free nucleons, discussed in Chapter 2. There, the correlations are introduced by a wave function describing the relative motion, which goes over into free or independent-particle motion when the nucleons are far away from each other. The optical model shows that, even though they are close to each other, the nucleons in a nucleus move quite far between collisions. Thus we can try to understand what happens when two nucleons scatter from each other inside the nucleus. For a more detailed discussion see deShalit and Feshbach, Chapter 3.

Brueckner argued that the scattering of two nucleons inside a nucleus should be rather similar to their scattering in free space, except for two main differences:

(1) Their asymptotic energy is bound;

(2) The Pauli principle prevents them from scattering into occupied quantum orbitals.

The modification of the scattering due to the nucleons' energies is straightforward to include. In fact, we have already seen how to do this by way of the $\Im$-matrix formalism, in which the scattering amplitude depends explicitly on energy. Including the Pauli principle is a little trickier.

Consider the equation for the $\Im$ matrix

$$\Im_E = V + \lim_{\epsilon \to 0} V \frac{1}{E - H_0 + i\epsilon} \Im_E. \tag{2.4.17}$$

Since we are trying to describe the scattering of two nucleons by their interaction V, we ought to choose H_0 so it describes their motion if they don't scatter. Thus we will expect H_0 to contain something like the mean-field potential $U(\vec{r})$. We will not specify precisely how to compute it until later; for now, we will suppose that H_0 is a known one-body operator,

$$H_0 = \sum_{i=1}^{A} H^0(\vec{r}_i, \vec{s}_i, \vec{t}_i) \tag{4.2.19}$$

and introduce a basis of its single-particle eigenstates $|\alpha\rangle$ with wave functions in coordinate, spin, isospin

$$\langle \vec{r} s_z t_3 | \alpha \rangle = \phi_\alpha(\vec{r}) \chi_\alpha(s_z, t_3) \tag{4.2.20}$$

and eigenvalues

$$H^0|\alpha\rangle = \varepsilon_\alpha |\alpha\rangle. \tag{4.2.21}$$

For the two-body states, analogous to the scattering states, we take product states $|\alpha\rangle|\beta\rangle$ as a basis (they aren't antisymmetric, but we can always find the matrix elements of antisymmetric states if we know the product-state matrix elements).

The Green function $(E-H_0 + i\epsilon)^{-1}$ of eq. (2.4.17) is diagonal in this basis, its matrix elements being

$$\left\langle \alpha\beta \left| \frac{1}{E - H_0 + i\epsilon} \right| \gamma\delta \right\rangle = \delta_{\alpha\gamma}\delta_{\beta\delta}(E - \varepsilon_\alpha - \varepsilon_\beta + i\epsilon)^{-1}. \qquad (4.2.22)$$

Because we are interested in bound states, instead of scattering states for which (2.4.17) was derived, we replace $\lim_{\epsilon\to 0}$ by a principal value. Thus eq. (2.4.17) becomes, in this basis,

$$\langle\alpha\beta|\Im_E|\gamma\delta\rangle = \langle\alpha\beta|V|\gamma\delta\rangle + \sum_{\mu\nu}\langle\alpha\beta|V|\mu\nu\rangle\frac{1}{E - \varepsilon_\mu - \varepsilon_\nu}\langle\mu\nu|\Im_E|\gamma\delta\rangle.$$

In the basis of 2-body states, we can see how to introduce the Pauli principle in the 2-body scattering: by requiring that the intermediate states $|\mu\rangle$ and $|\nu\rangle$ be unoccupied in the nucleus. Formally, we write

$$\langle\alpha\beta|\Im_E|\gamma\delta\rangle = \langle\alpha\beta|V|\gamma\delta\rangle + \sum_{\mu\nu}\langle\alpha\beta|V|\mu\nu\rangle\frac{\langle\mu\nu|Q|\mu\nu\rangle}{E - \varepsilon_\mu - \varepsilon_\nu}\langle\mu\nu|\Im_E|\gamma\delta\rangle \qquad (4.2.23)$$

where the Pauli operator Q is given in terms of the occupation probabilities n_μ, n_ν of states $|\mu\rangle, |\nu\rangle$ by

$$\langle\mu\nu|Q|\mu'\nu'\rangle = \delta_{\mu\mu'}\delta_{\nu\nu'}(1 - n_\mu)(1 - n_\nu). \qquad (4.2.24)$$

In a nucleus's ground state, $n_\mu=1$ if the state $|\mu\rangle$ is occupied, and $n_\mu=0$ if it is unoccupied.

The Brueckner equation (4.2.23) takes account of the Pauli exclusion principle in nuclei. It may be derived in a systematic way by introducing the basis of A-body antisymmetric wave functions corresponding to H_0, and then summing a subset of perturbation-series terms which correspond to the scattering of a pair of nucleons with the others as spectators [deShalit and Feshbach, Chapter 3]. While this derivation has the advantage of appearing systematic, the neglect of the other terms in the many-body perturbation theory can only be justified by the qualitative arguments we have used to deduce eq. (4.2.23).

We are left with the question of determining H_0. The idea here [Jeukenne, Lejeune and Mahaux] is borrowed from the Hartree-Fock theory. The Hartree-Fock estimate of this potential failed because the interaction matrix elements were evaluated with uncorrelated wave functions. As we learned in Chapter 2, the $\Im$ matrix is a kind of effective Hamiltonian whose matrix elements between uncorrelated states reproduce those of the two-body potential in the presence of correlations. Thus we evaluate H_0 by using the Hartree-Fock expression, but with the $\Im$-matrix effective interaction instead of the bare two-body interaction V. In our basis, we have

$$\langle\alpha|H^0|\beta\rangle = \left\langle \alpha \left| \frac{p^2}{2m_N} \right| \beta \right\rangle + \sum_{\gamma=1}^{A}(\langle\alpha\gamma|\Im_E|\beta\gamma\rangle - \langle\alpha\gamma|\Im_E|\gamma\beta\rangle) \qquad (4.2.25)$$

obtained from eq. (4.2.11) by multiplying from the left with ϕ_α^*, replacing V by $\mathfrak{I}$ and i by β, and integrating. The energy E to be used in evaluating $\mathfrak{I}_E$ is the sum of the single- particle energies of the nucleons which are interacting, $\mathfrak{I}_{\varepsilon_\alpha + \varepsilon_\gamma}$.

The set of equations (4.2.21, 23, 25) form a closed system of non-linear integrodifferential equations for the wave functions $|\alpha\rangle$, the mean field H^0, and the effective interaction $\mathfrak{I}_E$. They are known as the Brueckner-Hartree-Fock equations. They have never been solved for finite nuclei, except with uncontrolled numerical approximations. The solutions for nuclear matter, however, are easier to obtain, because the wave functions $|\alpha\rangle$ are plane waves. The results are promising: using Reid's phenomenological potential (Sect. 2.6) fit to two-body scattering data, the binding energy per particle of nuclear matter is found to be 14 MeV, at an equilibrium density only about 25% greater than the central density of heavy nuclei deduced from elastic electron scattering. The binding energy is comparable to the 8 MeV observed for most nuclei, eq. (4.1.2); we shall see in Sect. 4.3 that the computed value is even closer to the extrapolated value (15.6 MeV) for nuclear matter.

The reason that Brueckner-Hartree-Fock theory gives a better account of nuclear binding than the Hartree-Fock theory is clear: the $\mathfrak{I}$ matrix de-emphasizes the repulsive short-range interaction, by building up correlations that prevent the nucleons from coming too close. The saturation of nuclear binding at a reasonable density emphasizes another feature: as the nuclear density increases, the effects of the Pauli principle become more important. The exclusion of higher-momentum intermediate states in the Brueckner equation causes the $\mathfrak{I}$ matrix to become less attractive and more repulsive. To see this, notice that the second term on the right-hand side of (4.2.23) is negative,since $\mathfrak{I}_E$ and V mostly have the same sign while E is less than $\varepsilon_\mu + \varepsilon_\nu$. Indeed, the matrix elements of $\mathfrak{I}$ in nuclear matter show much weaker attraction for low relative momentum than those in free space, and more quickly become repulsive as the relative momentum increases. The greater the density of the matter, the more pronounced these effects. The presence of the other nucleons makes the effective interaction less attractive and more repulsive.

Unfortunately, calculations with the Reid soft core potential which take account of correlations between three and more nucleons give both saturation densities and binding energies which are substantially larger than the Brueckner-Hartree-Fock results. Independent computations in variational and extended Brueckner methods give, both for the Reid potential and the more accurate Paris and Bonn interactions, binding energies about 50% too large at saturation densities about twice the observed value. Evidently, more complicated multi-body interactions help to saturate nuclear matter, somehow cancelling the attraction from multiparticle correlations. Thus the simple Brueckner-Hartree-Fock picture with two-body interactions gives a reasonable account of nuclear binding.

4.2 D. The Local-Density Approximation

To make practical applications to real nuclei, it is necessary to use physical insight to simplify the Brueckner-Hartree-Fock theory. The main approximation is called the local-density approximation: the effective interaction between two nucleons in a nucleus depends on the medium around them only by way of the density in the neighborhood of the interacting nucleons. Thus the effective interaction $\Im$ can be computed in nuclear matter, as a function of the density, instead of being computed self-consistently in the nucleus being studied.

An additional approximation is to suppose that $\Im$ is "almost local", i.e. that its non-locality lies entirely in its dependence on spin and angular momentum. This is quite a good approximation in nuclear matter. To find the almost-local $\Im$ matrix that best approximates a nuclear-matter $\Im$ matrix, one demands that they give the same contribution to the average binding energy of nuclear matter. Such an effective interaction was first used by Negele, together with the local-density approximation, to estimate the Brueckner-Hartree-Fock structure of the magic nuclei. As expected from the nuclear matter results (the saturation density was 25% too big), the radii were too small; but tinkering with just one parameter, regulating the density dependence of the effective interaction, readily led to density distributions consistent with observations from electron scattering. At the same time, the binding energies were well reproduced.

The validity of the local-density approximation is not well tested. The intuitive idea is appealing: the nucleons, when scattered, are influenced mainly by the matter nearby. However, the single-particle density matrix in the nuclear surface is not very similar to that of nuclear matter at a corresponding density. We must regard the local density approximation as a plausible but uncontrolled approximation.

It is clear that the use of density-dependent effective interactions must be restricted to applications where the wave functions are uncorrelated, since the effects of correlations are already included in the effective interaction. Great care and physical insight are needed to avoid nonsense. As an example of the subtleties hidden in the seemingly innocent eq. (4.2.23), consider what happens when we try to use this effective interaction $\Im\left(\vec{r}-\vec{r}', n(\frac{\vec{r}+\vec{r}'}{2})\right)$ in Hartree-Fock-style variational theory. One expresses the mean energy by way of the densities (eq. (4.2.8)), and then varies the wave functions appearing in the density matrix. The result is **not** merely to replace V by $\Im_E$ in eq. (4.2.25) and in (4.2.11). Instead, an additional term appears, resulting from the variation of the density hidden in the effective interaction $\Im$:

$$U^D(\vec{r}) = \int d^3\vec{r}' \, \Im_E\left(\vec{r}-\vec{r}', n(\frac{\vec{r}+\vec{r}'}{2})\right) n(\vec{r}') + U^R(\vec{r}) \tag{4.2.26a}$$

$$U^R(\vec{r}) = \int d^3\vec{r}' \frac{\partial \Im_E(\vec{r}', n(\vec{r}))}{\partial n} \tag{4.2.26b}$$

$$\left[\rho\left(\vec{r}+\frac{\vec{r}'}{2}, \vec{r}+\frac{\vec{r}'}{2}\right)\rho\left(\vec{r}-\frac{\vec{r}'}{2}, \vec{r}-\frac{\vec{r}'}{2}\right) - \rho\left(\vec{r}+\frac{\vec{r}'}{2}, \vec{r}-\frac{\vec{r}'}{2}\right)^2\right].$$

The potential U^R is called the "rearrangement potential". It says that the force felt by a nucleon includes a piece due to its effect on the interactions among the other pairs of nucleons. A similar effect can be obtained by extending the Brueckner-Hartree-Fock theory to include terms in the many-body perturbation theory which are neglected in the Brueckner-Hartree-Fock approximation. Negele argues that the rearrangement potential of the density- dependent force provides a reasonable approximation of these many-body effects.

A model which includes the rearrangement potential by way of the effective interactions' density dependence is called Density-Dependent Hartree Fock. Some examples of the proton distributions and binding energies obtained (with a minor modification of the Brueckner nuclear-matter effective interaction) are shown in fig. 4.7 and table 4.1. The agreement with experiment seems quite satisfactory. The spectra of single- particle energies, on the other hand, do not agree well with observations (e.g. from (d,p) reactions), as fig. 4.8 shows. The calculated single-particle energies lie farther apart than the experimental ones. This fact reflects the limitations of the Brueckner-Hartree-Fock approximation.

4.2 E. Effective-Range Expansion and Skyrme Forces

To get to a truly simple and practical model of nuclei, an additional approximation is a great help. We saw in Chapter 2 that the scattering of nucleons with moderate relative momenta could be well described by the effective-range expansion of the effective interaction, requiring only a few parameters: singlet and triplet scattering lengths and effective ranges. When nucleons interact in the nuclear ground state, their relative momentum is not greater than twice the Fermi momentum, corresponding to scattering a beam of 140 MeV nucleons from a fixed target. Thus it is reasonable to hope that the effective interaction in nuclear matter may have a similar approximate expansion. Skyrme proposed [see Negele and Vautherin] that the plane-wave matrix elements of the almost-local density-dependent effective interaction in nuclear matter at density n could be expanded in a power series in the relative momenta of the initial and final 2-body states:

$$
\begin{aligned}
\langle \vec{p}_1', \vec{p}_2' \,|\, \Im(n) \,|\, \vec{p}_1 \vec{p}_2 \rangle \approx (2\pi\hbar)^{-3}\delta(\vec{p}_1 + \vec{p}_2 - \vec{p}_1' - \vec{p}_2')\Big\{ & \Im_0(n) \\
& + \frac{1}{8}\Im_1(n)[(\vec{p}_1 - \vec{p}_2)^2 + (\vec{p}_1' - \vec{p}_2')^2]/\hbar^2 \qquad (4.2.27) \\
& + \frac{1}{4}\Im_2(n)\,[(\vec{p}_1 - \vec{p}_2)\cdot(\vec{p}_1' - \vec{p}_2') + \ldots]/\hbar^2 \Big\}.
\end{aligned}
$$

Since the nuclear forces depend strongly on spin, the power-series coefficients $\Im_0$, $\Im_1$, and $\Im_2$ can be operators in the spin space:

$$
\Im_k(n) \rightarrow \Im_k(n)(1 + \chi_k(n)P_s) \qquad (4.2.28)
$$

where $P_s = \frac{1}{2}(1 + 4\vec{s}_1 \cdot \vec{s}_2)$ is -1 for singlet and $+1$ for spin-triplet states of relative motion. Often the density dependence of the coefficents χ_k, $\Im_1$, and $\Im_2$ is neglected.

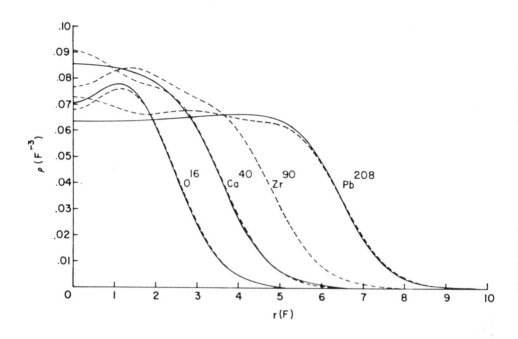

Figure 4.7   Theoretical (dashed lines) and empirical (solid lines) charge-density distributions, including proton size and c.m. motion corrections. The theoretical calculations are with a density-dependent effective interaction derived from the Brueckner theory of nuclear matter. [*from J. W. Negele, loc. cit.*]

Nucleus	^{16}O	^{40}Ca	^{43}Ca	^{90}Zr	^{208}Pb
E	-6.08	-7.28	-7.30	-7.77	-7.50
c.m. correction	-0.67	-0.21	-0.18	-0.08	-0.03
Theoretical B.E.	6.75	7.49	7.48	7.85	7.53
Experimental B.E.	7.98	8.55	8.67	8.71	7.87
Point proton rms radius	2.71	3.41	3.45	4.18	5.37
Point neutron rms radius	2.69	3.37	3.68	4.30	5.60

Table 4.1   Binding energies, in MeV, and rms radii, in fm, for closed-shell nuclei. [*from J. W. Negele, Phys. Rev.* **C1** *(1970) 225*]

We note that, in principle, all the coefficents depend on the energy of the interacting nucleons, but this energy dependence is usually ignored.

Skyrme's form of the density-dependent effective interaction is "almost local": it has the same matrix elements as the coordinate-space two-body operator

$$\Im(n,\vec{r}) = \Im_0(1+\chi_0 P_s)\delta(\vec{r}) - \Im_1(1+\chi_1 P_s)[\delta(\vec{r})\vec{\nabla}_r^{\,2} + \vec{\nabla}_r^{\,2}\delta(\vec{r})]/2$$
$$- \Im_2(1+\chi_2 P_s)\vec{\nabla}_r \cdot \delta(\vec{r})\vec{\nabla}_r \qquad (4.2.29)$$

where $\vec{r}$ is the relative coordinate between the nucleons. The δ-functions in eq.

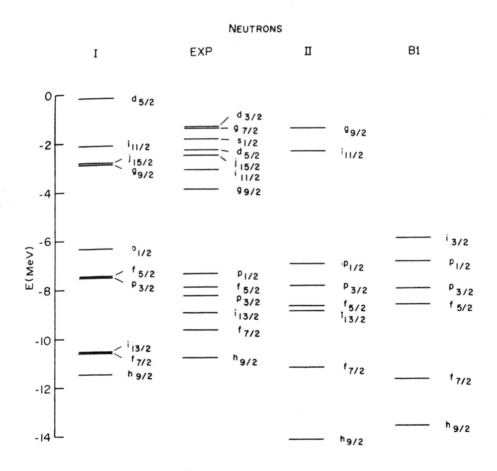

Figure 4.8a   Single-particle neutron states of ^{208}Pb near the Fermi level. The results obtained with various Skyrme interactions (I, II) are compared with the experimental values and with the spectrum calculated by Negele for a finite-range density-dependent effective interaction (B1). [*from J. Negele and D. Vautherin*]

(4.2.29) greatly simplify computations of matrix elements in coordinate representation, provided partial integrations are performed to avoid the derivatives of delta functions.

The effective-range expansion (4.2.29) leads to a very simple approximation for the energy of a Slater-determinant wave function. For example, if we take the same wave functions for neutrons and protons, spin up and spin down, and neglect the Coulomb interactions, then the energy may be written in terms of two functions of

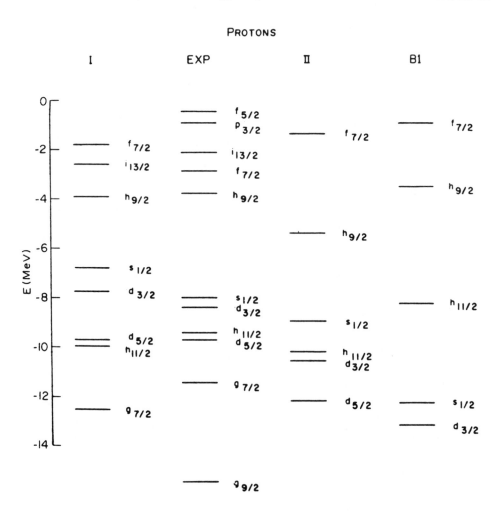

Figure 4.8b   Single-particle proton states of ^{208}Pb. See caption, fig. 4.8a.

$\vec{r}$, the diagonal density,

$$n(\vec{r}) = \rho(\vec{r}, \vec{r}) \tag{4.2.30}$$

and the kinetic-energy density,

$$\frac{\hbar^2}{2m_N} \tau(\vec{r}) = \frac{\hbar^2}{2m_N} \lim_{r \to r'} \vec{\nabla}_r \cdot \vec{\nabla}_{r'} \rho(\vec{r}, \vec{r}') \tag{4.2.31}$$

as follows

$$\langle \psi_{HF} | H | \psi_{HF} \rangle = \int d^3\vec{r}\, H(\vec{r}) \tag{4.2.32}$$

where the local energy density $H(\vec{r})$ is given by

$$H(\vec{r}) = \frac{\hbar^2}{2m_N} \tau + \frac{3}{8} \Im_0 n^2 + \frac{1}{16} \left[3\Im_1 + \Im_2(5 + 4\chi_2) \right] n\tau$$
$$+ \frac{1}{64} \left[9\Im_1 - \Im_2(5 + 4\chi_2) \right] \left| \vec{\nabla} n \right|^2 \tag{4.2.33}$$

(recall that $\Im_k$ may depend on n). In the interior of a large nucleus, n is approximately constant, and τ is related to n as in the Fermi gas (compare (4.2.15))

$$\tau = \frac{3}{5} k_F^2 n = \frac{3}{5} (3\pi^2 n/2)^{2/3} n. \tag{4.2.34}$$

Thus the first three terms in eq. (4.2.33) represent the energy density of nuclear matter at density n. The last term, proportional to $|\vec{\nabla} n|^2$, modifies the energy density in the surface. It is proportional to $\Im_1$ and $\Im_2$, the effective-range parameters, and represents the correction to the potential energy of a nucleon in the surface, due to the fact that its interactions sense the density a little distance away from it. Mean-field calculations with Skyrme forces can reproduce approximately the results with finite-range forces, as fig. 4.8 shows.

4.2 F.   Mean-Field Picture of Pion-Nucleus Scattering

We saw in Sect. 3.6 that pions with kinetic energies less than 100 MeV have reasonably long mean free paths in nuclei, comparable to those of nucleons. We might try, therefore, to apply a mean-field approach to low-energy pion-nucleus scattering.

Since we believe the pion-nucleon force to have a short range, we can look for an effective interaction $\Im$ which reproduces the form of low-energy pion nucleon scattering, eq. (2.7.7). Thus we look for an operator $\Im_E^{\pi N}$ in the relative motion which satisfies

$$\langle \Upsilon_{\vec{k}'} | \Im_E^{\pi N} | \Upsilon_{\vec{k}} \rangle = -\frac{2\pi(\hbar c)^2}{\varepsilon_\pi(\vec{k})} f_k^{\pi N}(\theta) \tag{4.2.35}$$

$$\approx -\frac{2\pi(\hbar c)^2}{\varepsilon_\pi(\vec{k})} \sum_{T=\frac{1}{2}}^{3/2} \Pi_T \left\{ f_{0\frac{1}{2}T}^{\pi N}(E) + \frac{\vec{k} \cdot \vec{k}'}{k^2} \left[f_{1\frac{1}{2}T}^{\pi N}(E) + 2f_{1\frac{3}{2}T}^{\pi N}(E) \right] \right.$$
$$\left. - \frac{2i}{\hbar k^2} \vec{s} \cdot \vec{k} \times \vec{k}' \left[f_{1\frac{1}{2}T}^{\pi N}(E) - f_{1\frac{3}{2}T}^{\pi N}(E) \right] \right\}$$

For low k, the scattering amplitudes $f_{LJT}^{\pi N}$ behave like k^{2L} (compare eq. 2.7.6) so if we replace the factors $\vec{k}$ and $\vec{k}'$ by momentum operators, their coefficients will remain finite for low energies. Thus we find an effective local pion-nucleon interaction analogous to the Skyrme force (4.2.29), which in the relative coordinate $\vec{r}$ of the πN system has the form

$$\Im_E^{\pi N}(\vec{r}) = \sum_{T=\frac{1}{2}}^{3/2} \Pi_T \left\{ \Im_{0\frac{1}{2}T}^{\pi N}(E)\delta(\vec{r}) - \left[\Im_{1\frac{1}{2}T}^{\pi N}(E) + 2\Im_{1\frac{3}{2}T}^{\pi N}(E) \right] \vec{\nabla} \cdot \delta(\vec{r})\vec{\nabla} \right. \qquad (4.2.36)$$

$$+ \frac{2i}{\hbar} \left[\Im_{1\frac{1}{2}T}^{\pi N}(E) - \Im_{1\frac{3}{2}T}^{\pi N}(E) \right] \vec{s} \cdot \vec{\nabla} \times \delta(\vec{r})\vec{\nabla} \Bigg\}$$

where the coefficients

$$\Im_{LJT}^{\pi N}(E) = -\frac{2\pi(\hbar c)^2}{k^{2L}\varepsilon_\pi(\vec{k})} f_{LJT}^{\pi N}(E) \qquad (4.2.37)$$

are analogous to the Skyrme parameters in the effective-range expansion of the two-nucleon force. We expect them to have a weak energy dependence for low energies, but depend markedly on E in the neighborhood of the Δ resonance.

The pions' mean field $V_{\pi A}$ is given by an expression analogous to eq. (4.2.25) but without the exchange term (because pions are distinguishable from nucleons!):

$$\langle \phi_\alpha | V_{\pi A} | \phi_\beta \rangle_{MF} = \int d^3\vec{r}\phi_\alpha^+(\vec{r}) \sum_{\gamma=1}^A \int d^3\vec{r}_N \Phi_\gamma^+(\vec{r}_N) \Im_E^{\pi N}(\vec{r} - \vec{r}_N) \Phi_\gamma(\vec{r}_N)\phi_\beta(\vec{r})$$

$$(4.2.38)$$

where the Φ_γ are nucleon spinor-isospinors. Using the expression (4.2.36) for $\Im_E^{\pi N}(\vec{r} - \vec{r}_N)$, we obtain after a few partial integrations, and assuming for simplicity that the nucleons' wave functions are independent of spin and isospin,

$$\langle \phi_\alpha | V_{\pi A} | \phi_\beta \rangle_{MF} = \delta_{\alpha\beta} \int d^3\vec{r}\phi_\alpha^+(\vec{r})\{a_0 n(\vec{r})$$

$$+ a_1(\vec{\nabla}^2 n(\vec{r}) + \tau(\vec{r}) + \vec{\nabla} \cdot n(\vec{r})\vec{\nabla})\}\phi_\alpha(\vec{r}) \quad (4.2.39)$$

where $n(\vec{r})$ is the density of the nucleons and $\tau(\vec{r})$ is their kinetic energy density, eq. (4.2.31). The constants a_0 and a_1 are

$$a_0 = \frac{2}{3}\Im_{0\frac{1}{2}\frac{3}{2}}^{\pi N}(E) + \frac{1}{3}\Im_{0\frac{1}{2}\frac{1}{2}}^{\pi N}(E) \qquad (4.2.40)$$

$$a_1 = \frac{2}{3}\Im_{1\frac{3}{2}\frac{3}{2}}^{\pi N}(E) + \frac{1}{3}\Im_{1\frac{3}{2}\frac{1}{2}}^{\pi N}(E) + \frac{1}{3}\Im_{1\frac{1}{2}\frac{3}{2}}^{\pi N}(E) + \frac{1}{6}\Im_{1\frac{1}{2}\frac{1}{2}}^{\pi N}(E);$$

they approach, for $k \to 0$, the average s- and p-wave scattering lengths respectively. Like the Skyrme parameters, we expect the nuclear matter to modify them compared to their free-space values for πN scattering. We might therefore expect them to depend on the nuclear density.

So far we have ignored the fact that the pion becomes relativistic at a modest energy E. This implies that we have to use the relativistic wave equation (2.8.25). We first look for stationary solutions of the form

$$\phi_\alpha(\vec{r}, t) = \phi_\alpha^{\text{Static}}(\vec{r}) + \phi_\alpha(\vec{r})e^{i\varepsilon_\pi t/\hbar} \qquad (4.2.41)$$

where $\phi_\alpha^{\text{Static}}(\vec{r})$ was discussed in Sect. 2.8. This leads to the time-independent wave equation for $\phi_\alpha(\vec{r})$

$$(-(\hbar c)^2\vec{\nabla}^2 + m_\pi^2 c^4)\phi_\alpha(\vec{r}) = \varepsilon_\pi^2\phi_\alpha(\vec{r}).$$

The conventional way to introduce the mean field is to assume that it is the zeroeth component of a 4-vector, as would be the case for the Coulomb potential $t_3 e\phi(\vec{r})$ which we also need to include. We thus make the replacement $\varepsilon_\pi \rightarrow (\varepsilon_\pi - V_{\pi A} - t_3 e\phi(\vec{r}))$; neglecting terms of second order in $(V_{\pi A} + e\phi(\vec{r}))/\varepsilon_\pi$, we arrive at the wave equation

$$\left[-k^2 - \vec{\nabla}^2 + a_0 n(\vec{r}) + a_1(\vec{\nabla}^2 n(\vec{r}) + \tau(\vec{r}) + \vec{\nabla} \cdot n(\vec{r})\vec{\nabla})\right]\phi_\alpha(\vec{r}) = 0 \qquad (4.2.42)$$

where $\hbar k$ is the momentum associated with the pion's kinetic energy $\varepsilon_\pi - m_\pi c^2$, and where the derivative operators in the last term on the right-hand side operate also on the pion wave function. This form of the pions' optical potential is known as the **Kisslinger potential**.

A simplified form of the Kisslinger potential, without the terms proportional to $\vec{\nabla}^2 n(\vec{r})$ and $\tau(\vec{r})$, was used to compute the theoretical angular distributions shown

$\varepsilon_\pi - m_\pi c^2$ (MeV)	a_0 (fm)	a_1 (fm^3)
30	$-1.16 + 0.82i$	$9.10 + 0.95i$
80	$-1.03 + 0.56i$	$9.00 + 3.60i$
120	$-0.87 + 0.42i$	$7.34 + 7.15i$
200	$-0.54 + 0.33i$	$-0.83 + 6.40i$
280	$-0.27 + 0.31i$	$-2.18 + 2.53i$

Table 4.2 Parameters of Kisslinger potential, eq. (4.2.42), for various pion kinetic energies. These values were used to give the computed angular distributions shown in fig. 3.5 [after M. M. Sternheim and R. R. Silbar, "Meson-Nucleus Scattering at Intermediate Energies", in Annual Review of Nuclear and Particle Science, Vol. 24, p. 249 (Academic Press, New York, 1974).]

in fig. 3.5. The parameters a_0 and a_1 were estimated from pion-nucleon scattering data, with an attempt to include the Fermi motion by an averaging procedure leading to the values in table 4.2. The free-space s-wave scattering lengths, eq. (2.7.8), give a very small value of $a_0 = -0.02$fm ≈ 0; it is not surprising that the estimated values of a_0 are small. The free-space value of a_1 would be $(6.5 + 1.8i)$fm^3 at a pion kinetic energy of 80 MeV. This implies that, at nuclear-matter density, the last term in eq. (4.2.42) cancels the pion's kinetic-energy term $-\vec{\nabla}^2$. With the estimated values of a_1, which are even larger, the pion's energy in nuclear matter would be a **decreasing** function of its momentum. This unreasonable behavior is known as the **Kisslinger anomaly**. It shows that a naive mean-field approach to pion-nucleus scattering leads to an unreasonable conclusion: pions of sufficiently high momentum would be arbitrarily deeply bound in nuclear matter!

Clearly the mean-field picture of pions in nuclei is unsatisfactory. A fully satisfactory account has yet to be made. The Kisslinger anomaly has been attributed to the way that the pion-nucleon coupling is inserted into the wave equation (4.2.42). The **delta-hole model** [see Oset, Toki and Weise] is an especially appealing way to avoid the difficulties. This model explicitly mixes wave-function components containing the pion with the nucleus in its ground state, described by ϕ_α above, with components of the same isospin, angular momentum, and parity in which there is no pion, but instead one of the nucleons is excited into a delta resonance. The eigenstates of energy, which determine the scattering, are constructed as superpositions of both kinds of states. Coupled equations for ϕ_α and the isobar-nucleus wave functions are obtained and solved with various approximations. The resulting picture of pions in nuclei is very far from the simple mean-field picture which so successfully describes the motion of nucleons in nuclei.

4.3   THE LIQUID-DROP MODEL

The picture of a large nucleus as a uniform region of nuclear matter, surrounded by a surface where the density falls rapidly, suggests a model for its binding energy

$$B \approx B_V V - B_S S - E_C \qquad (4.3.1)$$

where V is the volume of the nucleus, S is its surface area, and E_C is the Coulomb energy. The effective-range local-density approximation provides a microscopic justification for this view. The coefficient B_V, for example, would just be the first three terms of eq. (4.2.33), evaluated at the saturation density. The coefficient B_S represents the loss of energy due to formation of a surface. It is partly due to the range of the force, via the $|\vec{\nabla}n|^2$ term in (4.2.33), and partly to the fact that nucleons in the surface don't have as many other nucleons around to attract them, leading to a lower uniform-matter energy density.

If the densities of neutrons and protons are unequal then B_V and B_S will be modified. B_V will be decreased because the isospin unsymmetric nuclear matter has a higher energy than symmetric matter, for two reasons:

(1) To turn a proton into a neutron, a single-particle proton state with less than the Fermi energy μ has to be replaced by a neutron with energy greater than μ, on account of the Pauli principle; this costs kinetic energy, see Sect. 7.3.

(2) As explained in Chapter 3, unlike nucleons attract each other more than like nucleons, due to the differences in relative s-wave forces and wave functions. The number of unlike pairs is maximized when N=Z. To take these into account, the volume energy B_V has to contain a term proportional to $(N-Z)^2$ (it has an extremum when N=Z, so the first power is absent). Similar considerations lead to a corresponding correction to the surface energy.

If we assume that nuclei are uniform spheres, then V is proportional to A, S is proportional to $A^{2/3}$, and E_C to $Z^2/A^{1/3}$. Empirically, it is also found that nuclei with even numbers of neutrons or protons are more tightly bound than those with odd numbers. The Bethe-von Weiszacker liquid-drop mass formula, [see Bethe and Bacher], takes all these effects into account:

$$B_{LD}(N, Z) = b_V A \left[1 - K_V \left(\frac{N-Z}{A} \right)^2 \right] - b_S A^{2/3} \left[1 - K_S \left(\frac{N-Z}{A} \right)^2 \right]$$

$$-b_C \frac{Z^2}{A^{1/3}} + \delta[(-1)^N + (-1)^Z]A^{-1/2}. \tag{4.3.2}$$

It gives a remarkably good fit to the trends of nuclear binding energies for all nuclei with A > 12, see fig. 4.1. The constants used in the figure are

$$
\begin{aligned}
b_V &= 15.6\,\text{MeV}, \qquad b_S = 17.2\,\text{MeV} \\
b_C &= 0.70\,\text{MeV} \\
K_V &= 1.50, \qquad K_S = 0 \\
\delta &= 6\,\text{MeV}
\end{aligned}
\tag{4.3.3}
$$

The departures from the liquid-drop binding are mainly traced to the effects of closed shells near magic numbers of neutrons and protons. The combined effect of surface energy, largest for small A, and of symmetry energy ($\sim K_V$) and Coulomb energy, largest for large A, is to keep the average binding energy B/A remarkably constant throughout the periodic table. Its average value, 8 MeV, is about half the binding energy b_V of nuclear matter in the absence of Coulomb, surface, and symmetry energy.

4.4   THE OPTICAL MODEL BEYOND THE MEAN FIELD

We return now to the Optical Model, which provides the first motivation for the mean-field picture of nuclear structure. Can the Brueckner-Hartree-Fock theory give a quantitative explanation of the observed optical potentials? We consider first the imaginary and then the real potential.

4.4 A. The Imaginary Potential

We left Chapter 3 with a major riddle: the optical model predicts a much longer mean free path for nucleons in nuclei than would be expected from the 2-nucleon scattering cross sections. The reason for this is now apparent: it is the Pauli exclusion principle, due to the antisymmetrization of the many-body wave functions. The Pauli principle reduces the scattering cross sections in nuclear matter in two ways.

The most direct effect is called "Pauli blocking": the final states into which the nucleon may scatter are drastically limited in nuclear matter, because many of them are already occupied by other nucleons. The smaller the momentum and energy of the scattering nucleons, the more important this restriction becomes. In fact, as the nucleons' energies approach the Fermi energy μ, the number of available final states vanishes. This is the reason for the strong energy dependence of the imaginary optical potential.

Another, less direct Pauli effect also decreases the scattering cross sections in nuclear matter. Recall that the Pauli effect in the intermediate states of the Brueckner equation has the effect of emphasizing the repulsion and de-emphasizing the attraction, since the relative motion of the colliding nucleons can't accommodate the interactions as effectively in the presence of the other nucleons as in free space. The low-energy scattering of free nucleons is dominated by the attractive interactions; in the nucleus these are reduced, while the repulsive part of the interaction is felt more strongly and helps to cancel the attraction. The result is that the matrix elements of $\Im$ are much smaller than in free space. Once again, the scattering cross sections are reduced.

In fact, the combination of these effects results in such a strong inhibition of the scattering that the mean free paths are actually longer in the nuclear interior than in the surface. This effect is seen quantitatively in Brueckner computations of nuclear matter [Jeukenne, Lejeune and Mahaux], where the mean free path, indeed, attains a minimum value around half of the saturation density (fig. 4.9).

While the nuclear-matter picture could explain why the absorption is greater in the nuclear surface, it doesn't explain the observation (compare eq. 3.3.21b) that the extra surface absorption decreases with energy: in nuclear matter, the absorption increases with energy. The main part of the surface absorption has been traced instead due to the fact that a nucleon in the nuclear surface changes the shape of the mean field by pulling the other nucleons toward it. The shape of the nucleus cannot respond instantaneously to the extra nucleon (see Chapter 8) but changes in a characteristic time governed by how long it takes a nucleon with the Fermi energy to cross the nucleus. When the perturbing nucleon is moving at a velocity comparable to the Fermi velocity p_F/m_N, it can lose energy by exciting vibrational modes of the nuclear shape. A much faster nucleon loses less energy this way, because it moves on before the nucleus has time to change its shape.

4.4 B.   The Real Optical Potential

The phenomenological optical potential has a fairly strong energy dependence: the attraction decreases with increasing energy. This result is also found in the Brueckner-Hartree-Fock theory of nuclear matter, but the reason for it is different than the reason for the energy dependence of the imaginary potential. In fact, it is due to the effective range of the force. Qualitatively, this is easy to understand: as the momentum of the particle in the state $|\alpha\rangle$ (eq. (4.2.25)) increases, its momentum relative to the nucleons in the occupied states $|\gamma\rangle$ also increases, on the average. Thus the effective-range corrections reduce the $\mathfrak{S}$-matrix elements. In other words, the effect is not an energy dependence, but a momentum dependence; we say that the optical potential is non-local. Since the dependence is roughly proportional to the square of the momentum (and thus to the energy), one often says that a nucleon in nuclear matter has an "effective mass" which is smaller than in free space: its

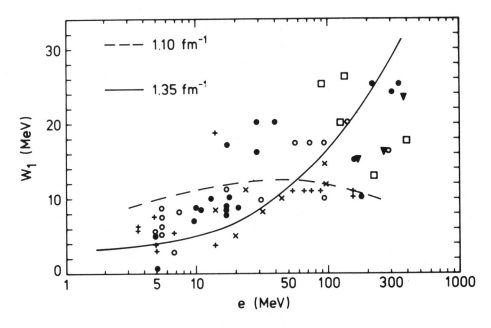

Figure 4.9   Dependence of $W_1(\varepsilon)$ (eqs. 3.3.19-21) on the single-particle energy ε, for $k_F = 1.35$ fm^{-1} (full curve) and $k_F = 1.10$ fm^{-1} (long dashes), respectively. The various points are taken from a compilation of empirical values. [*Jeukenne, Lejeune and Mahaux*].

energy and momentum are related by

$$\varepsilon(k) \approx -U_0 + U'k^2 + \frac{\hbar^2}{2m_N}k^2 = -U_0 + \frac{\hbar^2}{2m^*}k^2 \tag{4.4.1}$$

with $m^*/m_N \approx 0.7$ to 0.8. This is the origin of the energy-dependent term in the phenomenological optical model, eq. (3.3.21c): it is chosen to give the same relation between energy and wavelength as eq. (4.4.1).

The hardest part of the optical model to explain in Brueckner-Hartree-Fock theory is actually the spectrum of bound states near the Fermi energy μ. As we remarked in sect. 4.2.D, these states are observed closer together than predicted by Brueckner-Hartree-Fock theory; they can be fitted by an energy-independent potential with $m^* = m_N$. The increase of the effective mass near the Fermi surface has been understood by a detailed study of the rearrangement effects, which go beyond the mean-field theory. Like the surface absorption, this effective-mass enhancement is due to the nucleon's disturbing the density distribution of the other nucleons, which it does most effectively when its velocity is close to the Fermi velocity. The resulting added attraction approximately compensates for the momentum-dependent repulsion for nucleons moving with roughly the Fermi energy.

4.5 TRANSLATIONAL SYMMETRY AND THE LIMITATIONS OF THE MEAN FIELD

In describing the many fundamental successes of the self-consistent mean-field picture of nuclei, we have overlooked an equally deep and fundamental failing: its wave functions are not eigenstates of momentum. Since they are localized in space, they are wave packets containing many momenta. This is true, not only for the single-particle factors $\phi_i(\vec{r}_j)$, but also for the wave function $\psi(\vec{r}_1, \ldots \vec{r}_A)$. This is a serious drawback, because the fact that the Hamiltonian commutes with the total momentum operator $\vec{P} = \sum_i \vec{p}_i$ implies that the exact eigenstates of the Hamiltonian are to be found among the eigenstates of $\vec{P}$.

4.5 A. Translational Degeneracy in the Mean-Field Picture

The source of the trouble is not hard to identify: the mean fields $U^D(\vec{r})$ and $U^X(\vec{r}, \vec{r}')$ do not commute with $\vec{P}$, because they are not translationally invariant. This defect cannot be remedied by an improved treatment of two-body correlations: it lies at the conceptual heart of the mean-field picture. We say that the mean-field approximation "breaks" the translational symmetry of the Hamiltonian; the mean-field approximation to the ground state is a state of **broken symmetry**.

Of course, the mean-field wave function is only an approximation to the true ground state. It has a low energy because it keeps all the particles near each other, where they can enjoy the attractive interactions. The choice of a particular place for this cluster to form was quite arbitrary; what was important was the mutual proximity of the particles. Thus, given one solution, $\psi_0(\vec{r}_1, \ldots, \vec{r}_A)$ with its corresponding potentials $U_0^D(\vec{r})$ and $U_0^X(\vec{r}, \vec{r}')$, we can find a whole family of solutions, $\psi_{\vec{R}}$, each displaced by an arbitrary displacement $\vec{R}$ from the original one:

$$\psi_{\vec{R}}(\vec{r}_1, \vec{r}_2, \ldots, \vec{r}_A) = \psi_0(\vec{r}_1 - \vec{R}, \vec{r}_2 - \vec{R}, \ldots, \vec{r}_A - \vec{R}) \qquad (4.5.1)$$

with corresponding mean fields

$$U_{\vec{R}}^D(\vec{r}) = U_0^D(\vec{r} - \vec{R}) \qquad (4.5.2a)$$

$$U_{\vec{R}}^X(\vec{r}, \vec{r}') = U_0^X(\vec{r} - \vec{R}, \vec{r}' - \vec{R}). \qquad (4.5.2a)$$

Of course, each solution $\psi_{\vec{R}}$ has the same energy E_0 as the original solution ψ_0.

4.5 B.   Constructing Momentum Eigenstates

We can use this degenerate family of solutions to construct eigenfunctions of the momentum by forming a linear combination of them:

$$\begin{aligned}
\psi_{\vec{k}}(\vec{r}_1, \ldots, \vec{r}_A) &= \int \frac{d^3\vec{R}}{(2\pi)^{3/2}} \, e^{i\vec{k}\cdot\vec{R}} \, \psi_{\vec{R}}(\vec{r}_1, \ldots, \vec{r}_A) \\
&= \int \frac{d^3\vec{R}}{(2\pi)^{3/2}} \, e^{i\vec{k}\cdot\vec{R}} \, \psi_0(\vec{r}_1 - \vec{R}, \ldots, \vec{r}_A - \vec{R})
\end{aligned} \qquad (4.5.3)$$

It is easy to show that $\psi_{\vec{k}}$ is an eigenfunction of the total momentum $\vec{P} = \sum_i \vec{p}_i$ and the eigenvalue of $\vec{P}$ is $\hbar\vec{k}$ (problem 4.10).

Peierls and Yoccoz, who proposed the form (4.5.3), also pointed out an unsatisfactory property of $\psi_{\vec{k}}$: its energy,

$$E_k = \frac{\langle \psi_{\vec{k}} | H | \psi_{\vec{k}} \rangle}{\langle \psi_{\vec{k}} | \psi_{\vec{k}} \rangle}. \qquad (4.5.4)$$

While the energy depends on k, its dependence on k differs from the correct result

$$E(k) = \text{constant} + \frac{\hbar^2 k^2}{2M_A} \qquad (4.5.5)$$

where $M_A = Am_N$ is the total mass of the system (non-relativistically). (Eq. (4.5.5) follows from the fact that H separates into the sum of two terms, the translational kinetic energy $P^2/2M_A$ plus a part, including the interactions, which only depends on the relative motion of the nucleons.) Even for small k, where the approximation based on a stationary well ought to be reasonable, the term in the Taylor series of E_k in k,

$$E_k = \text{constant} + \alpha k^2 \tag{4.5.6}$$

has an incorrect value $\alpha \neq \hbar^2/2M_A$ (except for the case of a collection of coupled harmonic oscillators). Evidently there is something fundamentally wrong with the wave functions of the stationary well, even beyond their breaking of translational symmetry.

4.5 C.   Restoring Galilean Invariance

Indeed, there is another exact symmetry of non-relativistic mechanics, besides translational symmetry, which is also broken by the stationary well: Galilean relativity, the fact that an observer can't tell how fast he's travelling. We say that the mean-field approximation **breaks Galilean invariance as well as translational invariance**.

The key to restoring the broken translational symmetry was the construction of a set of degenerate wave functions $\psi_{\vec{R}}$. Similarly, following Peierls and Thouless, we can restore the Galilean symmetry by making a linear superposition of the set of wave functions

$$\psi_{\vec{k},\vec{k}'}(r_1,\ldots,r_A) = e^{i(\vec{k}-\vec{k}')\cdot\vec{r}_{cm}}\psi_{\vec{k}'}(r_1,\ldots,r_A) \tag{4.5.7}$$

where

$$\vec{r}_{cm} = \sum_i \frac{\vec{r}_i}{A}. \tag{4.5.7a}$$

Recalling that the factor $e^{i\vec{q}\cdot\vec{r}_{cm}}$ has the effect of boosting the momentum by an amount $\hbar\vec{q}$, we observe that every $\psi_{\vec{k},\vec{k}'}$ is an eigenstate of the total momentum $\vec{P}$ with eigenvalue $\hbar\vec{k}$, for each value of $\vec{k}'$. Thus every linear combination

$$\Psi_{\vec{k}}(\vec{r}_1,\ldots,\vec{r}_A) = \int \frac{d^3\vec{k}'}{(2\pi)^{3/2}} w_{\vec{k}}(\vec{k}')\psi_{\vec{k},\vec{k}'} \tag{4.5.8}$$

$$= e^{i\vec{k}\cdot\vec{r}_{cm}}\int \frac{d^3\vec{k}'d^3\vec{r}'}{(2\pi)^3} w_{\vec{k}}(\vec{k}')e^{i\vec{k}'\cdot(\vec{r}'-\vec{r}_{cm})}\psi_0(\vec{r}_1-\vec{r}',\ldots,\vec{r}_A-\vec{r}').$$

will be an eigenfunction of $\vec{P}$ with eigenvalue $\hbar\vec{k}$. To satisfy Galilean invariance, we have to choose the weight function $w_{\vec{k}}(\vec{k}') = w(\vec{k}')$ the same for all $\vec{k}$. Then

$$\Psi_{\vec{k}}(\vec{r}_1,\ldots,\vec{r}_A) = e^{i\vec{k}\cdot\vec{r}_{cm}}\Psi_w(\{\vec{r}_{ij}\}), \tag{4.5.9}$$

where $\vec{r}_{ij} = \vec{r}_i - \vec{r}_j$ and

$$\Psi_w(\{\vec{r}_{ij}\}) \equiv \int d^3\vec{r}\,\tilde{w}(\vec{r})\psi_0(\vec{r}_1-\vec{r}_{cm}-\vec{r},\ldots,\vec{r}_A-\vec{r}_{cm}-\vec{r}) \tag{4.5.9a}$$

$$\tilde{w}(\vec{r}) \equiv (2\pi)^{-3}\int d^3\vec{k}'e^{i\vec{k}'\cdot\vec{r}}w(\vec{k}'). \tag{4.5.9b}$$

(To derive eq. (4.5.9) substitute $\vec{r} = \vec{r}' - \vec{r}_{cm}$ in eq. (4.5.8), then note that $\vec{r}_i - \vec{r}_{cm} = \sum_j \vec{r}_{ij}/A$). Eq. (4.5.9) shows the Galilean invariance, because now when we boost $\Psi_{\vec{k}}$ by $e^{i\vec{q}\cdot\vec{r}_{cm}}$, we obtain $\Psi_{\vec{k}+\vec{q}}$ as we should. The translational invariance is also apparent because Ψ_w depends only on relative coordinates. This ensures that a translation of the origin would only multiply $\Psi_{\vec{k}}$ by an overall unobservable phase.

An especially simple choice of the weighting function $w(\vec{k}')$ would be to choose a constant, $w(\vec{k}) = 1$. For this choice, the wave function $\Psi_{\vec{k}}$ has the transparently simple form $e^{i\vec{k}\cdot\vec{r}_{cm}}\psi_0(\vec{r}_1 - \vec{r}_{cm}, \ldots, \vec{r}_A - \vec{r}_{cm})$. A better choice of w could be found by minimizing the energy, which now has the correct behavior:

$$E(k) = \frac{\langle \Psi_{\vec{k}}|H|\Psi_{\vec{k}}\rangle}{\langle \Psi_{\vec{k}}|\Psi_{\vec{k}}\rangle}$$
$$= \frac{\hbar^2 k^2}{2M_A} + \frac{\langle \Psi_{\vec{k}=0}|H|\Psi_{\vec{k}=0}\rangle}{\langle \Psi_{\vec{k}=0}|\Psi_{\vec{k}=0}\rangle}, \tag{4.5.10}$$

since the total momentum operator $\vec{P} = -i\hbar\vec{\nabla}_{r_{cm}}$ only affects the first factor in eq. (4.5.9). By studying the equations obtained from requiring that $E(\vec{k} = 0)$ be a minimum with respect to variations of w, Peierls and Thouless concluded that there is a unique w which minimizes E.

4.5D.   Broken Symmetry as an Artifact of the Mean-Field Approximation

In summary, we have found that the mean-field approximation leads to a degenerate set of states which breaks the translational and Galilean symmetries of the Hamiltonian. The mean-field ground state's broken symmetries are, like its degeneracy, an artifact of the mean-field approximation. A set of states with the symmetries restored can be found by making appropriate linear combinations. Among these linear combinations is a unique state of lowest energy, whose energy is lower than the mean-field states: the removal of the broken symmetries is accompanied by the removal of the corresponding degeneracies. Instead of the degenerate set of mean-field states, the linear combinations have a spectrum of excitations corresponding to collective motion of the degree of freedom whose symmetry was broken: the whole nucleus moves with a momentum $\vec{P} = \hbar\vec{k}$ and a corresponding kinetic energy $P^2/2M_A$.

PROBLEMS

4.1   Use a harmonic-oscillator potential $\hat{U}(r)$ in eq. (3.3.17):

$$\text{Re } H^{POM} = \frac{p^2}{2m_N} + \left(1 + U_{\ell s}^{HO} \, \vec{\ell} \cdot \vec{s} \frac{1}{r} \frac{\partial}{\partial r}\right) \hat{U}(r)$$

$$\hat{U}(r) = \frac{1}{2} m_N \Omega^2 r^2.$$

How many of the observed magic numbers can you reproduced by choosing $U_{\ell s}^{HO}$ properly? Estimate its value. Hint: make a graph of the eigenvalues (in units of $\hbar\Omega$) as a function of $U_{\ell s}^{HO}$.

4.2   Use the top and bottom figures in fig. 4.4 to estimate the single-particle energies for each of the states shown in the figure (neglect the peaks marked C, which are due to contaminant elements in the target). Do you find the same energies in both cases? If not, why not?

4.3   A beam of 20 MeV protons is shot at a target of ^{209}Pb, and the spectrum of emerging deuterons is detected at $\theta_{lab} = 34°$. This is called a pickup reaction.
  (a) Write the standard notation for the reaction, analogous to eq. (4.1.9).
  (b) The deuteron spectrum has a series of peaks, similar to those in fig. 4.4. Use the information in fig. 4.4 to predict the energy of the most-energetic peak.
  (c) Use information from fig. 4.2 (together with fig. 4.4) to estimate the energy of the next-most-energetic peak.

4.4   At a heavy-ion accelerator, a beam of 600 MeV nuclei of ^{208}Pb is scattered from a target of ^{209}Pb, and ^{209}Pb emerges in the forward direction ($\theta \approx 0$). Using the figures in this chapter, predict the energies of as many peaks in the cross section as you can.

4.5   Find the scattering lengths and effective ranges for the Skyrme interaction (4.2.29), for singlet and triplet states. Hint: first find the $\mathfrak{S}$-matrix elements between plane waves.

4.6   For a neutron in a large nucleus ($A \approx 100$), what is the average value of
  (a) the kinetic energy?
  (b) the potential energy?
  (c) the binding energy?
Reconcile your answers to parts (a), (b), and (c).

4.7   The equilibrium density of nuclear matter is $n_0 = 0.17$ fm^{-3}. In a collision of two heavy nuclei, matter is compressed to a density of 1.1 n_0. Estimate the change in

(a) the internal kinetic energy per nucleon of the matter.
(b) the potential energy per nucleon of the matter.

4.8   Find the ratio $(N-Z)/A$ which minimizes the energy in the liquid drop, neglecting pairing;
(a) for a given value of of A
(b) for a given value of of Z.

What is the most favorable isotope of calcium? of lead?

4.9   Give the most important reason why the imaginary part of the optical potential depends on energy. At what energy **must** it vanish?

4.10   Show that $\psi_{\vec{K}}$ (eq. (4.5.3)) is an eigenfuction of the total momentum $\vec{P}$ with eigenvalue $\hbar\vec{k}$.

4.11   Derive effective mass $m^*(\vec{r})$ and mean potential $U(\vec{r})$ from the interaction eq. (4.2.29). Use the variational principle in eq. (4.2.10) and the definitions

$$\left[-\vec{\nabla}\frac{\hbar^2}{2m*(\vec{r})}\vec{\nabla} + U(\vec{r})\right]\phi_i(\vec{r}) = \varepsilon_i\phi_i(\vec{r})$$

and density-independent parameters except $\Im_0$ replaced by $\Im_0 \to \Im_0 + \Im_3\frac{1}{6}n^\alpha$.

4.12   Assume all parameters in eq. (4.2.29) are density-independent except $\Im_0$ which is replaced by $\Im_0 \to \Im_0 + \frac{1}{6}\Im_3 n^\alpha$. Calculate the energy per nucleon E/A for nuclear matter. Determine constraints on the parameters from the requirements of
(i) $E/A = -15.6$ MeV;
(ii) $\partial(E/A)/\partial n = 0$;
(iii) $\partial^2(E/A)/\partial n^2 = \frac{1}{9}\cdot 200$ MeV/n^2 with the equilibrium density $n = 0.155$ nucleons/fm^3.
Use the effective mass expression (found in problem 4.11)

$$\frac{m_N}{m^*} = 1 + \frac{m_N}{\hbar^2}\frac{n}{8}\cdot(3\Im_1 + \Im_2(5 + 4\chi_2))$$

and the additional constraint $\frac{m^*}{m_N} = 0.75$ to obtain numerical values for the various parameter combinations.

4.13   Calculate the energy density (for nuclear matter) from the interaction in eq. (4.2.29) when the neutron and proton densities differ but the spin-up and spin-down densities are identical.

5

Static Deformations

In the previous chapters we have seen how an effective interaction between the nucleons in the nucleus arises from the nucleon-nucleon interaction. We have seen how the nucleus' Schrödinger equation with this interaction can be solved approximately by using the Hartree-Fock procedure. Its basic assumption is that the wave function is a Slater determinant of minimum energy.

Each nucleon moves in the one-body field created by all the other nucleons. This mean field is the same for all the various nucleons. It determines the motion of the nucleons which in turn determines the field. Thus the mean field and the Hartree-Fock solution are self-consistent. The density distribution for the nucleons therefore must resemble the mean field, and vice versa.

We have seen how the mean-field approximation breaks translational and Galilean invariance, giving a degenerate set of wave functions with different center-of-mass coordinates. We have also seen that these symmetries can be restored by constructing superpositions of the degenerate mean-field states representing collective translational motion, and that these superpositions are no longer degenerate but have a spectrum typical of the corresponding collective motion.

In this chapter, we shall learn that the mean-field picture can also lead to the breaking of rotational symmetry. We will explore this phenomenon within the mean-field picture, but will wait until Chapter 9 to study the restoration of the rotational symmetry and the corresponding collective motion.

5.1 THE INDEPENDENT-PARTICLE MODEL AS AN APPROXIMATION TO THE MEAN FIELD

The independent-particle model (ipm) may be viewed as an approximation to mean-field theory; the wave functions of the two models can be the same if the one-body field of the ipm is chosen to replicate the mean field, but the energies have to be corrected for the ipm's overcounting of the potential energy. A good approximation to the mean-field wave functions can be found by parameterizing the one-body field with just a few well-chosen variables, such as its depth, radius, and surface thickness. If these variables are choosen to minimize the mean value of the Hamiltonian in the resulting wavefunction, then the ipm's wave functions can be a very good approximation to those of mean-field theory.

In this chapter, we will study the mean-field theory using the independent-particle model with parametrized one-body fields. Solving for the single-particle wave functions in these fields, we can form Slater determinants to approximate the mean-field wave functions. The energies of these wave functions, evaluated using the appropriate two-body effective interactions, may be very close to the energies of the mean-field wave functions; however, the wave functions may not be fully consistent with the one-body potentials, since nothing has been done to guarantee that the one-body fields are indeed those that would arise from the nucleons' motion via the effective interactions. Instead, we can impose constraints to insist that the fields be approximately consistent with the resulting densities.

The parametrized independent-particle model has advantages and disadvantages when viewed as an approximation to mean-field theory. The self-consistency and minimum-energy requirements are at best only approximately fulfilled, and the direct connection to the (only approximately known) nuclear force is lost. On the other hand, we are much closer to experimental quantities, due to the significantly simpler calculations. The independent-particle model can in this way be viewed as a phenomenological intermediate step. It is adjusted to reproduce observations and subsequently used as the result required from a Hartree-Fock calculation, thereby giving information about the effective nuclear force.

5.2   SELF-CONSISTENCY IN THE SPHERICAL INDEPENDENT-PARTICLE MODEL

The saturation property of the nuclear force and the self-consistency condition tell us that the average potential (mean field) is flat and negative inside the nucleus and increases to zero outside the nuclear radius of $r_0 A^{1/3}$. The simplest, most popular and in fact very successful potential with these properties is the Woods-Saxon potential. The average potential, including the spin-orbit part and excluding the Coulomb potential for protons, is then approximated by

$$\hat{U}(r) = \left(U_0 + \vec{\ell} \cdot \vec{s} \, U_0^{\ell s} \frac{1}{r} \frac{d}{dr} \right) \frac{1}{1 + e^{(r-R)/a}} \qquad (5.2.1)$$

where $\vec{\ell}$ is the angular momentum and $\vec{s}$ the spin. The parameters can vary a little but they are always fairly similar to the real part of the Phenomenological Optical Model (eqs. (3.3.19, 20)). We use the hat to remind us that $\hat{U}$ is only an approximate parametrization of the self-consistent field.

The Schrödinger equation corresponding to eq. (5.2.1) cannot be solved analytically. This does not present any problem since numerical solutions can be found easily by using a computer. However, historically and pedagogically, simpler potentials are very useful. The harmonic oscillator, being especially simple, is especially useful.

At first sight the harmonic potential seems to be an incredibly bad approximation because of its infinitely large values at large distances, where the potential should vanish. This objection is devastating when the unbound continuum or scattering states are involved. On the other hand, when we confine our interest to radial distances smaller than about the nuclear radius, e.g. for nuclear structure studies, the harmonic-oscillator potential behaves reasonably.

The Woods-Saxon potential is therefore replaced by a parabola of curvature $m\Omega^2$ and with a minimum U_c at the center. The potential in eq. (5.2.1) then becomes

$$\hat{U}(r) = U_c + \tfrac{1}{2} m_N \Omega^2 r^2 + C\vec{\ell} \cdot \vec{s} \qquad (5.2.2)$$

where the spin-orbit strength C is given by

$$C = m_N \Omega^2 U_{\ell s}^{HO}. \qquad (5.2.3)$$

Ω is the frequency of the nucleon in the central part of the potential.

For eq. (5.2.2) to approximate eq. (5.2.1), we clearly must adjust parameters U_c, Ω and $U_{\ell s}^{HO}$. As the Woods-Saxon potential is flat at the center while the harmonic oscillator has a minimum, U_c must be somewhat smaller than U_0. Since U_c only translates the potential up and down by a constant, it enters the solutions only as a constant shift in the eigenvalue, and is quite uninteresting. The size of Ω can be determined by the requirement that the root mean square radius, on average over all nuclei, equals the observed value, i.e.

$$\langle r^2 \rangle = \frac{\sum_{i\ occ} \langle i | r^2 | i \rangle}{\sum_{i\ occ} 1} = \frac{3}{5}(1.2\,\text{fm}\ A^{1/3})^2 \qquad (5.2.4)$$

where the summations are over all occupied states i and the factor $\frac{3}{5}$ comes about by assuming a uniform density distribution. The degeneracy g(N) of an oscillator shell with principal quantum number N is

$$g(N) = 2(N+1)(N+2) \qquad (5.2.5)$$

when both neutrons and protons occupy the same states. The total number A of nucleons in the shells below $N=N_F$ (the Fermi level) is then

$$A = \sum_{N=0}^{N_F} g(N) = \frac{2}{3}(N_F + 1)(N_F + 2)(N_F + 3). \qquad (5.2.6)$$

Using the fact that the potential energy for each state is half of the total energy, i.e.

$$\langle N | \tfrac{1}{2} m_N \Omega^2 r^2 | N \rangle = \tfrac{1}{2} \hbar \Omega \left(N + \tfrac{3}{2}\right) \qquad (5.2.7)$$

we find

$$\langle r^2 \rangle = \frac{2\hbar}{Am_N\Omega} \sum_{N=0}^{N_F} (N + \tfrac{3}{2})(N + 1)(N + 2)$$

$$= \frac{1}{2}\frac{\hbar}{Am_N\Omega}(N_F + 1)(N_F + 2)^2(N_F + 3) \qquad (5.2.8)$$

$$= \frac{3}{4}\frac{\hbar}{m_N\Omega}(N_F + 2) \ .$$

and therefore

$$\hbar\Omega \approx \frac{5}{4}\frac{\hbar^2}{m_N(1.2\,\text{fm})^2}\left(\frac{3}{2}\right)^{1/3} A^{-1/3} \approx \frac{41\text{MeV}}{A^{1/3}} \ . \qquad (5.2.9)$$

Assuming that the harmonic oscillator and Woods-Saxon potentials cross each other at r=R we obtain for N=Z the estimate $U_c = -58$ MeV.

The spin-orbit potential $C\vec{\ell}\cdot\vec{s}$ does not change the spatial motion of the nucleons in the spherically symmetric oscillator potential. It merely couples the orbital and spin motion to obtain eigenstates of the total angular momentum. Thus eigenstates of the oscillator model may be written $|N\ell jm\rangle$ where N is the principle quantum number of the orbital motion, $\hbar^2\ell(\ell + 1)$ is the eigenvalue of $\vec{\ell}^2$, $\hbar^2 j(j + 1)$ is the eigenvalue of $\vec{j}^2$, and $m\hbar$ is the eigenvalue of $\vec{j}_z$. The corresponding energies are $\hbar\Omega(N + 3/2) + C/2[j(j + 1) - \ell(\ell + 1) - 3/4]$. A reasonable fit to the observed spin-orbit splitting determines C [Bohr and Mottelson vol.1]. Summarizing:

$$\hbar\Omega = \frac{41\text{MeV}}{A^{1/3}}$$

$$U_c = -58\text{MeV} \qquad (5.2.10)$$

$$C = -\frac{20\text{MeV}}{A^{2/3}}$$

These parameters give a reasonable representation of the single-particle energies and wave functions for bound states of nuclei with $12 \le A \le 60$ nucleons (see fig. 5.1). For heavier nuclei, it is necessary to rely on computers for numerical solutions to the Schrödinger equation in Woods-Saxon potentials. The quantum numbers of the single-particle states in spherical nuclei correspond, however, to those of the oscillator model. j, ℓ, and m retain exactly the same significance for the angular and spin dependence, while the principle quantum number N retains its meaning as the number of nodes in the wave functions,

$$N = 2n_r + \ell \qquad (5.2.11)$$

where n_r is the number of nodes in the radial wave function, counting neither the node at r=∞ nor at r=0. If analytic approximations are needed for heavy nuclei, it is better to base them on the Fermi-gas model of nucleons in a box of constant internal potential, than on the harmonic oscillator which provides a good description of light nuclei.

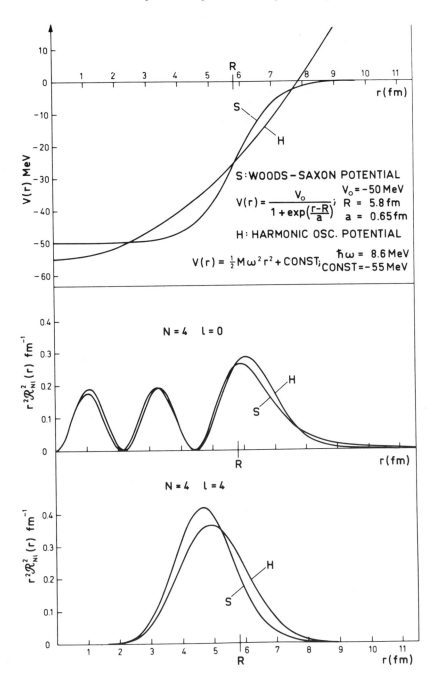

Figure 5.1   The square of the wave function times r^2 for the harmonic oscillator and the Woods-Saxon potential are plotted in units of fm^{-1}. [*from Bohr and Mottelson, vol. 1*]

5.3   THE DEFORMED INDEPENDENT-PARTICLE MODEL

In general there is no reason to believe that the Hartree-Fock mean field should be spherically symmetric. In order to study this possibility, the independent-particle model may be generalized to have a deformed potential. The philosophy is still that the independent-particle model potential is an approximation to the Hartree-Fock mean field.

A deformed Woods-Saxon potential can be constructed from eq. (5.2.1) in several ways. One can for example [Bohr and Mottelson vol.1] assume an angular dependence of the radius

$$R = R(\theta, \phi) = R(A) \left[1 + \sum_{\ell m} \alpha_{\ell m} Y_\ell^m(\theta, \phi) \right]. \qquad (5.3.1)$$

If the deformation parameters $\alpha_{\ell m}$ were allowed to vary completely without constraints, the resulting average nucleon density could depend significantly on deformation. This is not allowed, if we want to approximate calculations with a saturating nuclear force, which leads to almost constant nuclear density for all nuclei.

Therefore we require a constant nuclear volume independent of deformation. Assuming furthermore self-consistency between nuclear density and the average potential, we can instead require volume conservation of the equipotential surface corresponding to the nuclear surface. Usually this surface is approximated by $\hat{U}(\vec{r}) = \frac{1}{2} U_0$ for $|\vec{r}| = R(\theta, \phi)$. The condition would be different, if another equipotential surface had been chosen.

Another problem is that the nuclear center of mass becomes a function of deformation. To fix the position of the center of mass, another constraint on the deformation parameters is needed. In case of axial symmetry, i.e. $\alpha_{\ell m} = \delta_{0m} \alpha_{\ell 0}$, these two constraints can easily be shown to determine α_{00} and α_{10} by

$$\alpha_{00} = -\frac{1}{\sqrt{4\pi}} \sum_{\ell=1} \alpha_{\ell 0}^2 \qquad (5.3.2)$$

$$\alpha_{10} = -\frac{3}{2} \sqrt{\frac{3}{\pi}} \sum_{\ell=2} \frac{(\ell+1)\alpha_{\ell 0}\alpha_{\ell+1,0}}{(2\ell+1)(2\ell+3)}, \qquad (5.3.3)$$

through second order in $\alpha_{\ell 0}$. Eq. (5.3.2) is the volume conservation condition, and eq. (5.3.3) ensures that the center of mass remains at r=0 for all deformations.

As was the case for a spherical potential, it is very instructive to use a harmonic oscillator instead of the Woods-Saxon potential. The obvious generalization of eq. (5.2.2) is to take different frequencies Ω_x, Ω_y and Ω_z in the x, y and z-directions. Ignoring the angular-momentum-dependent terms, we have a very simple volume conservation condition

$$\Omega_x \Omega_y \Omega_z \equiv \Omega_0^3 \approx \frac{(41\text{MeV}/\hbar)^3}{A} \qquad (5.3.4)$$

which in this case is independent of the chosen equipotential surface.

In this simple model we can, following Bohr and Mottelson, study the self-consistency condition in a little more detail. The equipotential surfaces are ellipsoids of half-axes a_x, a_y and a_z inversely proportional to the corresponding frequencies, i.e.

$$a_x \Omega_x = a_y \Omega_y = a_z \Omega_z \tag{5.3.5}$$

The squares of the corresponding variances for the density distribution are given by

$$\begin{aligned}
\langle x^2 \rangle &= -\frac{\sum_{i \text{ occ}} \langle i|x^2|i \rangle}{\sum_{i \text{ occ}} 1} \\
&= \frac{1}{A} \frac{\hbar}{m_N \Omega_x} \sum_{i \text{ occ}} \left[n_x(i) + \tfrac{1}{2} \right] \\
&\equiv \frac{1}{A} \frac{\hbar}{m_N \Omega_x} \Sigma_x,
\end{aligned} \tag{5.3.6}$$

with analogous expressions for y and z. If the density were uniformly distributed inside an ellipsoid, then the variances of (5.3.6) would be proportional to the half-axes of the ellipsoid. If the axes of the potential and density are proportional, as may be expected from the self-consistency requirement, we obtain by combining (5.3.5) and (5.3.6)

$$\Omega_x \Sigma_x = \Omega_y \Sigma_y = \Omega_z \Sigma_z. \tag{5.3.7}$$

Thus there must be a correlation between the frequency in a given direction and the quanta of the occupied states, e.g. large Ω_x (steep oscillator) means small values of the quantum numbers n_x of the occupied orbits.

In the following section, we will discuss the model of a non-spherical oscillator potential without spin-orbit coupling. To make a quantitative model including spin-orbit coupling, we would solve the single-particle wave equation on a computer.

5.4 DEFORMATION ENERGY IN THE OSCILLATOR MODEL

In the previous section we described parametrizations of the deformed mean field. Now we address the question of how the energy of the system depends upon the deformation parameters: how to calculate it, how large is it and where are the stable points?

For a strict independent-particle model the energy is simply the sum of occupied single-particle energies. In our case, the average potential on each nucleon is created from the effect of all the other nucleons. The sum of single-particle energies would then count two-particle interactions twice, three-particle interactions three times, etc. as we discussed in Chapter 4. Assuming only two-body interactions, we can express the total energy in terms of the single-particle energies ε_i and the expectation values in the corresponding states of either the kinetic energy τ_i or potential

U_i operators, i.e.

$$
\begin{aligned}
\langle\psi|H|\psi\rangle = \sum_{i \text{ occ}}(\tau_i + \tfrac{1}{2}U_i) = \tfrac{1}{2}\sum_{i \text{ occ}}(\varepsilon_i + \tau_i) = \tfrac{1}{2}\sum_{i \text{ occ}}(2\varepsilon_i - U_i) \\
= \tfrac{3}{4}\sum_{i \text{ occ}}\varepsilon_i - \tfrac{1}{4}\sum_{i \text{ occ}}(U_i - \tau_i).
\end{aligned}
\tag{5.4.1}
$$

In actual numerical calculations, any of these expressions could easily be used to give the energy. For a harmonic oscillator potential, $\hat{U}_i = \hat{\tau}_i$ and the last expression of eq. (5.4.1) would be simple. The energy would be proportional to, and therefore its stationary points identical with those of, the sum of single-particle energies. For a realistic self-consistent potential, the influence of the $U_i - \tau_i$ term is small for small nuclei; for large nuclei the qualitative picture remains similar.

The sum of single-particle energies, $\hat{\mathcal{E}}$, for an axially symmetric harmonic oscillator of frequencies $\Omega_\perp$ and Ω_z is

$$
\begin{aligned}
\hat{\mathcal{E}} = \sum_{i \text{ occ}}\hat{\varepsilon}_i = \sum_{i \text{ occ}}\left\{\hbar\Omega_\perp[n_x(i) + n_y(i) + 1] + \hbar\Omega_z[n_z(i) + \tfrac{1}{2}]\right\} \\
= \hbar\Omega_\perp(\Sigma_x + \Sigma_y) + \hbar\Omega_z\Sigma_z
\end{aligned}
\tag{5.4.2}
$$

where $\vec{\Sigma} = (\Sigma_x, \Sigma_y, \Sigma_z)$ is given in eq. (5.3.6). Defining q as the ratio of frequencies, we have by using the volume conservation condition

$$
\begin{aligned}
q = \frac{\Omega_\perp}{\Omega_z}, \qquad \Omega_z\Omega_\perp^2 = \Omega_0^3 \\
\Omega_z = \Omega_0 q^{-2/3}, \qquad \Omega_\perp = \Omega_0 q^{1/3}.
\end{aligned}
\tag{5.4.3}
$$

The degenerate spherical harmonic oscillator shell of principal quantum N splits up when q deviates from unity. When q is larger than one, the levels within the shell are ordered by decreasing number of quanta in the z-direction or equivalently by increasing total number of quanta in the perpendicular directions. For q less than one, the level order is reversed. If the deformation is big enough, the lowest levels of the shell N+1 will be pulled down below the highest levels of the shell N. The first level crossing for q>1 between levels from oscillator shells labeled N and N+1 will occur when

$$
q^{-2/3}\left(N + \frac{3}{2}\right) + q^{1/3} = q^{-2/3}\frac{1}{2} + q^{1/3}(N + 1)
\tag{5.4.4}
$$

or

$$
q_1 = 1 + \frac{1}{N}.
\tag{5.4.5}
$$

The same level from the shell N+1 continues downwards and crosses the intermediate level of quantum number $(n_z, n_\perp) = (N/2, N/2)$ at the deformation

$$
q_2 = 1 + \frac{2}{N}.
\tag{5.4.6}
$$

The occupied states are found by filling the appropriate number of nucleons in the lowest possible single particle levels. This lowest configuration changes as a function of deformation, due to level crossings. The quantities $\vec{\Sigma}$ change discontinuously at level crossings but remain constant otherwise. We can then, between crossings, find the minimum of the energy (rewritten from eqs. (5.4.2) and (5.4.3))

$$\hat{\mathcal{E}} = \hbar\Omega_0 \left[q^{1/3}(\Sigma_x + \Sigma_y) + q^{-2/3}\Sigma_z \right] \tag{5.4.7}$$

to be

$$q_{eq} = \frac{2\Sigma_z}{\Sigma_x + \Sigma_y}. \tag{5.4.8}$$

Using this value of q, together with its definition (5.4.3), we see that the self-consistency condition (5.3.7) is fulfilled. Thus we have found that the same deformation minimizes the energy, and leads to self-consistency between potential and density shapes. This is quite satisfying because it shows that the nucleons' density follows the potential in the same way that the potential follows the density.

We can find explicit expressions for q_{eq} as a function of the number of nucleons. We assume that we fill the shell N_F-1 completely and occupy the lowest levels of the shell N_F up to (including) the level of quantum number $(n_z, n_\perp) = (N_F-n, n)$. Each level contains four nucleons (neutrons and protons with spin up and down). Besides this fourfold degeneracy, there is also the degeneracy $n_\perp + 1$ associated with the two dimensions of the perpendicular motion.

We can then calculate $\vec{\Sigma}$ as a sum of contributions from the core of completely-full lowest-lying shells $\vec{\Sigma}^c$ and the partially-filled valence shell $\vec{\Sigma}^v$. We obtain easily (see eqs. (5.2.5) and (5.2.8))

$$\Sigma_x^c = \Sigma_y^c = \Sigma_z^c = \tfrac{1}{3} \sum_{N=0}^{N_F-1} \left(N + \tfrac{3}{2}\right) g(N) = \frac{A_c}{4}(N_F + 1) \tag{5.4.9}$$

where A_c is the number of nucleons in these lowest shells. The calculation of $\vec{\Sigma}^v$ is also straight forward. We find for $q > 1$

$$\Sigma_x^v + \Sigma_y^v = 4 \sum_{n_\perp=0}^{n} (n_\perp + 1)^2 = \frac{4}{3}(n+1)(n+2)\left(n + \tfrac{3}{2}\right) \tag{5.4.10}$$

$$\Sigma_z^v = 4 \sum_{n_z=N_F}^{N_F-n} (n_z + \tfrac{1}{2})(N_F - n_z + 1)$$

$$= \tfrac{2}{3}(n+1)(n+2)\left(3N_F - 2n + \tfrac{3}{2}\right). \tag{5.4.11}$$

The equilibrium deformation of eq. (5.4.8) is now seen to be

$$q_{eq} = \frac{1 + \Sigma_z^v/\Sigma_z^c}{1 + (\Sigma_x^v + \Sigma_y^v)/2\Sigma_z^c}. \tag{5.4.12}$$

Thus $q_{eq} = 1$ for $n=N_F$ and $n = -1$, i.e. when the shell N_F is either completely filled or completely empty. In other words the full-shell configurations are always spherical.

When the valence contribution is small compared to the core contribution, we have

$$q_{eq} \approx 1 + \frac{2}{\Sigma_z^c}(n+1)(n+2)(N_F - n).$$

(5.4.13)

The maximum deformation occurs when n has the value

$$n_{max} = \tfrac{1}{3}N_F - 1 + \tfrac{1}{3}\sqrt{N_F^2 + 3N_F + 3} \approx \tfrac{2}{3}(N_F - 1)$$

(5.4.14)

which results in (see eqs. (5.2.6), (5.4.9), (5.4.13) and (5.4.14))

$$q_{eq} = 1 + \frac{16}{9}\frac{(N_F + \tfrac{1}{2})(N_F + 2)}{N_F(N_F + 1)^2}.$$

(5.4.15)

If we have $n=0$, i.e. four valence nucleons, we obtain

$$q_{eq}(n = 0) = 1 + \frac{24}{(N_F + 1)^2(N_F + 2)}.$$

(5.4.16)

The numerical results for these deformations are shown in table 5.1. As was already apparent from the analytical results, we obtain non-spherical equilibrium deformations. This qualitative finding is general and not a peculiarity of the harmonic oscillator. It has many consequences for the understanding of the nucleus.

We notice the general decrease of q_{eq} with nucleon number. The first level crossing occurs at a larger deformation than q_{eq} $(n=0)$, and the calculation is therefore consistent. Since the maximum deformation occurs at about q_2, the occupation numbers used may not be quite consistent with filling the levels from the bottom. However, the general picture is apparent.

The shape of the nucleus can be characterized by the ratio of axes in different directions of the matter distribution. By means of eqs. (5.3.6), (5.4.3) and (5.4.8) we obtain immediately

$$\sqrt{\frac{\langle 2z^2 \rangle}{\langle x^2 + y^2 \rangle}} = \sqrt{\frac{2\Sigma_z}{\Omega_z}\frac{\Omega_\perp}{\Sigma_x + \Sigma_y}} = q_{eq}.$$

(5.4.17)

Thus q_{eq} is simply the ratio of the half axes of the nuclear matter distribution, in harmony with the self-consistency condition eq. (5.3.7) as we noted above.

It is also easy to calculate the quadrupole moment Q_2 of the density distribution. Using eqs. (5.3.6), (5.4.3), (5.2.6) and (5.4.8) we get

$$Q_2 = \langle gs|2z^2 - x^2 - y^2|gs \rangle = \frac{\hbar}{m_N}\left[\frac{2\Sigma_z}{\Omega_z} - \frac{\Sigma_x + \Sigma_y}{\Omega_\perp}\right]$$

$$= \frac{\hbar}{m_N\Omega_0}\frac{1}{3}\frac{q_{eq}^2 - 1}{q_{eq}^{1/3}}\left[N_F(N_F + 1)^2(N_F + 2) + 4(n+1)(n+2)\left(n + \tfrac{3}{2}\right)\right].$$

(5.4.18)

As an example we obtain for ^{20}Ne with $\hbar\Omega_0$ from eq. (5.2.9)

$$Q_2(^{20}\text{Ne}) = \frac{\hbar}{m_N\Omega_0}\frac{7 \times 2^6}{9(1 + \frac{4}{7})^{1/3}} = 117\text{fm}^2 \qquad (5.4.19)$$

which compares favorably with the experimental value of 94 fm^2.

Many nuclei are deformed in their ground state, but the two largest groups of deformed nuclei have total nucleon number A in the intervals 150<A<188 and 230<A. These are the rare earths (lanthanides) and the actinides, respectively. From table 5.1 we have found axis ratios of about 1.3 for these nuclei. This is in very good agreement with experiments, where the tendency towards smaller deformations for the actinides also is present.

When the equilibrium deformations are calculated systematically in a more realistic model, they can be compared with corresponding experimental values. Furthermore the model also predicts the ground state spin of the nucleus. It is zero for even-even nuclei due to a special coupling between pairs of nucleons (see Chapter 6). For an odd-even system, viewed as an even-even system plus the last unpaired nucleon, the spin is consequently equal to that of the last nucleon. The last nucleon can quite easily be lifted to close but higher-lying single-particle orbits without changing anything else. The spins of these low-lying excited states are therefore also predicted by the model. These properties and others are remarkably well reproduced by the deformed independent particle model. Its introduction was a tremendous success.

The independent-particle approximation to the mean-field picture gives a good account of the shapes of nuclear ground states, which are only moderately deformed from the spherical shape. The approximation fails badly, however, when applied to more severe distortions of the shape. To understand the limitations of the independent-particle approximation, we turn to another, complementary approximation to the self-consistent mean field picture: the liquid-drop model.

N_F	A_c	$q_{eq}(n=0)$	$q_{eq}(n_{max})$	q_1	q_2
1	4	2.0	3.0	∞	3.0
2	16	1.57	1.99	2.0	2.00
3	40	1.29	1.65	1.50	1.67
4	80	1.16	1.48	1.33	1.50
5	140	1.094	1.38	1.25	1.40
6	224	1.061	1.31	1.20	1.33
7	336	1.042	1.27	1.17	1.29

Table 5.1. Equilibrium deformations as function of the quantum number N_F of the partially filled last shell. $q_{eq}(n=0)$ is calculated from eq. (5.4.12) but $q_{eq}(n_{max})$ is from eq. (5.4.15). The number of core particles A_c and the relevant level-crossing deformations q_1 and q_2 of eqs. (5.4.5) (N=$N_F - 1$) and (5.4.6) (N=N_F) are also given.

5.5   LIQUID-DROP DEFORMATION ENERGY

The basic ideas of the liquid drop model are described in Chapter 4. The binding energy is given by

$$B_{LD}(N, Z) = b_V A \left[1 - K_V \left(\frac{N - Z}{A} \right)^2 \right] - b_S A^{2/3} \left[1 - K_S \left(\frac{N - Z}{A} \right)^2 \right]$$

$$- b_C \frac{Z^2}{A^{1/3}} + \delta \left[(-1)^N + (-1)^Z \right] A^{-1/2}. \qquad (4.3.2)$$

for a spherical nucleus. When the nucleus assumes a deformed shape, its surface area must be larger, since the sphere has the smallest area for a given volume. Thus the surface-energy term proportional to $A^{2/3}$ will increase by a factor $S_0 > 1$, the ratio of the area of the deformed nucleus to that of the sphere. Similarly, a deformed nucleus will have a smaller Coulomb energy, because it is less compact and the charges lie farther from one another; thus the Coulomb energy will decrease by a factor $C_0 \leq 1$ compared to the spherical shape. Including these factors, we find

$$B_{LD} = b_V A \left[1 - K_V \left(\frac{N - Z}{A} \right)^2 \right] - b_S A^{2/3} \left[1 - K_S \left(\frac{N - Z}{A} \right)^2 \right] S_0$$

$$- b_C \frac{Z^2}{A^{1/3}} C_0 + \delta \left[(-1)^N + (-1)^Z \right] A^{-1/2}. \qquad (5.5.1)$$

The volume term is assumed unchanged, because the nucleus is nearly incompressible.

The factors S_0 and C_0 may be easily found using the surface defined by eq. (5.3.1) with all $m=0$ (i.e. axial symmetry). The surface area is

$$S = 2\pi \int_0^\pi R^2(\theta) \sin\theta \sqrt{1 + \frac{1}{R^2} \left(\frac{dR}{d\theta} \right)^2} \, d\theta \qquad (5.5.2)$$

which to second order in the deformation parameters $\alpha_{\ell m}$ results in [Bohr and Wheeler]

$$S = 4\pi R^2(A) S_0 = 4\pi R^2(A) \left[1 + \frac{1}{8\pi} \sum_{\ell=2} (\ell - 1)(\ell + 2) \alpha_{\ell 0}^2 \right]. \qquad (5.5.3)$$

The Coulomb energy of a uniformly charged drop of density n is

$$E_C = \frac{1}{2} n^2 \int\int_{vol} \frac{d^3\vec{r}_1 d^3\vec{r}_2}{|\vec{r}_1 - \vec{r}_2|} \qquad (5.5.4)$$

which for the shape in eq. (5.3.1) can be calculated to second order in the $\alpha_{\ell m}$. We find

$$E_C = \frac{3}{5} \frac{Z^2 e^2}{R(A)} C_0 = \frac{3}{5} \frac{Z^2 e^2}{R(A)} \left[1 - \frac{5}{4} \pi \sum_{\ell=2} \frac{\ell - 1}{2\ell + 1} \alpha_{\ell 0}^2 \right]. \qquad (5.5.5)$$

The deformation energy E_D (opposite sign of the binding energy!) measured relative to the spherical configuration, in units of the spherical surface energy, is

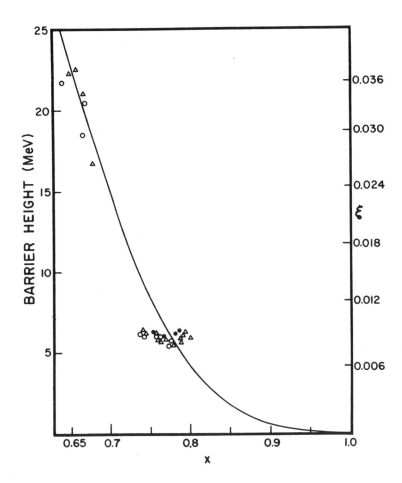

Figure 5.2   The simple liquid-drop model prediction of the fission barrier height as a function of the fissility parameter x. In order to present the results in MeV, the ordinate corresponds at each x value to typical values of Z and A for known nuclei, using the surface energy constant of eq. (4.3.3). The experimental barrier heights were plotted at x values assuming $(Z^2/A)_c$ is equal to 50.13. The ratio of the barrier height to the surface energy of a sphere is indicated by the slightly non- linear scale at the right. ○ Even-even, △ odd-A, ● odd-odd. [*from R. Vandenbosch and J. R. Huizenga, Nuclear Fission, Academic Press, New York 1973*]

then

$$E_D = \sum_{\ell=2} C_\ell \alpha_{\ell 0}^2 \equiv \sum_{\ell=2} \frac{(\ell-1)}{8\pi} \alpha_{\ell 0}^2 \left[\ell + 2 - \frac{20x}{2\ell+1}\right] \qquad (5.5.6)$$

where the fissility parameter x is defined by

$$x = \frac{b_C}{2b_S} \frac{Z^2}{A\left[1 - K_S\left(\frac{N-Z}{A}\right)^2\right]} \equiv \frac{Z^2/A}{(Z^2/A)_c}. \qquad (5.5.7)$$

When C_ℓ is positive (negative), the nucleus is stable (unstable) against deformations of the type $\alpha_{\ell 0}$. Thus the nucleus is least stable against deformations of smallest multipole order. In fact $C_2 > 0$ if and only if $x < 1$. This is the reason for the definition of $(Z^2/A)_c$, where c means critical.

The conclusion is that the liquid drop model predicts spherical equilibrium shapes for all nuclei of $x < 1$. These shapes are stable with respect to small deviations from equilibrium. For large deviations, we have to parametrize the shape better. For example, S_0 and C_0 can be calculated exactly for an ellipsoid. The result is that, if the charge is greater than $Z=34$, the spherical shape is only a local minimum. For these heavy nuclei, the liquid-drop energy reaches a maximum at a certain deformation, and then falls off for still larger deformations, where the increase in surface energy cannot keep up with the decrease in Coulomb energy. The maximum in the potential energy, at intermediate deformations, is called the **fission barrier**. Its height, measured with respect to the energy of the spherical configuration, is shown in fig. 5.2 as a function of x (for $K_S=0$) and compared to corresponding experimental numbers. The result is correct on average but wrong in details by up to about 10 MeV.

5.6 STRUTINSKY UNIFICATION OF THE LIQUID-DROP AND INDEPENDENT-PARTICLE MODELS

We conclude from the discussion of the previous two sections that the mean-field picture is capable of describing important features of both small and large deformations of the nuclear shape. However, the simplified approximations of the independent-particle model and the liquid-drop model are only useful for small or large deformations respectively. One way to unify the picture is to proceed with direct computations of the Density-Dependent Hartree-Fock model. While the model as described in Chapter 4 only has solutions at stable deformations, it can readily be extended by introducing an external one-body field adjusted to make other shapes stable. This procedure leads to some surprising results, including stable shapes of very great deformation which could not have been predicted by either the liquid-drop or independent-particle models. To understand how these new, highly deformed configurations — the **fission isomers** — come about, and to unify the two simple models, we follow arguments due to Strutinsky, who predicted the fission isomers before the advent of Density-Dependent Hartree-Fock — and also before they were discovered experimentally [see the review by Brack].

First, we have to understand why the independent-particle model gives wrong energies for large deformations. The problem can be traced to the very approximate treatment of the self-consistency condition.

The difficulty is easy to understand qualitatively. Volume conservation was introduced as a consequence of the saturation property of the nuclear force. This is a relevant requirement only for the bulk part of the nuclear density distribution. The leading order (volume term) of the nuclear energy is then assumed independent of deformation. A small mistake in such an approximation can clearly affect the total energy significantly. On top of this comes the fact that the variation of surface and Coulomb energies nearly cancel each other (for heavy nuclei) for many shapes, for example at the fission barrier. Thus we infer that, for the purpose of deformation energy calculations, self-consistency must be included much better than by the volume conservation condition.

While the independent-particle picture has an insufficient treatment of self-consistency, the liquid-drop model is not good enough either. The energy differences responsible for ground-state deformations in the independent-particle model are quite small, and depend on the details of which single-particle orbits are occupied. Indeed, simply by moving particles among states degenerate in the spherical nucleus, we can obtain oblate or prolate deformations with almost the same energy. The liquid-drop model, like the Fermi-gas model, does not take account of these details. Thus we need a better understanding of how the liquid-drop model originates in the self-consistent mean field. Strutinsky found a method for calculating the deviations from the liquid-drop energy, using the mean-field picture. He found the result that the deviations of the energy from the liquid drop can be approximated using the single-particle energies of the independent-particle model. Strutinsky's approximation has been tested by Density-Dependent Hartree-Fock computations, and gives the deviations from the liquid drop within about 20%. The approximation would be almost exact in ordinary Hartree-Fock theory, as we now show.

Let us therefore start again from the Hartree-Fock approximation and assume an effective density-independent nucleon-nucleon interaction V. The total energy $\langle \psi_{HF} | H | \psi_{HF} \rangle$ in the Slater determinental state ψ formed by the set $\{\phi_k\}$ of single particle wave functions is stationary with respect to variations of ϕ_k; the resulting Hartree-Fock equation may be written

$$\left(\frac{\vec{p}^{\,2}}{2m_N} + U \right) \phi_k = \varepsilon_k \phi_k \tag{5.6.1}$$

where we have introduced the average potential U given by (see eq. (4.2.11)).

$$U(\vec{r})\phi_i(\vec{r}) = \int d^3\vec{r}' V(\vec{r} - \vec{r}') \left[\rho(\vec{r}', \vec{r}')\phi_i(\vec{r}) - \rho(\vec{r}', \vec{r})\phi_i(\vec{r}') \right] \tag{5.6.2}$$

in terms of the density matrix $\rho(\vec{r}, \vec{r}')$.

We now suppose that the density matrix can be separated into two parts

$$\rho = \tilde{\rho} + \delta\rho. \tag{5.6.3}$$

$\tilde{\rho}$ is called the smooth part, and corresponds to the liquid-drop model. We will not need to find an exact functional form for $\tilde{\rho}$, but only suppose that it exists, and is close to the exact density matrix, so that

$$|\delta\rho| \ll \rho. \tag{5.6.4}$$

Using $\tilde{\rho}$ everywhere in eq. (5.6.2) instead of ρ, we obtain a smooth single-particle potential $\tilde{U}$. Its eigenvalues $\hat{\varepsilon}_k$ and eigenfunctions $\hat{\phi}_k$ are solutions to an equation completely analogous to eq. (5.6.1). From the eigenfunctions, $\hat{\phi}_k$, the corresponding density $\hat{\rho}$ is defined analogous to ρ. The difference $\delta U \equiv U - \tilde{U}$ between the two potentials is of the same order, i.e. $\delta\rho$, as the difference between the related densities.

To make a corresponding separation of the energy, we expand eq. (4.2.12) to first order in $\delta\rho$:

$$
\begin{aligned}
\langle \psi_{HF} | H | \psi_{HF} \rangle = \sum_{i=1}^{A} \varepsilon_i - \frac{1}{2} \int\int d^3\vec{r}\, d^3\vec{r}'\, V(\vec{r} - \vec{r}')[\tilde{\rho}(\vec{r},\vec{r})\tilde{\rho}(\vec{r}',\vec{r}') - \tilde{\rho}(\vec{r}',\vec{r})^2 \\
+ 2\tilde{\rho}(\vec{r},\vec{r})\delta\rho(\vec{r}',\vec{r}') - 2\tilde{\rho}(\vec{r}',\vec{r})\delta\rho(\vec{r}',\vec{r})] + \mathcal{O}(\delta\rho^2) \\
= \sum_{i=1}^{A} \varepsilon_i - \frac{1}{2} \int d^3\vec{r}\, \tilde{U}^D(\vec{r})\tilde{\rho}(\vec{r}) + \frac{1}{2} \int d^3\vec{r}\, d^3\vec{r}'\, \tilde{U}^X(\vec{r},\vec{r}')\tilde{\rho}(\vec{r}',\vec{r}) \\
- \int d^3\vec{r}\, \delta U^D(\vec{r})\tilde{\rho}(\vec{r}) + \int d^3\vec{r}\, d^3\vec{r}'\, \delta U^X(\vec{r},\vec{r}')\tilde{\rho}(\vec{r}',\vec{r}) + \mathcal{O}(\delta\rho^2).
\end{aligned}
\tag{5.6.5}
$$

To obtain the corresponding expansion of the sum of the eigenvalues, we use perturbation theory: the difference between the eigenvalues ε_i of U and the eigenvalues $\hat{\varepsilon}_i$ of $\tilde{U}$ is just the expectation value of δU with the unperturbed wave functions $\hat{\phi}_i$ of the smooth potential. Summing this relation over the single-particle states, we have

$$\sum_{i=1}^{A} \varepsilon_i = \hat{\mathcal{E}} + \int d^3\vec{r}\, \delta\tilde{U}^D(\vec{r})\tilde{\rho}(\vec{r}) - \int d^3\vec{r}\, d^3\vec{r}'\, \delta\tilde{U}^X(\vec{r},\vec{r}')\tilde{\rho}(\vec{r}',\vec{r}) + \mathcal{O}(\delta\rho^2). \tag{5.6.6}$$

where

$$\hat{\mathcal{E}} = \sum_{i=1}^{A} \hat{\varepsilon}_i. \tag{5.6.7}$$

Inserting this in eq. (5.6.5) we find

$$
\begin{aligned}
\langle \psi_{HF} | H | \psi_{HF} \rangle = \hat{\mathcal{E}} - \frac{1}{2} \int\int d^3\vec{r}\, d^3\vec{r}'\, V(\vec{r} - \vec{r}')\,[\tilde{\rho}(\vec{r},\vec{r})\tilde{\rho}(\vec{r}',\vec{r}') - \tilde{\rho}(\vec{r}',\vec{r})^2] \\
+ \mathcal{O}(\delta\rho^2)
\end{aligned}
\tag{5.6.8}
$$

where the integral involves only "smooth" quantities. This is the main result, called Strutinsky's energy theorem. It states that all first order (in $\delta\rho$) fluctuations in the Hartree-Fock energy are contained in the sum of the single-particle energies $\hat{\mathcal{E}}$. In actual calculations we identify $\tilde{U}$ with the independent-particle model potential, for example the deformed Woods-Saxon potential which was approximated by oscillators in sect. 5.4.

The final step is to relate the energy to the liquid drop model. We can identify the liquid-drop energy as the smooth part of eq. (5.6.8). The last item is already "smooth", so we only need to find the smooth part of $\hat{\mathcal{E}}$. This can be done in several equivalent ways which all amount to averaging the sum of single-particle energies over various parameters such as nucleon number, deformation or single-particle energy. Here we shall use the temperature averaging procedure, in which the single-particle energy is calculated with Fermi functions as occupation numbers, i.e.

$$\mathcal{E}_T = \sum_{k=1}^{\infty} \hat{\varepsilon}_k n_k(T) \tag{5.6.9}$$

$$n_k(T) = \frac{1}{1 + \exp[(\hat{\varepsilon}_k - \mu)/T]}. \tag{5.6.10}$$

The Fermi energy μ for each temperature T is determined by the requirement

$$A = \sum_{k=1}^{\infty} n_k(T). \tag{5.6.11}$$

If T is chosen larger than the spacing between levels $\hat{\varepsilon}_k$, but smaller than μ, then $\mathcal{E}_T$ is quadratic in T, since the excitation energy is equal to the number of nucleons excited times their average energy. The first factor is proportional to the number of single-particle levels per unit energy around μ, times the contributing energy interval, which is proportional to T. The last factor is also proportional to T, since each excited nucleon on average is lifted about T in energy.

For a smearing interval T which is bigger than the spacing between levels, $\mathcal{E}_T$ does not contain information about the detailed structure at T=0. Therefore the behavior of $\mathcal{E}_T$ for these values of T leads to the desired single-particle energy average by extrapolation down to T=0, i.e.

$$\mathcal{E}_T \underset{\text{T large}}{\approx} \tilde{\mathcal{E}} + a\,T^2 \tag{5.6.12}$$

where $\tilde{\mathcal{E}}$ is the smooth part of the single-particle energy $\hat{\mathcal{E}}$. In Chapter 10 we will prove eq. (5.6.12) and find an expression for a.

We are now able to identify the liquid-drop energy $\tilde{E}$:

$$\tilde{E} = \tilde{\mathcal{E}} - \frac{1}{2} \int \tilde{\rho}\tilde{U}d^3\vec{r} \approx -B_{LD}. \tag{5.6.13}$$

The difference between the mean-field energy $\langle\psi_{HF}|H|\psi_{HF}\rangle$ and the liquid-drop energy is known as the **shell correction** $\delta\mathcal{E}$. From eqs. (5.6.8) and (5.6.13) we have

$$\langle\psi_{HF}|H|\psi_{HF}\rangle = \widetilde{E} + \delta\mathcal{E} \tag{5.6.14}$$

where

$$\delta\mathcal{E} = \widehat{\mathcal{E}} - \widetilde{\mathcal{E}} \tag{5.6.15}$$

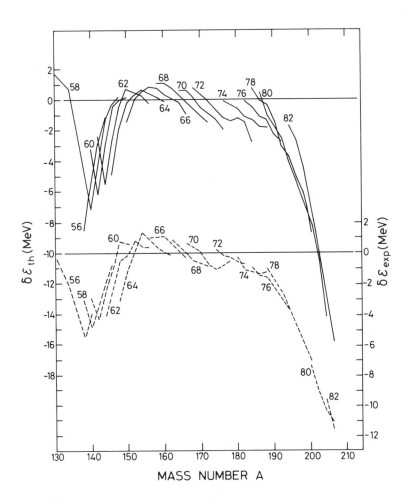

Figure 5.3   Theoretical and experimental shell corrections for even-even nuclei. The curves are labelled by their proton number. [*from H. C. Pauli and T. Ledergerber, Proc. Symp. Physics and Chemistry of Fission, IAEA, Rochester, NY, 1973, p. 463*].

depends only on the spectrum of single-particle energies in the smooth potential $\tilde{U}$.

The Strutinsky method thus allows us to estimate the energy of the nucleus, as a function of its deformation, without solving the self-consistent mean-field equations. Instead, we only need a good approximation to the "smoothed" mean field $\tilde{U}$ and its eigenvalues, and a good estimate of the liquid-drop energy. The Strutinsky method can also be applied to the case of density-dependent effective interactions. In that case, the rearrangement terms in the single-particle potential lead to corrections to $\delta\mathcal{E}$ which are of the same order as $\delta\mathcal{E}$. Numerical comparisons with density-dependent mean-field computations show that the rearrangement corrections reduce $\delta\mathcal{E}$ by 20 to 30% for realistic effective interactions.

We can also try out the Strutinsky method directly against experimental observations. For example, it is designed to give better binding energies than the liquid-drop model, which fits the known nuclear binding energies within a deviation of about ± 10 MeV. Fig. 5.3 compares the shell correction calculated for a Woods-Saxon po-

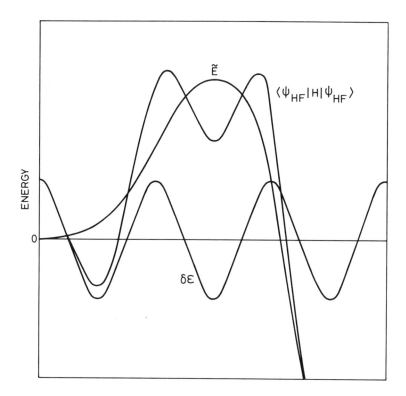

Figure 5.4   Schematic plot of the liquid-drop, shell correction and total energy as a function of deformation, i.e. a degree of freedom leading from the spherical shape to fission.

tential, to the difference between experimental and liquid-drop binding energies. These are nicely reproduced except for the exaggerated shell closure at ^{208}Pb. The shell-correction method also gives a good description of the ground state deformations.

It is not possible to mention the Strutinsky method without explaining about its immediate and greatest triumph. The shell correction oscillates as a function of deformation, while the liquid drop energy is more smooth. Fig. 5.4 shows schematically what they look like for a nucleus like ^{240}Pu. The first minimum in the total energy is the deformed ground state. The second minimum comes about because the liquid drop is flat in this region while the shell correction changes rapidly. Between the two minima is a barrier, and outside the second minimum is another barrier separating this equilibrium deformation from sliding downhill to fission (separation of ^{240}Pu in two almost equal halves). Thus a new stable deformation has appeared. States corresponding to this second minimum are called **fission** or **shape isomers**. Many of these have been identified in the actinide nuclei.

PROBLEMS

5.1   Use the harmonic-oscillator model to estimate the energy in MeV of the most energetic deuterons from the reaction ^{17}O(p,d) ^{16}O
(a)   without spin-orbit force
(b)   with spin-orbit constant C from eq. (5.2.10).

5.2   Consider an axially symmetric oscillator, $\Omega_x = \Omega_y = \Omega_\perp \neq \Omega_z$, as a model of ^{20}Ne, which has 2 neutrons and 2 protons in the N=2 shell. Neglect spin-orbit effects.
(a)   Find linear combinations of the cartesian oscillator wave functions, which are eigenfunctions of ℓ_z, for all states with N=2 in the spherical nucleus.
(b)   Show that the states of part (a) have axially symmetric densities.
(c)   For each of the wave functions of part (a), construct a Slater determinant with 2 neutrons and 2 protons in the same valence (N=2) orbital, and with nucleons in all the closed shells N=0, 1. Using the volume conservation condition, calculate the energy of the nucleus as a function of the deformation for each determinant. Plot, as a function of deformation.
(d)   Use the self-consistency condition on the shapes to find the deformation corresponding to each orbital. What is the ground state?

5.3   Which of the following nuclei would you expect to have the most deformed ground state? And why?

$$^{40}\text{Ca}, \quad ^{42}\text{Ca}, \quad ^{173}\text{Yb}, \quad ^{204}\text{Hg}, \quad ^{208}\text{Pb}$$

5.4 The nucleus ^{24}Mg has a typical, highly-deformed ground state. What do you expect its quadrupole moment Q_2 to be? $Q_2 \equiv \langle\psi|2z^2 - x^2 - y^2|\psi\rangle$, express your answer in fm^2.

5.5 What would the ground state deformation of ^{24}Mg be in the liquid drop model?

5.6 Calculate for given Ω_0 and q>1 the contribution to the quadrupole moment from four particles in the N=2 oscillator shell. Compare to the contribution from all the core nucleons in the same potential given by Ω_0 and q.

5.7 Calculate $\Sigma_x^v + \Sigma_y^v$ and Σ_z^v for oblate deformations (q<1). Find the equilibrium value q_{eq} of q for $4(N_F+1)$ nucleons in the valence shell. Compute also the minimum value of q_{eq}. Compare the energies of the prolate and oblate minima for 12 valence nucleons (6 protons and 6 neutrons) in the N=2 oscillator shell.

5.8 Calculate the deformation energy of ^{16}O with axially symmetric and volume conserving harmonic oscillator potential wavefunctions (both q≤1 and q≥1) for the interaction in eq. (4.2.29) where

$$\Im_0 \rightarrow \Im_0 + \frac{1}{6}\Im_3 \cdot n$$

and all other parameters are density independent. Use $\Im_0 = -1087$ MeV $\cdot$ fm^3; $\Im_3 = 12200$ MeV fm^6; $3\Im_1 + \Im_2(5+4\chi_2) = 710$ MeV fm^3; $9\Im_1 - \Im_2(5+4\chi_2) = 364$ MeV fm^5.

5.9 Calculate the energy of ^{16}O with axially symmetric harmonic oscillator wavefunctions of given q as function of Ω_0 for the interaction defined in problem 5.8.

5.10 Repeat the procedure described in problem 5.8 for ^{28}Si.

6

Pairing

6.1 INTERACTIONS BEYOND THE MEAN FIELD

We have seen that the mean-field picture can explain many of the basic observations about nuclei: their sizes and shapes, their cross sections for elastic scattering, many of the excited states of odd-mass nuclei, and most of the liquid-drop binding energy formula. Now we have to go beyond the mean-field picture. Our first step in this direction will explain the origin of the term in the mass formula (4.3.2) which describes the striking difference in binding energies of odd and even numbers of neutrons and protons: the pairing term

$$B_P = \frac{\left[(-1)^N + (-1)^Z\right]\delta}{A^{1/2}}, \tag{6.1.1}$$

which favors nuclei with even numbers of neutrons and protons. The need for such a term is apparent from fig. 6.1.

The pairing term in the binding energy is only one of several trends that show that the ground states of even-even nuclei (i.e. N and Z even) are unusual. We saw in Sect. 4.1.C that the odd-A nuclei have excited states corresponding to a nucleon being in a single-particle orbit other than the lowest energy state, with an excitation energy E^* equal to the difference of the single-particle eigenvalues. Such states would also be expected in even-even nuclei; they are often seen, but only when their exitation energy E^* is more than twice B_P. We are led to conclude that the ground states of even-even nuclei are especially tightly bound. It is also remarkable that the ground states of all even-even nuclei have zero angular momentum.

The explanation of these and many other observations has to be sought in correlations beyond independent-particle motion. A convenient way to study these correlations is by considering them as a perturbation on the mean-field picture. We write

$$H = H_{MF} + H_R \tag{6.1.2}$$

where H_{MF} contains the kinetic energy and the one-body field whose eigenstates are the Slater determinants of Chapter 4 (to avoid double-counting the potential energy, a constant has to be added as in the Strutinsky method). H_R is called the residual interaction.

The most straightforward approach is to try to diagonalize H_R in the basis of eigenstates of H_{HF}. This is only possible with huge computers and drastic approximations, and is known as the Interacting Shell Model (the name Shell Model was formerly used for the independent-particle model).

We will discuss somewhat more intuitive methods to approximate the effects of H_R. The choice of the method depends on the properties of H_R. Very crudely, we can divide H_R into long-range and short-range parts. The long-range part is responsible for collective motion, and is conveniently studied by extending the mean-field picture in Chapters 8 and 9. In this present chapter, we study the effect of the short-range part.

6.2   THE δ-FORCE

Even if H_R is very small compared to H_{MF}, it will have a decisive influence when H_{MF} has degenerate states. We begin by considering such a case. Let us assume a spherical H_{MF} and a number of closed (full) angular momentum subshells. We add two identical nucleons in the next empty subshell of angular momentum j. The

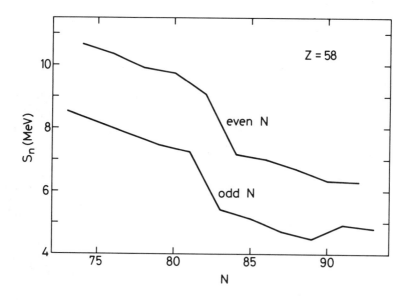

Figure 6.1   The neutron separation energy, eq. (4.1.8) as function of neutron number for the cerium isotopes (Z=58). The lines connect points of the same neutron-number parity. The even-neutron nuclei clearly bind the neutrons more strongly than the odd-neutron nuclei. The steep decrease is due to the shell closure at N=82. [*Data from A. H. Wapstra and G. Audi, Nucl. Phys.* **A432** *(1985) 55*]

two-nucleon wave function of angular momentum J and projection M is then

$$\psi_{\text{JM}}^{n\ell j}(1,2) = \sum_{m_1 m_2} \langle jm_1 \, jm_2 | \text{JM} \rangle \frac{\Phi_{n\ell jm_1}(1)\Phi_{n\ell jm_2}(2) - \Phi_{n\ell jm_1}(2)\Phi_{n\ell jm_2}(1)}{\sqrt{2}}$$

(6.2.1)

where $\langle jm_1 jm_2 | \text{JM} \rangle$ is the vector coupling or Clebsch-Gordon coefficient and Φ is given by

$$\Phi_{n\ell jm_1} = \frac{1}{r} u_{n\ell j}(r) \sum_{m,s} \langle \ell m \tfrac{1}{2} s | jm_1 \rangle Y_{\ell}^{m}(\theta,\phi)\chi_s$$

(6.2.2)

in terms of the radial wave function $u_{n\ell j}$ of principal quantum number n for the subshell, the spherical harmonic Y_{ℓ}^{m} and the spin wave function χ_s of spin 1/2 and projection s. Then eq. (6.2.2) defines the single-particle wave function Φ of angular momentum j obtained by coupling the orbital (ℓ) and spin (1/2) angular momenta, and eq. (6.2.1) defines the two-particle wave function of angular momentum J obtained by coupling the two angular momenta j of the nucleons. Note that J has to be even because of the antisymmetry of ψ, together with the fact that $\langle jm_1 jm_2 | \text{JM} \rangle = (-1)^{2j-J} \langle jm_2 jm_1 | \text{JM} \rangle$.

The shortest possible range of H_R is that of a δ-force, i.e.

$$H_R = V_0 \delta(\vec{r}_1 - \vec{r}_2)$$

(6.2.3)

which leads to the energy spectrum E_J

$$E_J = V_0 \int \psi_{\text{JM}}^{*} \delta(\vec{r}_1 - \vec{r}_2)\psi_{\text{JM}} \, d^3\vec{r}_1 \, d^3\vec{r}_2.$$

(6.2.4)

Here and in Sect. 6.3 we ignore the mean-field energy, since it is the same for all states within the degenerate j-shell. Using eqs. (6.2.1) and (6.2.2) we easily obtain our equation for E_J in terms of the radial wave function and a multiple sum of products of Clebsch-Gordon coefficents. By extensive use of orthogonality relations between these vector coupling coefficents we then arrive at the simple result

$$E_J = \frac{V_0 \left[1 + (-1)^J\right] (2j+1)^2}{32\pi(2J+1)} |\langle j, \tfrac{1}{2}, j, -\tfrac{1}{2} | J \, 0 \rangle|^2 \int_0^{\infty} r^{-2} u_{n\ell j}^4(r) \, dr.$$

(6.2.5)

The vanishing of E_J for odd values of J reflects the fact that two identical Fermi particles in the same j-shell can only exist in even angular-momentum states.

The only J-dependence of E_J is the $(2J+1)$-factor and the Clebsch-Gordon coefficent. Since E_J increases with J for an attractive force ($V_0 < 0$) the lowest state has J=0 and the first excited state has J=2. For all $j > \frac{3}{2}$, the ratio E_2/E_0 is approximately $\frac{1}{4}$ and the relative gap between first excited and ground state is about $|(E_2 - E_0)/E_0| \approx \frac{3}{4}$. Thus the zero-angular-momentum state is separated from the almost-degenerate excited spectrum by a large gap. In this J=0 state, the two nucleons are placed in time-reversed single-particle states with equal probability. This is seen from eq. (6.2.1) and the expression for the Clebsch-Gordon coefficient

$$\langle jm\,j -m|00\rangle = \frac{(-1)^{j-m}}{\sqrt{2j+1}}.$$ (6.2.6)

The two nucleons have their largest spatial overlap in this state, and the attractive δ-interaction consequently leads to the depression of its energy (see fig. 6.2).

Figure 6.2   The energy spectrum for our attractive zero-range interaction, without mean field, for a system of two nucleons in a j=15/2 subshell. The limiting value of vanishing interaction energy is approached for increasing angular momentum (shown on the right-hand side of the levels).

6.3   THE DEGENERATE PAIRING MODEL

Even with a force as simple as the δ-force, computations for states with more than two particles outside a closed shell become tedious. We therefore introduce a further simplificaton, based on the experience with the δ-force in the two-particle system.

The essential feature of the δ-force is maintained in the pairing force V. It only has non-vanishing matrix elements between time-reversed states, for which they all are identical. In the model of two nucleons in one j-shell we have

$$\langle jm_1 \overline{jm_1} |V| jm_2 \overline{jm_2} \rangle \equiv -G \tag{6.3.1}$$

where the bar above the quantum numbers denotes "time reverse". Our phase choice is such that

$$|\overline{jm}\rangle = (-1)^{j-m}|j-m\rangle. \tag{6.3.2}$$

Then the J=0 state in eq. (6.2.1) is a simple sum (without phase factors) of time-reversed states $|jm\,\overline{jm}\rangle$ of equal weight (see eq. (6.2.6)).

Let us choose those $j+\frac{1}{2} \equiv \Omega$ states as basis vectors defining a subspace. Clearly, the space of two-particle states is much larger. Since all matrix elements outside pairs of time-reversed states are zero, those others states, which obviously have zero energy, don't couple to our subspace. The Schrodinger equation is then conveniently expressed in matrix notation within the subspace of paired states:

$$-G\begin{pmatrix} 1 & 1 & \cdots & 1 \\ 1 & & & 1 \\ \vdots & & \ddots & \vdots \\ 1 & & \cdots & 1 \end{pmatrix} \begin{pmatrix} x_1 \\ x_2 \\ \vdots \\ x_\Omega \end{pmatrix} = E \begin{pmatrix} x_1 \\ x_2 \\ \vdots \\ x_\Omega \end{pmatrix} \tag{6.3.3}$$

where $\vec{x} = (x_1 \ldots x_\Omega)$ is the eigenvector of energy E. It is equivalent to

$$-G(x_1 + \ldots + x_\Omega) = Ex_1 = Ex_2 = \ldots = Ex_\Omega \tag{6.3.4}$$

with the solutions

$$E = -G\Omega, \qquad \vec{x} = \frac{1}{\sqrt{\Omega}}(1,1,\ldots,1) \tag{6.3.5}$$

and

$$E = 0, \qquad x_1 + x_2 + \ldots + x_\Omega = 0. \tag{6.3.6}$$

The state in eq. (6.3.5) is that of J=0 from eq. (6.2.1), and the degenerate states of eq. (6.3.6) therefore can be chosen as non-zero angular momentum states. The δ-force spectrum in fig. 6.2 is then modified such that all J>0 states coincide.

Let us now assume n particles in the j-shell ($n \leq 2\Omega$) and furthermore that we have p pairs of particle, i.e. states of the type in eq. (6.3.5). The energy E(n,p) of this state can be found by the following argument. The energy of one pair equals minus G times the number of available paired states (see eq. (6.3.5)). Since some of the states are blocked by other pairs or particles we find the energy per pair

$$\frac{E(n,p)}{p} = -G[\Omega - (n - 2p) - p + 1].$$ (6.3.7)

therefore we obtain

$$E(n,p) = -G\, p(\Omega - n + p + 1).$$ (6.3.8)

This can be expressed by the number of unpaired nucleons S=n−2p called the seniority S, i.e.

$$E(n,S) = -\frac{G}{4}(n - S)(2\Omega - n - S + 2).$$ (6.3.9)

When all nucleons are paired (S=0), the energy is, as expected, lowest, i.e.

$$E(n,0) = -\tfrac{1}{4}Gn(2\Omega - n + 2)$$ (6.3.10)

which equals $-G\Omega$ for n=2 as it should from eq. (6.3.5). When the subshell is full, $E(2\Omega,0) = -Gn/2$ which is what we would expect from the mean field, since the particles no longer have any freedom to correlate.

Thus the pairing force will create as many pairs of nucleons as possible. Each pair has zero angular momentum, and the force acts only between identical nucleons. Therefore the total ground state spin of even-even nuclei is zero. For odd nuclear masses, the ground-state spin is determined by the spin of the unpaired nucleon. This general prediction is remarkably well fulfilled throughout the periodic table.

Another piece of experimental evidence is the systematically smaller binding energy of odd-mass nuclei compared to even-even nuclei. The degenerate model predicts this energy difference to be

$$E(2p + 1, p) - E(2p, p) = Gp$$ (6.3.11)

We can also get useful information about the low-lying excited states in this model. For odd-mass nuclei, the lowest-lying excited states may show up at non-systematic and perhaps very low energy, depending on the mean-field states. For even-even nuclei, the lowest excited state in the model is that of one broken pair. Its energy is therefore

$$E(2p, p - 1) - E(2p, p) = G\,\Omega.$$ (6.3.12)

Thus an energy gap of $G\Omega$ in the excited spectrum of even-even nuclei should be a systematic feature. This is indeed found throughout the periodic table.

As a non-trivial example we can consider ^{75}As where the observed ground-state configuration of the unpaired nucleon is a $p_{3/2}$ state. The last three occupied mean-field levels for the protons are shown in fig. 6.3. The lowest-energy configuration seems offhand to be two pairs in the $p_{3/2}$ orbit and the last unpaired proton in the $f_{5/2}$ level. The energy obtained from eq. (6.3.8) is $\varepsilon_1 - 2G$ while the energy of the configuration with one pair in the $f_{5/2}$ level is $2\varepsilon_2 - 4G$. Thus the latter energy is lowest when the energy loss by lifting one proton up into the $f_{5/2}$ orbit is more than compensated by the extra pairing of this configuration. This is the case when $\varepsilon < 2G$, i.e. for close-lying mean-field levels. Other cases of similar nature exist, e.g. ^{61}Ni (see problem 6.3).

Another subtle effect of pairing is the existence of many spherical nuclei which are not closed shells. In the mean-field approximation, most non-closed-shell nuclei are deformed, though the deformations are small if they are close to closed shells. For such nuclei, the pairing energy favors spherical shapes, because then the shells have large degeneracy Ω. Deformation reduces the degeneracy and thus the pairing energy. The pairing energy lost by deformation for nuclei with only a few nucleons in (or missing from) closed shells can be greater than the mean-field energy gained by deformation. Thus the transition from spherical to deformed nuclei near magic nuclei may be viewed as a competition between the mean field and the pair correlations.

6.4   GENERAL PAIRING THEORY

In the previous section, we considered the influence of pairing within a subspace of degenerate levels of the mean-field model, on the assumption that H_R was small compared to H_{MF}. This allowed us to use degenerate perturbation theory and diagonalize H_R within the degenerate subspace. The criterion for the validity of degenerate perturbation theory is that the resulting energy shifts should be small compared to the energy spacing of the unperturbed levels. A first estimate would have been $G \ll \varepsilon_i - \varepsilon_j$, which is usually fulfilled. However, the total energy shifts turn out to be not G but $G\Omega$ which is much larger; thus the requirement for lowest-order degenerate perturbation theory is usually not met. Instead, we have to consider the effects of H_R within a larger subspace of single-particle levels near the Fermi energy. Our treatment of this case is inspired by the BCS theory of superconductivity [see Schrieffer].

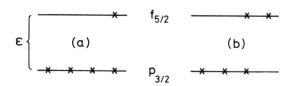

Figure 6.3   The last two occupied proton independent-particle model levels for ^{75}As. The energy difference between them is ε, and there are 28 protons in the orbits below. Ground-state configurations are shown (a) without pairing, (b) with pairing.

6.4 A.   The Model Hamiltonian

We further generalize the pairing force such that it will maintain the essential features of the δ-force. In second quantization our model interaction is taken to be

$$H_R = -G \sum_{i,k}{}' a_k^\dagger a_{\bar{k}}^\dagger a_{\bar{i}} a_i \qquad (6.4.1)$$

where $a_k^\dagger$ and a_k are creation and annihilation operators of the single-particle level
with quantum numbers k. A bar above the index means the time-reversed state.
Eq. (6.4.1) is an attractive interaction between pairs of nucleons, where one pair
is defined as two nucleons in time-reversed orbits. The approximation in (6.4.1)
assumes constant matrix elements independent of the state. Then the infinite sum
cannot be maintained, because a constant interaction matrix between **all** states
necessarily leads to infinite energy. Instead the summation is restricted to an interval
$|\varepsilon_k - \mu| \leq S$ around the Fermi energy. This restriction is indicated by the prime on
the summation sign.

   Since we can no longer assume $H_R \ll H_{MF}$, we must find another division of the
Hamiltonian. We choose

$$H = H_0 + \delta H \qquad (6.4.2)$$

where

$$H_0 = \sum_k \varepsilon_k \left(a_k^\dagger a_k + a_{\bar{k}}^\dagger a_{\bar{k}} \right) - \Delta \sum_k{}' \left(a_{\bar{k}}^\dagger a_k^\dagger + a_k a_{\bar{k}} \right) \qquad (6.4.3a)$$

$$\delta H = -G \sum_{i,k}{}' a_k^\dagger a_{\bar{k}}^\dagger a_{\bar{i}} a_i + \Delta \sum_k{}' \left(a_{\bar{k}}^\dagger a_k^\dagger + a_k a_{\bar{k}} \right). \qquad (6.4.3b)$$

The terms proportional to Δ are added and substracted in order to make δH small,
so that it can be treated as a perturbation. The reason for the choice of this form
is not yet obvious.

   The first term of H_0, the Hartree-Fock term, is diagonal in the chosen representation. We have clearly assumed degeneracy of time-reversed states. The number
operator N is also diagonal:

$$N = \sum_k \left(a_k^\dagger a_k + a_{\bar{k}}^\dagger a_{\bar{k}} \right). \qquad (6.4.4)$$

Since we have a definite number of particles we want to chose eigenfunctions of both
N and H. This is only possible when N and H commute.

   A direct computation show that this is the case with the expressions in eqs. (6.4.1)
and (6.4.4). The structure of H could have told us this directly, since annihilation
always is followed by creation of the same number of particles. However, neglecting

δH in (6.4.2) destroys this structure. Therefore we can at most require a given average number of particles. The Hamiltonian is then modified to

$$H' = H - \mu N = H'_0 + \delta H, \qquad H'_0 = H_0 - \mu N \tag{6.4.5}$$

where μ is a Lagrangian multiplier. It will, later on, be determined by the requirement of a given expectation value of the operator N. The abandonment of number conservation is quite counterintuitive for most physicists: it seems to be a move in the wrong direction. Luckily, Bardeen, Cooper and Schrieffer were cleverer than most; that may be why they earned the Nobel prize.

Minimization of the mean value of H'_0 for fixed μ leads to

$$\delta\langle H'_0 \rangle = 0 = \delta\langle H_0 \rangle - \mu \delta\langle N \rangle \tag{6.4.6a}$$

or

$$\mu = \frac{\delta\langle H_0 \rangle}{\delta\langle N \rangle} \tag{6.4.6b}$$

which shows that μ is the energy needed to remove one nucleon, i.e. the chemical potential.

6.4 B.   Solving the Unperturbed Hamiltonian

An exact diagonalization of H'_0 is possible. Since the interaction vanishes outside the interval $|\varepsilon_k - \mu| \leq S$, H'_0 is already diagonal there. Inside this interval, we introduce a new set of creation and annihilation operators $\alpha^\dagger_k$ and α_k given by

$$\alpha_k = u_k a_k + v_k a^\dagger_{\bar{k}} \qquad \alpha_{\bar{k}} = u_k a_{\bar{k}} - v_k a^\dagger_k. \tag{6.4.7}$$

This change of variables is known as the Bogolyubov transformation. The numbers u_k and v_k are at our disposal to minimize $\langle H'_0 \rangle$ or, equivently, to diagonalize H'_0. However, we require normalization

$$u^2_k + v^2_k = 1 \tag{6.4.8}$$

which ensures the same commutation relations between the α operators as those of the a operators, i.e.

$$\left\{ \alpha_i, \alpha_k \right\} = 0$$
$$\left\{ \alpha^\dagger_i, \alpha^\dagger_k \right\} = 0 \tag{6.4.9}$$
$$\left\{ \alpha_i, \alpha^\dagger_k \right\} = \delta_{ik}$$

where {A,B} means anticommutator of A and B.

The transformation of the a operators is

$$a_k = u_k \alpha_k - v_k \alpha_{\bar{k}}^\dagger, \qquad a_{\bar{k}} = u_k \alpha_{\bar{k}} + v_k \alpha_k^\dagger. \qquad (6.4.10)$$

Introducing this transformation (eq. 6.4.10) in eq. (6.4.5), we obtain

$$H_0' = \Omega_{gs} + \sum_k{}' H_k^{(1)} \left(\alpha_k^\dagger \alpha_k + \alpha_{\bar{k}}^\dagger \alpha_{\bar{k}} \right) + \sum_k{}' H_k^{(2)} \left(\alpha_{\bar{k}}^\dagger \alpha_k^\dagger + \alpha_k \alpha_{\bar{k}} \right) \qquad (6.4.11)$$

where

$$\Omega_{gs} = 2 \sum_k{}' \left[(\varepsilon_k - \mu) v_k^2 - \Delta u_k v_k \right] \qquad (6.4.12)$$

$$H_k^{(1)} = (\varepsilon_k - \mu)(u_k^2 - v_k^2) + 2u_k v_k \Delta \qquad (6.4.13)$$

$$H_k^{(2)} = 2(\varepsilon_k - \mu) u_k v_k - (u_k^2 - v_k^2)\Delta \qquad (6.4.14)$$

and where we have introduced the notations $u_k = 1$, $v_k = 0$ for $\varepsilon_k - \mu > S$ and $u_k = 0$, $v_k = 1$ for $\varepsilon_k - \mu < -S$, to make the equations compact.

If H_0' then is required to be diagonal in the α operators we must have vanishing $H_k^{(2)}$ for all k inside the interval $|\varepsilon_k - \mu| \leq S$, i.e.

$$2(\varepsilon_k - \mu) u_k v_k = (u_k^2 - v_k^2)\Delta, \qquad (6.4.15)$$

with the solutions

$$v_k^2 = \frac{1}{2} \left(1 - \frac{\varepsilon_k - \mu}{\epsilon_k} \right) \leq 1 \qquad (6.4.16a)$$

$$u_k^2 = \frac{1}{2} \left(1 + \frac{\varepsilon_k - \mu}{\epsilon_k} \right) \leq 1 \qquad (6.4.16b)$$

$$\epsilon_k = \sqrt{(\varepsilon_k - \mu)^2 + \Delta^2} \qquad (6.4.17)$$

where u_k, v_k, and Δ are chosen to be positive. Outside the interval $|\varepsilon_k - \mu| \leq S$, we take $\epsilon_k = |\varepsilon_k - \mu|$. Then H_0' reduces to

$$H_0' = \Omega_{gs} + \sum_k \epsilon_k \left(\alpha_k^\dagger \alpha_k + \alpha_{\bar{k}}^\dagger \alpha_{\bar{k}} \right). \qquad (6.4.18)$$

The ground state |gs⟩ is defined by

$$\alpha_k |gs\rangle = \alpha_{\bar{k}} |gs\rangle = 0. \qquad (6.4.19)$$

It can be related to the vacuum state |vac⟩ which obeys

$$a_k |vac\rangle = a_{\bar{k}} |vac\rangle = 0 \qquad (6.4.20)$$

by the expression (the BCS wave function)

$$|\text{gs}\rangle = \prod_k \left(u_k + v_k a_k^\dagger a_{\bar{k}}^\dagger \right) |\text{vac}\rangle \tag{6.4.21}$$

satisfying eq. (6.4.19). This solution does **not** describe a system with a definite number of particles. The average number of particles N_0 is

$$N_0 = \langle \text{gs}|N|\text{gs}\rangle = 2\sum_k v_k^2 = \sum_k \left(1 - \frac{\varepsilon_k - \mu}{\epsilon_k} \right) \tag{6.4.22}$$

and the fluctuation in the number of particles is measured by

$$\sigma^2 = \langle \text{gs}|N^2|\text{gs}\rangle - (\langle \text{gs}|N|\text{gs}\rangle)^2 = 4\sum_k u_k^2 v_k^2 = \Delta^2 {\sum_k}' \frac{1}{\epsilon_k^2}. \tag{6.4.23}$$

The energy gain compared to the normal solution is

$$\begin{aligned}
\Delta E &= \langle \text{gs}|H_0' + \mu N|\text{gs}\rangle - 2\sum_{\varepsilon_k < \mu} \varepsilon_k \\
&= 2\sum_k \varepsilon_k v_k^2 - \Delta^2 {\sum_k}' \frac{1}{\epsilon_k} - 2\sum_{\varepsilon_k < \mu} \varepsilon_k.
\end{aligned} \tag{6.4.24}$$

From eq. (6.4.21) we see that v_k^2 is the probability of finding the original level ε_k occupied by one pair. This probability decreases from unity below μ to zero above μ, as ε_k passes through the interval $|\varepsilon_k - \mu| < \Delta$ around the Fermi level μ. The system is now described in terms of new "particles", called quasi-particles, of energies ϵ_k (see eq. (6.4.17)). They are created by the operators $\alpha_k^\dagger$ and $\alpha_{\bar{k}}^\dagger$ and therefore are linear combinations of an "old" hole and an "old" particle (see eq. (6.4.7)). The transformation to quasiparticles is called the Bogolyubov transformation.

So far all quantities are expressed in terms of the undetermined parameter Δ. An estimate can be obtained by minimizing the expectation value of $H - \mu N$ in the ground state $|\text{gs}\rangle$ with respect to Δ. Thus

$$\begin{aligned}
E_\Delta &\equiv \langle \text{gs}|H - \mu N|\text{gs}\rangle \\
&= 2\sum (\varepsilon_k - \mu) v_k^2 - G \left({\sum}' u_k v_k \right)^2
\end{aligned} \tag{6.4.25}$$

and its derivative (assuming μ independent of Δ)

$$\frac{\partial E_\Delta}{\partial \Delta} = \Delta \left(1 - \frac{1}{2} G {\sum}' \frac{1}{\epsilon_k} \right) {\sum}' \frac{(\varepsilon_k - \mu)^2}{\epsilon_k^3} \tag{6.4.26}$$

vanishes when

$$0 = \Delta \left(1 - \frac{1}{2} G {\sum}' \frac{1}{\epsilon_k} \right) \tag{6.4.27}$$

(the same value of Δ is obtained if μ is allowed to vary with Δ). One solution is always $\Delta = 0$, which implies that all the u_k and v_k are either zero or one so that the ground state in eq. (6.4.21) is nothing but the original Hartree-Fock solution in the absence of pairing. For non-vanishing Δ we have

$$\frac{2}{G} = \sideset{}{'}\sum \frac{1}{\epsilon_k}. \tag{6.4.28}$$

The two quantities Δ and μ are solutions to eqs. (6.4.22) and (6.4.28). For given values of G and μ we have

$$\frac{2}{G} = \sideset{}{'}\sum \frac{1}{\epsilon_k} \le \sum \frac{1}{|\varepsilon_k - \mu|} \equiv \frac{2}{G_c}. \tag{6.4.29}$$

Thus to obtain non-trivial solutions we must have $G \ge G_c$.

6.4 C.   Justifying the Neglect of δH

So far, we have constructed the solutions to H_0, eq. (6.4.2), ignoring δH. We now are in a position to see that δH is unimportant. To see this, we compare its expectation value in the ground state $|gs\rangle$ of H_0. Using eqs. (6.4.3b) and (6.4.19), we find after repeated use of (6.4.9)

$$\langle gs|\delta H|gs\rangle = \left(\Delta - G\sum_k u_k v_k\right)\sum_k u_k v_k - G\sideset{}{'}\sum_k v_k^4. \tag{6.4.30}$$

Using eqs. (6.4.16), (6.4.17), (6.4.18) we see that eq. (6.4.28) is equivalent to

$$\Delta = G\sum_k u_k v_k \tag{6.4.31}$$

and therefore

$$\langle gs|\delta H|gs\rangle = -G\sideset{}{'}\sum_k v_k^4. \tag{6.4.32}$$

Using eqs. (6.4.8), (6.4.22), and (6.4.23) we find

$$\langle gs|\delta H|gs\rangle = \frac{G}{2}(N_0 - \tfrac{1}{2}\sigma^2). \tag{6.4.33}$$

If we had included variations of $\langle gs|\delta H|gs\rangle$ instead of just $\langle gs|H_0|gs\rangle$ in determining the ground state, the first term would have produced no changes, since N_0 was held fixed in the variation (formally, it would redefine the Lagrange parameter, but not the chemical potential). The second term, due to the lack of particle-number conservation, ought to be small if the approach is to be physically reasonable.

6.5   THE UNIFORM MODEL

Let us furthermore assume a large Δ compared to the single-particle level spacing at the Fermi energy. Then we can consider the single-particle level density g as continuous. The gap eq. (6.4.28) becomes

$$\frac{2}{G} = \int_{\mu-S}^{\mu+S} \frac{\frac{1}{2}g(\varepsilon)d\varepsilon}{\sqrt{(\varepsilon-\mu)^2+\Delta^2}} \approx g(\mu) \int_0^S \frac{d\varepsilon}{\sqrt{\varepsilon^2+\Delta^2}}$$

$$= g(\mu)\ln\left[\frac{S}{\Delta} + \sqrt{1+\left(\frac{S}{\Delta}\right)^2}\right] \approx g(\mu)\ln\left(\frac{2S}{\Delta}\right) \qquad (6.5.1)$$

where we used a large effective interval $S \gg \Delta$ and an approximately constant $g(\varepsilon)$ around μ. (The factor $\frac{1}{2}$ is needed because each term in each state sum in this chapter includes contributions from two degenerate, time-reversed single-particle states, both of which are counted in $g(\varepsilon)$.) Thus this model, the uniform model, results in the Δ-value

$$\Delta = 2S\, e^{-2/Gg(\mu)} \qquad (6.5.2)$$

which for constant G is proportional to S. Since G is an average matrix element in the interval 2S around μ, it is clear that S and G must be related. As we shall see later, Δ is the physically relevant quantity, and the reasonable constraint would be to fix Δ and $g(\mu)$ from experiments and relate S and G by eq. (6.5.1).

The energy gain in eq. (6.4.24) is, in the same approximations, estimated as

$$\Delta E = \frac{1}{2}\int_{-S}^{S} g(\varepsilon)\varepsilon\left(1 - \frac{\varepsilon}{\sqrt{\varepsilon^2+\Delta^2}}\right)d\varepsilon - \frac{\Delta^2}{G} - \int_{-S}^{0} \varepsilon\, g(\varepsilon)d\varepsilon$$

$$\approx \frac{1}{2}g(\mu)S^2\left[1 - \sqrt{1+\frac{\Delta^2}{S^2}}\right] \qquad (6.5.3)$$

$$\approx -\frac{1}{4}g(\mu)\Delta^2$$

and the width of the particle number distribution in eq. (6.4.23) becomes

$$\sigma^2 \approx \frac{1}{2}\Delta^2 g(\mu)\int_{-S}^{S}\frac{d\varepsilon}{\varepsilon^2+\Delta^2} = g(\mu)\Delta\arctan\left(\frac{S}{\Delta}\right) \approx \frac{\pi}{2}g(\mu)\Delta. \qquad (6.5.4)$$

6.6   RELATION TO EXPERIMENTAL INFORMATION

The single particle levels have, for spherical nuclei, a definite angular momentum. Each pair has zero total angular momentum, as in the degenerate model. Here, more shells are involved, but still in such a way that time-reversed orbits of the same j-shell enter with equal probability. Thus spherical even-even nuclei have total angular momentum equal to zero and spherical odd mass nuclei have total angular momentum equal to that of the unpaired nucleon. The same holds for deformed nuclei but for a very different reason which we shall deal with in Chapter 9.

The excited states of the Hamiltonian eq. (6.4.18) have energies of the form $\sum_k \epsilon_k$. The lowest of these, for an even-even nucleus, is the lowest possible two-quasiparticle excitation of energy

$$2\epsilon_k = 2\sqrt{(\varepsilon_k - \mu)^2 + \Delta^2} \geq 2\Delta. \qquad (6.6.1)$$

It has to correspond to a particle-hole excitation, so that two quasiparticles are needed. This excitation corresponds to the breaking of one pair.

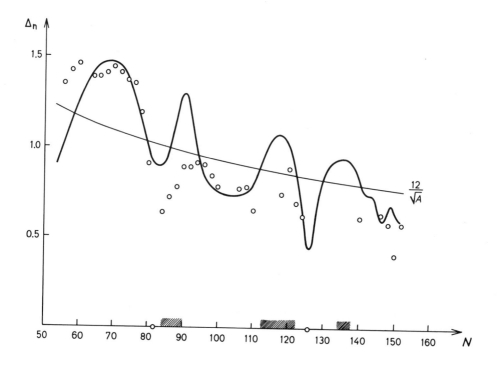

Figure 6.4a   Neutron pairing gaps as a function of the neutron number N. The points are calculations with a realistic single particle model [from T. Dossing and A. S. Jensen, Nucl. Phys. A222 (1974) 493]. The oscillating solid curve is a smooth extraction of the experimental gaps. [P. E. Nemirovsky and Y. V. Adamchuk, Nucl. Phys. 39 (1962) 551.] At the shaded regions the deformations change unusually fast.

If an odd nucleus is described approximately as an even-even core plus an extra quasiparticle, then eq. (6.4.18) can also be assumed to apply for an odd system. The ground state energy is then $\Omega_{gs} + \epsilon_{k_0}$, where k_0 denotes the state of lowest ϵ_k. The lowest excited states have excitation energies $\epsilon_k - \epsilon_{k_0}$ which can be very small depending on the underlying single-particle structure.

Thus odd nuclei don't have an energy gap in their excited spectrum, and even-even nuclei have an energy gap of 2Δ between the ground state and the first excited two-quasi-particle state.

In the same approximation, the binding energy of an odd mass nucleus is systematically $\epsilon_{k_0} \approx \Delta$ larger than for the corresponding even-even system. This quantity, which is known experimentally from the mass formula, is in this way closely related to the energy gap of even-even nuclei. The value is roughly (compare eq. 4.3.2)

$$\Delta = \frac{2\delta}{\sqrt{A}} = \frac{12 \mathrm{Mev}}{\sqrt{A}} \tag{6.6.2}$$

where A is the total number of nucleons. Thus Δ decreases from 2 MeV at A=36

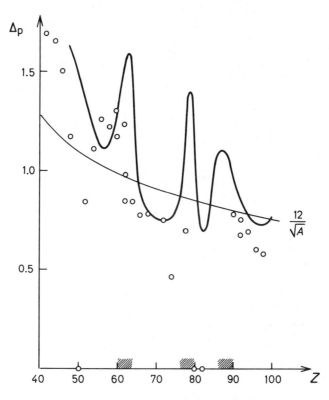

Figure 6.4b   The same as fig. 6.4.a for protons.

to 0.75 MeV at A=256. The energy gap in the spectrum of an intermediate even-even nucleus is then expected to be about 1 MeV. Figs. 6.4 compare smoothed experimental values with those of a realistic mean field model. Rather strong shell effects are clearly seen.

The various model predictions are remarkably well in agreement with the very extensive body of similar experimental results. It is therefore a safe conclusion that the nuclear force contains a residual two-body interaction (in addition to the mean field) which somehow resembles the pairing force or the δ-force.

The correlations from the residual interactions represented by the pairing picture affect not only the energies, but also the spatial distributions of the nucleons. This implies, for example, that when a nucleon is removed from the nucleus, the change in the nuclear density is a smoother function of position than one would estimate

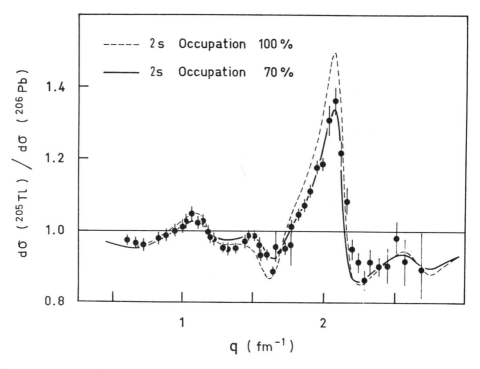

Figure 6.5   Ratio of differential cross sections for scattering of electrons from targets of ^{206}Pb and ^{205}Tl. The electron energies ranged from 200 to 500 MeV. The full line gives the prediction of the independent-particle model in which the nuclei differ by a $2s_{1/2}$ proton; the dashed line shows this prediction reduced by a factor 0.7. [*from B. Frois in Nuclear Structure 1985, R. Broglia, G. Hagemann, and B. Herskind, eds., North-Holland, Amsterdam (1985) p. 25*]

from the independent-particle model, because the quasiparticle is made up of various single-particle states whose densities have their nodes and maxima at different radii. Fig. 6.5 shows a comparison of electron scattering from ^{205}Tl and ^{206}Pb. In the independent-particle model, these nuclei differ by a proton in a $2s_{1/2}$ orbit, whose density distribution has 3 maxima and thus leads to an electron-scattering peak at a momentum transfer (eq. 3.5.1) of $|\vec{k} - \vec{k}'| = \hbar q$ where $q \approx 2$ fm$^{-1} \approx 3/R(A)$. The observed cross-section enhancement at this momentum transfer is only about 70% of that predicted by the ipm, which shows that the residual interactions lead to a smoother charge density for the last proton. In fact, the total nuclear charge density is also smoothed by the residual interactions. Fig. 6.6 compares the observed charge density of ^{140}Ce with a computation with a density-dependent mean field, which gives a distribution much less uniform than the observed one. Inclusion of pairing by a self-consistent computation using density-dependent forces with Bogolyubov

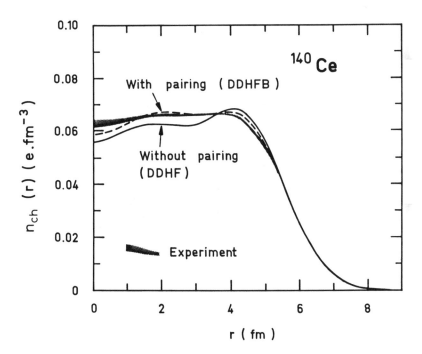

Figure 6.6   Charge density distribution of ^{140}Ce. The shaded curve shows the result of analysis of electron scattering experiments; its width indicates the experimental uncertainty. The full curve is from a mean field computation with a density-dependent force; the dashed curve is a self-consistent computation with the same force but with Bogolyubov quasiparticles instead of single-particle orbits. [*from B. Frois and C. Papanicolas in Annual Review of Nuclear and Particle Science, Palo Alto, to be published*]

quasiparticles gives better agreement with experiment. Of course these experiments cannot show whether the smoothing is due to correlations of the pairing type or to other collective many-body correlations. More detailed information can be extracted by comparing reactions where one or two nucleons are transferred between target and projectile. These are analyzed in the DWBA to provide additional evidence for pair correlations.

Let us now conclude this chapter by giving numerical estimates of the relevant quantities. The single-particle level density at the Fermi energy is on average for the neutrons

$$g(\mu) = \frac{N}{22\text{MeV}} \tag{6.6.3}$$

where $N \approx \frac{1}{2}A$ is the number of neutrons. The gain in energy from the normal to the paired solution is then estimated from eq. (6.5.3) to be 0.8 MeV for one kind of particles. Adding the contributions for both neutrons and protons we get approximately 1.6 MeV.

By assuming an effective pairing interval S slightly larger than the shell distance, i.e.

$$S \approx 1.1\hbar\omega = 1.1\frac{\text{MeV}}{A^{1/3}} = \frac{45\text{MeV}}{A^{1/3}}, \tag{6.6.4}$$

we can estimate the corresponding pairing matrix element G from eq. (6.5.2). This gives

$$G = \frac{1}{A}\frac{88\text{MeV}}{2.0 + \frac{1}{6}\ln A} \tag{6.6.5}$$

which decreases from about 0.3 MeV at A=100 to 0.12 MeV at A=250.

As described, the pairing Hamiltonian only conserves the average number of particles. The width of the distribution is obtained in eq. (6.5.4), and with the actual numbers we get

$$\sigma = 0.8N^{1/4} < 3. \tag{6.6.6}$$

Thus the fluctuation in particle number is fairly small, but significant admixtures of neighboring nuclei are present.

PROBLEMS

6.1   Show that (6.4.7) and (6.4.8), together with the canonical commutation relations for the a_i and α_k, imply (6.4.9).

6.2   Show that (6.4.21) satisfies (6.4.19).

6.3   Use the levels of the harmonic oscillator as in Sect. 5.2, together with the numerical estimates in Sect. 6.6, to predict the ground-state angular momentum and parity of ^{61}Ni.

6.4   Calculate u_k and v_k by minimizing the expectation value of H_0' (eq. 6.4.11) in the ground state $|gs\rangle$.

6.5   Express the ground state $\alpha_{k_0}^+|gs\rangle$ of an odd nucleus in terms of u_k, v_k, particle creation operators and the vacuum state.

6.6   Express the two-quasiparticle excited state $\alpha_{\underset{k}{\pm}}^+\alpha_k^+|gs\rangle$ in terms of the quantities u_k, v_k, particle creation operators, and the vacuum state.

6.7   Assume that the matrix element G_{ik} in eq. (6.4.1) depends on the state quantum numbers i and k. Use a k-dependent Δ in eqs. (6.4.3) and solve the corresponding unperturbed Hamiltonian in terms of this Δ_k. Derive the equations for Δ_k. Show that δH can be neglected as for constant matrix element. What is now

(i) the ground state energy of an odd nucleus?

(ii) The lowest possible two-quasiparticle excitation energy of an even-even nucleus?

(iii) The odd-even mass difference and the excitation energy gap for even-even nuclei?

6.8   Assume a level scheme within a given energy interval between ε_1 and $\varepsilon_2(\varepsilon_1 < \varepsilon_2)$. Levels outside this interval are far away and can be ignored. Approximate the gap and number equations in the continuous limit of the uniform model and evaluate the integrals. Show that the solutions are

$$\Delta^2 = 4D^2 f(1-f)C/(C-1)^2$$
$$\mu = \varepsilon_1 + D[f(C+1)-1]/(C-1)$$

where f is the fraction of occupied states and

$$D = \varepsilon_2 - \varepsilon_1 \; ; \quad C = \exp\left(4/\left(Gg(\mu)\right)\right)$$

Discuss the dependence of Δ^2 and μ as functions of $f\epsilon[0,1]$.

6.9   Evaluate the quantity

$$Q_m(N,Z) = -\frac{1}{2}[B(N-1,Z) - 2B(N,Z) + B(N+1,Z)]$$

for an even-even nucleus (N,Z) by using the liquid-drop approximation to the nuclear binding energy $B(N,Z)$ and expanding $B_{LD}(N+\eta,Z)$ to second order in η. Estimate the various terms. Repeat the procedure for the quantity

$$\Delta_m(N,Z) \equiv \frac{1}{4}[B(N-2,Z) - 3B(N-1,Z) + 3B(N,Z) - B(N+1,Z)]$$

Discuss how to extract the odd-even mass difference parameter δ from a given table of measured binding energies.

7

The Ground State: Stability and Decay

We are now in a position to understand the properties of all nuclear ground states. These can be described either by discrete quantum numbers, e.g. number of neutrons and protons, angular momentum, parity and isospin, or by continuous variables like radius, binding energy and electric and magnetic moments. All these characteristics of the nucleus vary systematically through the periodic table. We shall first describe a few of these quantities for stable nuclei. Then we shall formulate the basic limits for stability to find the boundaries of the nuclear world.

7.1 ANGULAR MOMENTUM AND PARITY

The spherically-symmetric single-particle potential leads to wave functions which can be chosen as eigenvectors for the z-component and square of the angular-momentum operator. The corresponding quantum numbers are j and m. The eigenvalues do not depend on m, due to the rotational symmetry. They have a degeneracy of $2j + 1$ equal to the number of possible m values.

If one j-shell is filled up with $2j+1$ particles, the individual angular momenta are equal and point in all possible directions. They therefore add up to zero, which is the total angular momentum of such a state. This is the result for the product wave function of the independent-particle model. Since unitary transformations between occupied states lead to the same total wave function, we conclude that all possible ways of filling $2j + 1$ particles into a j-shell give zero total angular momentum. In particular this is the case for the BCS type of wave function.

Let us now consider an even number of particles in one partially-filled j-shell. The angular momentum of the corresponding independent-particle wave function depends on the occupied single-particle states. In the BCS wave function, eq. (6.4.22), all available orbits of a given j-shell are occupied with the same probability, because the eigenvalue ε_k is degenerate and the u_k and v_k depend only on ε_k. Thus the total angular momentum is zero, as for a completely filled shell.

For an odd number of particles the lowest energy configuration is that of an even system plus one extra quasiparticle. The even system has zero angular momentum and the total odd system consequently has the same angular momentum as the unpaired nucleon.

The crucial assumption in the above arguments is the equal occupation probability of the j-shell states with different z-components of the single-particle angular momentum. The conclusion is then that an even-even (spherical) nucleus has total angular momentum zero and an odd-mass nucleus has angular momentum equal to that of the unpaired nucleon. For an odd-odd nucleus, the angular momenta of the two unpaired nucleons can in principle couple to give any result consistent with the vector addition rule.

For a non-spherical single-particle potential, the total Hamiltonian (including the pairing force) is time-reversal invariant. The resulting single-particle energies are therefore (at least) pairwise degenerate, since each state has the same energy as its time-reverse. The z-component of the total angular momentum for an even system is therefore zero. The system is deformed and quantum-mechanically able to rotate. This means that the different orientations have to be weighted in the total wave-function. The lowest energy then corresponds to an equal weighting, resulting in a spherically-symmetric state of zero total angular momentum. In Chapter 9 we shall systematically discuss these rotational states.

For an odd system the z component of the total angular momentum is equal to that of the unpaired nucleon. The necessary weighting of the different orientations then results in the lowest possible total angular momentum, i.e. equal to that of the z component.

For non-spherical intrinsic systems we then have a total angular momentum equal to that of its z component, i.e. zero for even-even nuclei and the unpaired nucleon value for odd-mass nuclei.

These results are found experimentally without exception for even-even nuclei. For other systems the theoretical result depends somewhat on the chosen model. It therefore provides constraints on the single-particle potential to be used.

The single-particle orbits of a spherical j-shell all have the same parity. The total nuclear parity of a spherical nucleus is therefore positive for an even-even system, that of the unpaired nucleon for an odd-mass nucleus and the product of the two unpaired parities for an odd-odd nucleus.

The symmetry of the shapes of the known deformed nuclei are such that the pairwise-degenerate single-particle orbits have the same parity. Thus the conclusions about parity for deformed nuclei are the same as for spherical nuclei.

7.2   MAGNETIC DIPOLE MOMENTS

The magnetic dipole $\vec{\mu}$ arises from the orbital motion of the nucleons inside the nucleus, the intrinsic spins of the nucleons and the meson degrees of freedom. Neglecting the latter we can write

$$\vec{\mu} = \sum_{k=1}^{A} \left(g_k^{(\ell)} \vec{\ell}_k + g_k^{(s)} \vec{s}_k \right) e/2m_p c. \tag{7.2.1}$$

where the values of the gyromagnetic ratios, different for neutrons and protons, are given by

$$\begin{aligned} g_p^{(\ell)} &= 1; & g_n^{(\ell)} &= 0 \\ g_p^{(s)} &= 5.5856; & g_n^{(s)} &= -3.8263. \end{aligned} \tag{7.2.2}$$

The magnetic dipole moment μ is defined as the expectation value

$$\mu \equiv \langle J\ M = J | \mu_z | J\ M = J \rangle \tag{7.2.3}$$

where the z component M of the angular momentum has its maximum value (equal to J). The matrix elements of any vector operator between states of the type $|J\ M\rangle$ are proportional to those of the angular momentum. Thus μ is proportional to M=J and given by the nuclear g-factor as

$$\mu \equiv gJ\mu_N \tag{7.2.4}$$

where $\mu_N = e\hbar/2m_p c$ is the nuclear magneton. The immediate consequence is that all even-even nuclei have zero magnetic dipole moment in their ground states.

For odd-mass nuclei, where the total angular momentum is due to the unpaired nucleon, we can estimate μ from the contribution of that nucleon. Then we have

$$\mu_{sp} = \langle j\ m = j | \mu_z | j\ m = j \rangle \tag{7.2.5}$$

which can be calculated by noting that it is equal to the expectation value of the projection on the z component of the angular momentum. This means

$$\mu_{sp} = \left\langle j\ m = j \left| \frac{\vec{\mu} \cdot \vec{j}}{\vec{j}^2} j_z \right| j\ m = j \right\rangle \tag{7.2.6}$$

which by use of the vector-coupling result

$$\langle \vec{j} \cdot \vec{s} \rangle = \frac{1}{2} \langle \vec{j}^2 + \vec{s}^2 - \vec{\ell}^2 \rangle \tag{7.2.7}$$

leads to

$$\mu_{sp} = j \left[g^{(\ell)} \pm \left(g^{(s)} - g^{(\ell)} \right) \frac{1}{2\ell+1} \right] \mu_N \tag{7.2.8}$$

for the possible values of $j = \ell \pm \frac{1}{2}$.

The extreme single particle estimates μ_{sp}, viewed as functions of j, are called the Schmidt lines. They are shown in fig. 7.1 together with the experimental values of

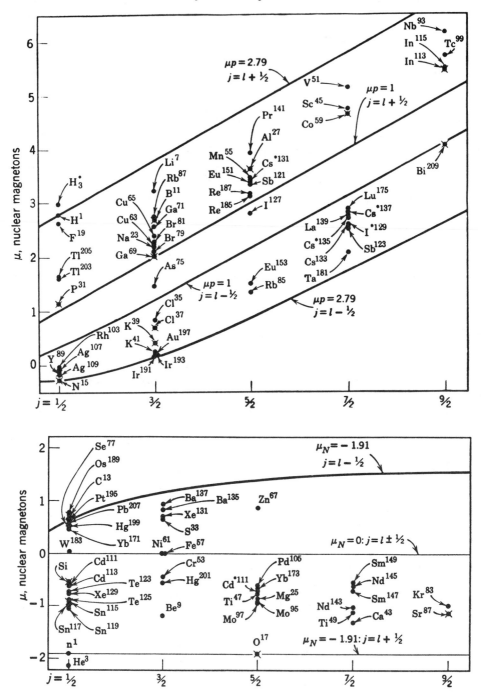

Figure 7.1   Magnetic moments of odd nuclei plotted against the spin. Top: odd Z, Bottom: odd N. [*from M. Mayer and H. Jensen*]

μ. They are certainly not accurate but only order of magnitude estimates. Deviations arise from the neglected remaining particles either through collective effects (deformed nuclei) or from the influence of the valence particle on the other nucleons.

7.3   ISOSPIN OF A NUCLEUS

The isospin of one nucleon $\vec{t}$ has magnitude $\frac{1}{2}$, and the total isospin of a nucleus is then

$$\vec{T} = \sum_{k=1}^{A} \vec{t}_k. \tag{7.3.1}$$

The 3-component, the sum of individual contributions of $-1/2$ for neutrons and $+1/2$ for protons, is

$$T_3 = -\tfrac{1}{2}(N - Z). \tag{7.3.2}$$

Let us first consider only charge-independent interactions (the Coulomb force is for the moment neglected). In that case, nuclei with the same A but different N and Z have states which are closely related to each other, called **analog states**. They are obtained by interchanging some neutrons and protons but otherwise maintaining all other characteristics of the wave functions. These states have identical energies and belong to different nuclei with the same total number of nucleons. The set is therefore called an **isobaric multiplet**.

The simplest of these are T=1/2 multiplets of the A=1 system. They each consist of two states which are identical states occupied by either a neutron or a proton. This is shown schematically in fig. 7.2

The second-simplest isobaric multiplet is that of T=0 for the A=2 system. The deuteron is a familiar example, see sect. 2.3. As mentioned in Chapter 2, the two-nucleon system can also form a T=1 state as illustrated in fig. 7.3. The $T_3 = 0$ state is obtained either by acting on the $T_3 = -1$ state with the raising operator $T_+ = \sum_{k=1}^{A} [t_1(k) + it_2(k)]$ or on the T=+1 state with T_-. As the di-neutron

Figure 7.2   The T=1/2 multiplet for the A=1 system. The line illustrates a specific state with a set of known quantum numbers. The cross and the circle indicate occupied and empty states, respectively.

and di-proton states only are observed in the elastic-scattering cross sections, we infer that the analogous np state of $T_3 = 0$ has an energy above that of the T=0 np state, which consequently must be the bound state of the deuteron. Thus the isospin-antisymmetric (T=0) or space-spin symmetric state of the deuteron has a lower energy than the isospin-symmetric T=1 state.

The maximum number of states in an isobaric multiplet would be A+1, obtained by counting the number of neutron-proton changes necessary to change A neutrons into A protons. Many of these A+1 states do not exist because they violate the exclusion principle. This is illustrated in fig. 7.4 for the T=1/2 multiplet of the A=3 system. The lowest two particles cannot be changed due to the exclusion principle. Only two analog states exist in this case.

The existence of isobaric-analog multiplets is not limited to nuclear ground states. In principle, each state with T> 0 has an analog in another nucleus. As an example, consider the case of the nuclei with A=14. Only ^{14}N has a T=0 state, because it has

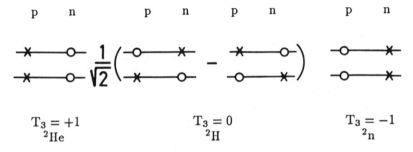

Figure 7.3   The T=1 multiplet of the two nucleon system. The notation is that defined in fig. 7.2. The $T_3 = 0$ state is the normalized linear combination of the two states indicated. The minus sign comes from the antisymmetry of the wave functions, which is not shown explicitly, see problem 7.2.

Figure 7.4   The T=1/2 multiplet of the A=3 system. The notation is defined in fig. 7.2.

N=Z; as in the case of the deuteron, it therefore ha J=1. The ground states of ^{14}O and ^{14}C have T=1, since they have $T_3 = \pm 1$. In the independent-particle model, these states are constructed from single-particle states as shown in fig. 7.5. The linear combination of determinants for T=1, T_3=0 is obtained by operating either with T_- on the ground state of ^{14}O, or with T_+ on the ground state of ^{14}C. The Pauli principle forbids all other states made up of the independent-particle wave functions in the N=0 and N=1 oscillator shells shown. The T=1 state of ^{14}N has a higher energy than the T=0 state, because the nuclear force is more attractive for lower isospin, as we saw in Chapter 2 for the deuteron.

The lowest half-dozen excited states of ^{14}O (see fig. 7.6) have the same angular momenta and almost the same excitation energies as the lowest excited states of ^{14}C, and correspond to a similar series of states in ^{14}N which however lie at similar excitation energies measured from its lowest T=1 state, Thus we see many isobaric multiplets with T=1. The T=0 states of ^{14}N, of course, have no analogs in the neighboring nuclei. On the other hand, at still higher excitation energies, all three nuclei have excited states which have larger isospins, and thus may be seen in other nuclei with A=14, such as ^{14}B and ^{14}F, which don't have analogs of the T=0 and 1 states because their values of T_3 are too large.

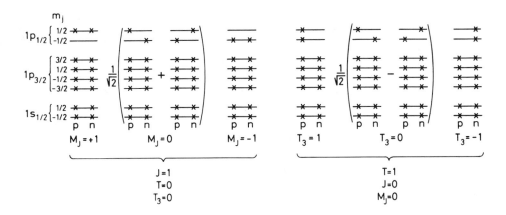

Figure 7.5   Independent-particle-model ground states of ^{14}O, ^{14}N, and ^{14}C, with lowest T=1 state in ^{14}N. The degenerate single-particle states with various projections m_j are drawn separately to emphasize the distinctness of their wave functions. Each column represents a Slater determinant of the single-particle states marked with an x.

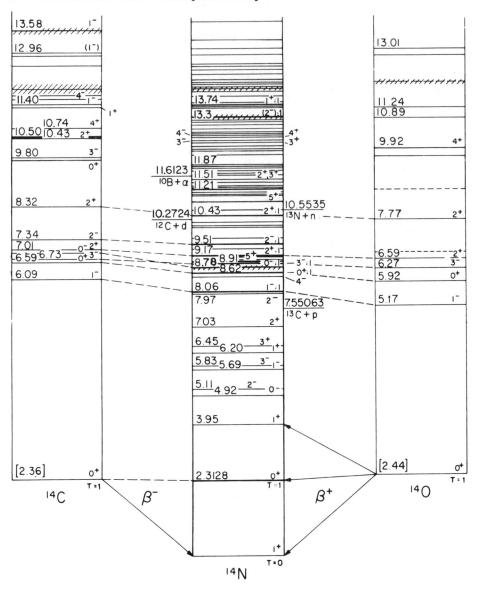

Figure 7.6   Lowest energy levels of three nuclei with A=14. The excitation energy is shown at left of each line, the spin and parity at right. The diagrams for individual isobars have been shifted vertically to eliminate the neutron-proton mass difference and the Coulomb energy, taken as $E_C = 0.60Z(Z-1)/A^{1/3}$. Energies in square brackets represent the (approximate) nuclear energy, $E_N = M(Z, A)c^2 - ZM(H)c^2 - NM(n)c^2 - E_C$, minus the corresponding quantity for ^{14}N; here Mc^2 represents the atomic mass in MeV. Levels which are presumed to be isospin multiplets are connected with dashed lines. [*from F. Ajzenberg-Selove, Nucl. Phys.* **A449** *(1986) 1*]

The states of different isospin for a given even total nucleon number A can be shown in a plot like fig. 7.7. The maximum T value is A/2 where all isospins are lined up in one direction. The ordinate shows T, but could as well have illustrated the energy of the system, since the nuclear force favors small T (see next section).

The total wave function, including ordinary space, spin and isospin coordinates, is required to be fully antisymmetric. The degree of symmetry of the space and spin wave function therefore must be reflected in the isospin wave function, and can be seen in its quantum numbers. For example, fig. 7.5 shows completely antisymmetric space and spin states for which the isospin assumes its maximum value T=1. The ground state of ^{14}N has the highest space-spin symmetry between identical nucleons, so its isospin value is minimum, T=0. Thus the isospin quantum number is a measure of the symmetry of the space-spin wave function.

It was clear from the beginning that isospin could only be an approximate quantum number due to the symmetry-violating Coulomb potential. For a uniform charge distribution inside the nuclear radius R, the potential is given as

$$U_C(r) = \begin{cases} Ze^2/R\left[3/2 - \left(r^2/2R^2\right)\right] & \text{for } r<R \\ Ze^2/r & \text{for } r>R. \end{cases} \tag{7.3.3}$$

Thus it bends over inside the nucleus and is relatively flat. The not-so-obvious but still correct consequence is that U_C acts essentially as an extra added constant in the energy. Even when the Coulomb potential is large, its gradient inside the nucleus is small: the Coulomb force is largest outside the nucleus. The Coulomb potential has very small non-diagonal matrix elements in the basis where U_C is neglected. Thus U_C does not crucially violate isospin conservation. It does, however, change the relative energies of the members of a multiplet. Using U_C as a perturbation we can estimate its effects on the energy difference by

$$\begin{aligned} \Delta E_C &\approx \int U_C(\vec{r})|\phi_p(\vec{r})|^2 d^3\vec{r} \\ &\approx \frac{\int d^3\vec{r}\, U_C(\vec{r})n(\vec{r})}{\int d^3\vec{r}\, n(\vec{r})} \\ &\approx \frac{6}{5}\frac{Ze^2}{R}. \end{aligned} \tag{7.3.4}$$

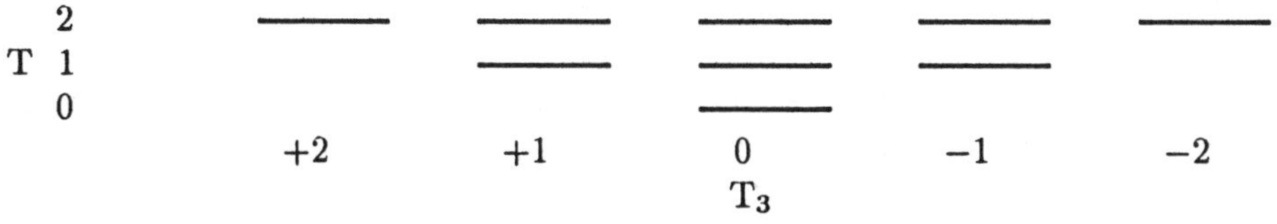

Figure 7.7 The various isospins for a given even A as function of the neutron excess or T_3.

Therefore members of a multiplet have displaced energy scales given roughly by eq. (7.3.4), compare fig. 7.6. For a heavy nucleus $\Delta E_C \approx 20$ MeV. In spite of this the individual levels are observed at approximately the same relative positions measured in fractions of an MeV. For a more complete discussion of isospin symmetry and isobaric analog states, see Bohr and Mottelson, vol. 1.

7.4   ISOSPIN POTENTIAL AND SYMMETRY ENERGY

Although the force between nucleons is charge-independent and therefore independent of the isospin projection T_3, we saw in Chapter 2 that it does depend on the total isospin T. Similarly, the effective nucleon-nucleon interaction in a nucleus has a term $\Im_t$ proportional to the scalar product of the two nucleons isospin

$$\Im_t(\vec{r}_{12}, \vec{t}_1, \vec{t}_2) = \vec{t}_1 \cdot \vec{t}_2 \Im^t(r_{12}). \tag{7.4.1}$$

This is a consequence of the requirement that the interaction must be a scalar in isospin space, together with the fact that any power of eq. (7.4.1) is of the same form, except for an additive isospin-independent term.

The direct (Hartree) term U^D of the mean field then has a component

$$U^t(\vec{r}, \vec{t}) = \vec{t} \cdot \sum_{i=1}^{A} \langle i | \vec{t}_i \Im^t(\vec{r}_i - \vec{r}) | i \rangle \tag{7.4.2}$$

where $\vec{t}$ is the isospin of the nucleon. When the single-particle states $|i\rangle$ are either neutrons or protons (have a definite value of the 3-component of the isospin), we have, for a zero-range two-body force of strength $\Im^t$

$$U^t(\vec{r}, t_3) = \Im^t t_3 \left(n_p(\vec{r}) - n_n(\vec{r}) \right) / 2. \tag{7.4.3}$$

The radial dependence $n(\vec{r})$ may be approximated by the Woods-Saxon shape, and the strength U_0^t can be found by comparison to eq. (3.3.21),

$$\hat{U}^t = -t_3 \frac{N-Z}{2} U_0^t f(r), \quad U_0^t = \frac{96 \text{ MeV}}{A} = n_0 \Im^t / A. \tag{7.4.4}$$

The expectation value of the corresponding two-body interaction, eq. (7.4.1), is simply given for a mean-field determinant $|gs\rangle$ by

$$\langle gs | \Im_t | gs \rangle = \frac{1}{2} U_0^t \left(\frac{N-Z}{2} \right)^2 = \frac{(N-Z)^2}{A} 12 \text{MeV} \tag{7.4.5}$$

where the factor $\frac{1}{2}$ is due to $\Im_t$ being a two-body force.

The force is repulsive (U_0^t is positive) and its effect is of volume nature, i.e. proportional to the nuclear volume or the number of nucleons. By doubling the number of neutrons and protons the energy is doubled. These properties follow from the structure of U_0 in eqs. (3.3.21) which in turn was a consequence of adjusting the potential to experimental information about single-particle motion.

The energy in eq. (7.4.5) has the same form as the symmetry energy of the semi-empirical mass formula. Such a term can therefore be explained by the force in eq. (7.4.1), except that it is too small. The remaining part is due to kinetic energy arising because of the exclusion principle. It can be estimated using the Fermi-gas model. Assuming that the neutrons and protons move in potential wells of the same size, we can follow sect. 4.1 to introduce separate Fermi momenta p_n and p_p for neutrons and protons, with

$$N = 2 \cdot \frac{4}{3}\pi R(A)^3 \cdot \frac{4}{3}\pi \frac{p_n^3}{h^3} \equiv \xi p_n^3.$$
$$Z = \xi p_p^3. \tag{7.4.6}$$

The same arguments that led to eq. (4.2.4) give for the total kinetic energy

$$\langle gs|KE|gs \rangle = \frac{3}{5}\frac{p_n^2}{2m_N}N + \frac{3}{5}\frac{p_p^2}{2m_N}Z. \tag{7.4.7}$$

Eliminating p_n and p_p by using eq. (7.4.6) we find

$$\langle gs|KE|gs \rangle = \frac{3}{10\,m_N\xi^{2/3}}\left(N^{5/3} + Z^{5/3}\right). \tag{7.4.8}$$

After changing variables from N and Z to A and $(N-Z)/A$, we can expand eq. (7.4.8) in a Taylor series in $(N-Z)/A$ to obtain

$$\langle gs|KE|gs \rangle = 33\,\text{MeV}\left(\frac{3}{5}A + \frac{1}{3}\frac{(N-Z)^2}{A}\right) \tag{7.4.9}$$

where ξ has been evaluated as in eq. (4.1.6). The kinetic energy thus contributes $\frac{1}{3} \cdot 33$ MeV = 11 MeV to the symmetry-energy coefficent.

We notice that the origin of this energy is the higher kinetic energy of an asymmetric ($N \neq Z$) nucleus compared to that of the symmetric system. The reason is that the Pauli principle prohibits more than one identical particle in each quantum state. Thus changing a proton to a neutron involves increasing its single-particle energy.

By adding the results of potential (eq. (7.4.5)) and kinetic energy (eq. (7.4.9)) we find a total symmetry energy coefficient (of $(N-Z)^2/A$) of 23 MeV. Considering the simplicity of the estimates we have obtained an excellent agreement with the mass-formula value, eq. (4.3.2), of $K_V b_V = 23.4$ MeV.

We have so far tacitly assumed that the wave function is one Slater determinant built of the lowest single-particle wave functions. In other words we have only considered the ground state. Let us now calculate the interaction energy of the two-body potential in eq. (7.4.1) with the strength U_0^t, i.e.

$$\mathfrak{H}_t^{(2)} = \frac{U_0^t}{2} \sum_{i \neq k} \vec{t}_i \cdot \vec{t}_k \tag{7.4.10}$$

for a state of given isospin T and 3-projection $T_3 = \frac{1}{2}(Z-N)$. This gives immediately

$$\left\langle TT_3 \left| \mathfrak{H}_t^{(2)} \right| TT_3 \right\rangle = \frac{U_0^t}{2} \sum_i \left\langle TT_3 \left| \vec{t}_i \cdot \left(\sum_k \vec{t}_k - \vec{t}_i \right) \right| TT_3 \right\rangle$$

$$= \frac{U_0^t}{2} \left[T(T+1) - \frac{1}{2} \left(\frac{1}{2} + 1 \right) \sum_{i=1}^A 1 \right] \tag{7.4.11}$$

$$\approx \frac{48}{A} \text{MeV } T(T+1) - 36 \text{ MeV}.$$

Thus the lowest energy is clearly obtained with the lowest possible value of T, viz. $T = \frac{1}{2}|N-Z|$. In other words the force prefers the lowest possible isospin, which means that the nuclear force leads to the lowest energy in the symmetric space-spin state.

The interaction in eq. (7.4.10) leads to the one-nucleon average potential

$$U^t = U_0^t \vec{t} \cdot \left\langle TT_3 \left| \sum_{k=1}^{A-1} \vec{t}_k \right| TT_3 \right\rangle \equiv U_0^t \vec{t} \cdot \langle \vec{T}^{A-1} \rangle. \tag{7.4.12}$$

This can be written in terms of raising and lowering operators $t_\pm$ and $T_\pm$

$$U^t = U_0^t \left[t_3 \left\langle T_3^{A-1} \right\rangle + \frac{1}{2} \left(t_+ \left\langle T_-^{A-1} \right\rangle + t_- \left\langle T_+^{A-1} \right\rangle \right) \right]. \tag{7.4.13}$$

The last term with $t_\pm$ clearly changes neutrons into protons or vice versa. It is therefore relevant under scattering conditions where charge exchange can take place. Analysis of such experimental results therefore also leads to estimates of the strength U_0^t. These two apparently different pieces of information can consistently be collected in an interaction like eq. (7.4.10).

7.5   BETA STABILITY

The saturation of the nuclear density, the large incompressibility and the roughly constant binding energy per unit volume are characteristics of a fluid. The semi-empirical liquid-drop-model mass formula is an extremely simple result of exploiting this analogy. The only less-intuitive term is the symmetry energy discussed in the previous subsection. It is, of course, of crucial importance for the stability of the nuclei. Its absence would make it favorable to change nearly all protons into neutrons by β-decay, and thereby avoid the expensive Coulomb energy.

An average estimate of the neutron-proton composition for the β-stable nuclei is easily obtained from the requirement of minimum energy for fixed total nucleon number. With the liquid-drop estimate, eq. (4.3.2), we find from eq. (4.6.1) the condition

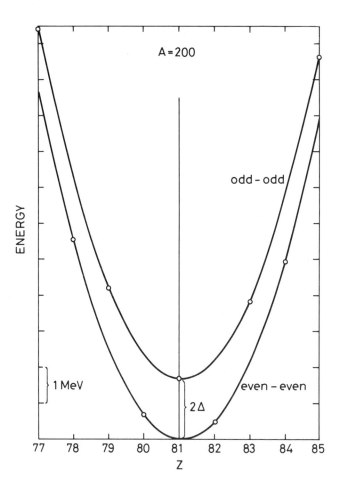

Figure 7.8   Energies of nuclear ground states for A=200.

$$\frac{\partial}{\partial Z} M(N, Z)c^2 \bigg|_A = (m_p - m_n) c^2 - 4b_V K_V \frac{(A - 2Z)}{A}$$

$$+ 2b_C \frac{Z}{A^{1/3}} + 4b_S K_S \frac{(A - 2Z)}{A^{4/3}} = 0. \tag{7.5.1}$$

which gives the neutron excess as function of mass number

$$(N - Z)_{\beta-\text{stable}} = A^{5/3} \frac{b_c}{4b_V K_V} \frac{1 - A^{-2/3}(m_n c^2 - m_p c^2)/b_c}{1 - (b_S K_S/4b_V K_V) A^{-1/3} + (b_c/4b_V K_V) A^{2/3}}$$

$$\approx \left(\frac{A}{18.9}\right)^{5/3} \frac{1 - 1.8A^{-2/3}}{1 + 0.0075A^{2/3}}. \tag{7.5.2}$$

For small A, $N \approx Z$, and when A exceeds 40, significant neutron excess is preferred. Eq. (7.5.1) is an average estimate where shell effects and odd-even effects in the binding are ignored. The shell effects can only lead to small local (in A) deviations from this estimate.

$$\frac{94\text{MeV}}{A} + \frac{0.7\text{MeV}}{A^{1/3}} = \frac{4b_V K_V}{A} + \frac{bc}{A^{1/3}} < 2\Delta = \frac{24\text{MeV}}{\sqrt{A}} \tag{7.5.3}$$

or equivalently $A > 18$. For larger mass numbers we can consequently expect that no odd-odd nuclei are β-stable.

For odd-A nuclei, where the δ-term vanishes both for (N,Z)=(odd,even) and (even,odd), only one nucleus is β-stable for given A. For even-A nuclei the situation is slightly more complicated, since β-decay means jumping 2Δ in binding energy. The curvature of the energy curve for fixed A is roughly $8b_V K_V/A + 2b_C A^{-1/3}$. The change in energy by a one-unit change of Z is therefore less than 2Δ when

For even-even nuclei the energy change caused by a change of n units of Z is larger than that of a n−1 unit change plus 2Δ when (see fig. 7.8)

$$n > \frac{1}{2} + \frac{\Delta \cdot A}{4b_v K_v + b_c A^{2/3}} \approx \frac{1}{2} + \frac{12\sqrt{A}}{94 + 0.7A^{2/3}}. \tag{7.5.4}$$

Thus for $A \leq 1500$ (corresponding to n=3) we can expect at the most three β-stable even-A nuclei for a fixed A.

This does not mean that the number of β-stable isotopes is limited to three. Especially for the heavier nuclei, where the neutron excess is significant, a larger number are possible and in fact often exist in nature (see fig. 7.9).

7.6   DRIP LINES

The valley of β stability, eq. (7.5.2), represent nuclei of especially high nuclear binding energy. The opposite extreme can also be investigated by use of the mass formula. The nucleon separation energy S is defined as the energy required to remove one nucleon from the nucleus. It can therefore be expressed in terms of the binding energy as

$$S_n(N, Z) \equiv B(N, Z) - B(N - 1, Z) \approx \left. \frac{\partial B}{\partial N} \right|_Z \qquad (7.6.1)$$

$$S_p(N, Z) \equiv B(N, Z) - B(N, Z - 1) \approx \left. \frac{\partial B}{\partial Z} \right|_N.$$

By adding more and more nucleons, the separation energy decreases, and eventually we reach a point where an extra nucleon is not bound any more. It does not stick to the nucleus, it falls off or drips off by itself. This point is characterized by

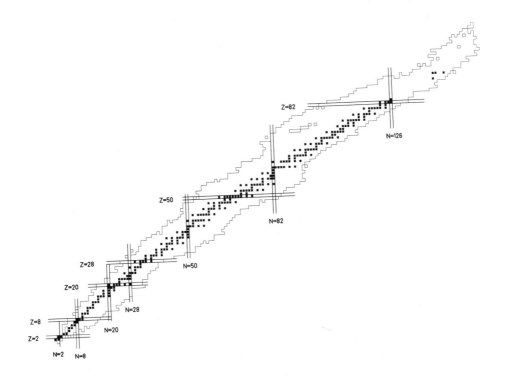

Figure 7.9   Nuclear species as a function of N(horizontal axis) and Z (vertical axis). The nuclei found in nature are indicated by black squares; the thin line indicates the boundary of other observed nuclei. The magic numbers are shown as double lines [*courtesy of P. Armbruster and K. H. Schmidt*].

zero separation energy. The resulting curves in a neutron-proton diagram are called drip lines.

Analytic expressions can easily be derived by using B_{LD} in eq. (7.6.1). They are shown in fig. 7.10. Within the drip lines are the stable nuclei, with the valley of β stability in the middle. Nuclei do not exist outside these lines which therefore surround the territory of nuclear physics.

7.7   FISSION

The nuclear binding energy per nucleon along the valley of β stability is a fairly well defined function of A. It has, as seen in fig. 4.1, a maximum around A=50. The structure is qualitatively easy to understand. For large A the fall-off is caused by the dominating repulsive Coulomb energy. For smaller A, the repulsive surface energy becomes more important. The surface region constitutes an appreciable and increasing fraction of the total nuclear volume and the binding energy consequently decreases.

The maximum can easily be found if the system has N=Z as predicted in eq.

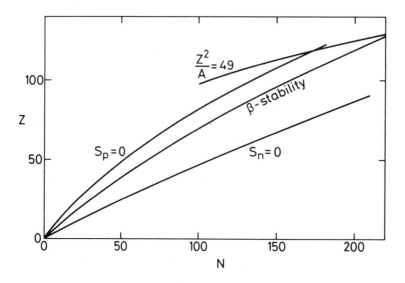

Figure 7.10   The limits of nuclear stability. The β-stability line and the proton and neutron drip-lines are calculated from the semi-empirical mass formula, eq. (4.3.2). The limiting curve $Z^2/A = 49$ of vanishing fission barrier is also shown.

(7.5.2) for small A. Then

$$\frac{\partial(B_{LD}/A)}{\partial A} = \frac{b_C}{6} A^{-4/3} \left(\frac{2b_S}{b_C} - A \right)$$ (7.7.1)

which vanishes for $A \approx 49$. Eq. (7.7.1) clearly shows the origin of the maximum as the result of a competition between Coulomb and surface effects.

This structure of the binding-energy function immediately tells us that we can gain energy either by dividing a heavy system (fission) or putting light systems together (fusion). The criterion is only that we end up closer to the maximum binding energy.

The energy Q gained by division of the nucleus (N,Z) into equal pieces is

$$Q = B(N, Z) - 2B \left(\frac{N}{2}, \frac{Z}{2} \right)$$
$$\approx 2b_S A^{2/3} \left[1 - K_S \left(\frac{N - Z}{A} \right)^2 \right] \left(1 - 2^{-2/3} \right) \left(x + \frac{1}{2} \frac{1 - 2^{1/3}}{1 - 2^{-2/3}} \right)$$ (7.7.2)

where we have introduced the liquid drop model and the fissility parameter x is given by (see sect. 5.5)

$$x \equiv \frac{b_C}{2b_S} \frac{Z^2}{A \left[1 - K_S \left(\frac{N-Z}{A} \right)^2 \right]} \equiv \frac{Z^2/A}{(Z^2/A)_c}.$$ (5.5.7)

The condition for gaining energy is then

$$x > 0.35 \quad \text{or} \quad \frac{Z^2}{A} > 0.35 \left(\frac{Z^2}{A} \right)_c .$$ (7.7.3)

By ignoring the neutron excess we have $(Z^2/A)_c \approx 49$ and therefore $Z^2/A > 17$ or (for N=Z) $A > 68$.

Thus it is energetically favorable either to divide systems of nucleon number larger than about 68 or combine two identical systems each of A < 34.

The energy released by such a symmetric fission of one ^{238}U nucleus is obtained from eq. (7.7.2) as about 200 MeV. This estimate is very realistic and in fact the basis for the interest in industrial-scale nuclear power production.

The fission instability of nuclei heavier than $A \approx 68$ is in apparent contradiction to our knowledge of the natural occurrence of nuclei as heavy as ^{238}U. The explanation is that the above estimate gives the energy release for each fission event. It does not say anything about the probability for the occurrence of the fission process: whether a huge barrier hinders the process resulting in a lifetime of cosmological size, or whether perhaps no barrier exists leading to an instantaneous decay.

This problem was already investigated in Chapter 5 where the barrier was found to vanish for x=1. Thus nuclei of 0.35 < x < 1 will fission spontaneously by a quantum-mechanical barrier-penetration process. The lifetime depends sensitively on the barrier height and thickness.

Nuclei of x > 1 have no barrier and fission immediately due to the Coulomb repulsion. Two equal fragments touching each other have a mutual potential Coulomb energy of

$$E_C \approx \left(\frac{Z}{2}\right)^2 e^2/(2R). \tag{7.7.4}$$

If the fragments start at rest a typical observed relative kinetic energy is half of this maximum value and the corresponding velocity v is

$$v \approx \sqrt{\frac{Z^2 e^2}{2 \cdot 8R\frac{1}{2}\frac{1}{8}m_N A}} = \sqrt{\frac{Z^2 e^2}{m_N AR}}. \tag{7.7.5}$$

The time it takes to move a distance R is then a measure of the fission time τ_f

$$\tau_f \approx \frac{R}{v} \approx \frac{m_N AR^2}{Z^2 e^2} \approx \frac{A}{Z} \frac{m_N r_0^3}{e^2} \approx 2 \times 10^{-22} \text{sec}. \tag{7.7.6}$$

Thus nuclei of x > 1 can exist only for an extremely short time.

The line x=1 is shown in fig. 7.10. Nuclei above the line decay by fission in about 10^{-22} sec. Thus the nuclear territory is limited above by this theoretical line. In practice, nuclear lifetimes become so short that it has not been possible to observe nuclei with Z greater than 109.

It is at this point worth emphasizing that all the estimates in this chapter are based on the semi-empirical mass formula. Only average properties are therefore expected to be correct, but important local modifications of this broad picture are possible. For example, it is widely believed that the shell structure will lead to magic nuclei with Z=114 to 120, which could have much longer lifetimes than those already found. This new closed shell is said to be an "island of stability" in the Z−N plane. Unfortunately, these nuclei have not been observed despite heroic attempts to make them.

PROBLEMS

7.1   What angular momentum J (where the eigenvalue of $\vec{J}^2$ is $J(J+1)\hbar^2$) do the following have in their ground states?
(a) deuteron
(b) ^{4}He
(c) ^{16}O
(d) ^{17}O
(e) ^{20}Ne

7.2   Write the T=1 state of two neutrons at rest as a determinant of single-particle states with spin up and down. Apply the isospin raising operator to show that the T=1 np state is as shown in fig. 7.3. What happens if you apply T_+ once more?

7.3   The reaction (π^+, π^0) changes a neutron into a proton in such a way that the total isospin of the target remains unchanged. A beam of 200 MeV (kinetic energy) π^+ from Los Alamos' LAMPF facility is directed onto a target of ^{208}Pb. Estimate the maximum energy of π^0 emerging from the target. What would you predict about the reaction (π^-, π^0)?

7.4   Make a computation showing why ^{23}Ne (Z=10) is unstable. What do you expect to be the decay products? How much energy can be released?

7.5   About half the energy released in the symmetric fission of ^{238}U appears as kinetic energy of the fission fragments; the rest is internal excitation energy of the two fragments (equally divided).
  (a) Suppose all the fragments' excitation energy is used to boil off nucleons. How many nucleons can be emitted per fragment? (Allow for the dependence of the ground state energy on N and Z.)
  (b) Suppose all the fragments' excitation energy were used to emit gamma rays. After this had happened, how many electrons or positrons could be emitted, per fragment?
  (c) Which do you think most likely happens, (a) or (b)? Why?

7.6   The nucleus ^{18}O (Z=8) has a spherical ground state, and an excited state corresponding to a prolate deformation.
  (a) What are the angular momentum, parity, and isospin of the ground state?
  (b) What are the angular momentum, parity, and isospin of the excited state?
  (c) Use an oscillator model with a reasonably realistic spin-orbit potential to estimate the pairing energy of the ground state, ΔE_{pair}=?
  (d) Use an oscillator model **without** a spin-orbit potential to estimate the energy gained by deforming the nucleus to a prolate deformation, E_D=? You need only consider one configuration for the valence neutrons, but you should motivate your choice of the configuration.
  (e) Use your answers to (c) and (d) to help explain why the ground state is not deformed.
  (f) What is the most stable nuclide with A=18? Give a quantitative argument why. Predict the maximum energy of a positron or electron emitted in the beta decay of ^{18}F (Z=9).

7.7   Calculate the magnetic dipole moment of the deuteron when the L=2 component of the relative motion is ignored. The measured value is 0.86 μ_N.

7.8   Assume the ground state (angular momentum $\frac{1}{2}$) of ^{3}He is a deuteron bound to a proton in a relative s-state. Calculate the magnetic dipole moment of

^{3}He in this state from those of the deuteron (0.86 μ_N) and the proton. Compare to the Schmidt value for the odd neutron in an $s_{1/2}$ state and to the measured value of -2.13 μ_N.

7.9   Which states in ^{14}N cannot be populated in the reaction

$$^4\text{He} + {}^{12}\text{C} \rightarrow {}^{14}\text{N} + {}^2\text{H}$$

when isospin is an exact quantum number?

7.10   Consider the reactions

$$^3\text{He} + {}^{12}\text{C} \rightarrow {}^{14}\text{N} + \text{p}$$

$$^3\text{He} + {}^{12}\text{C} \rightarrow {}^{14}\text{O} + \text{n}$$

populating isobaric analog states of T=1 in ^{14}N and ^{14}O. What is the ratio between the corresponding differential cross sections?

7.11   What is the isospin of ^{209}Pb in states consisting of a neutron added to the ground state of ^{208}Pb? What is the isospin of ^{209}Bi in states consisting of the ground state of ^{208}Pb plus a proton in a single particle orbit which (i) already is occupied by a neutron? (ii) is empty?

7.12   Show that eq. (7.4.5) differs from eq. (7.4.11) for the ground state (T $= \frac{1}{2}|N - Z|$) due to the neglected exchange potential. Eq. (7.4.11) predicts a term proportional to $|N - Z|$ in the nuclear binding energy. There is some experimental evidence for this so-called Wigner term. Why is it difficult to extract from measurements?

7.13   A power station is able to utilize all possible fission energy from 1 gram of uranium within 24 hours. How many megawatts can the station produce?

7.14   Calculate the energy gained by division of a given nucleus into two fragments as function of the nucleon number for one fragment on the β-stability line. Use ^{238}U as an example.

8

Nuclear Collective Motion

The collective model of nuclei is an attempt to explain certain striking properties of nuclear spectra and reactions in terms of only a few degrees of freedom, chosen to represent gross properties of the nucleus such as its shape and orientation. This possibility is suggested by analogy with macroscopic systems, such as rigid bodies or drops of liquid, whose behavior in most practical conditions can be satisfactorily described and predicted in terms of similar average properties such as shape, orientation, flow patterns, temperature, etc. In these macroscopic systems, it is seldom necessary to describe in detail the motion of the many constituent atoms and electrons, even though these ultimately make up the motion and so determine it. Instead, the macroscopic variables seem to be enough not only to describe but also to predict their own motion.

The laws governing macroscopic motion were discovered heuristically, and in general they take the form of equations involving a few, empirically determined coefficients such as moments of inertia, viscosity coefficients and the like. These quantities governing the macroscopic motion are generally known as transport coefficients, and may themselves be functions of the macroscopic dynamical variables - for example, a moment of inertia may depend on the speed of rotation, a viscosity will depend on temperature and density. The transport coefficients contain a summary of all the information about the system that is important for the collective motion, and are therefore enough to determine it. However, the actual evaluation of these transport coefficients often requires a detailed understanding of the dynamics of the many microscopic degrees of freedom whose average properties are represented by the collective variables. For example, prediction of the elastic modulus of compressibility of a solid or liquid requires a detailed knowledge of the correlations and interactions of its constituent atoms and molecules. In most macroscopic systems, this connection is conveniently formulated in the linear-response theory, which we will follow here.

A priori it is by no means clear that nuclear dynamics should be amenable to the same approach, since nuclei are composed of not more than a few hundred constituent particles. Nevertheless, the collective model of nuclei as pioneered by Bohr and Mottelson has been very fruitful, leading to a quantitative understanding of nuclear rotational and vibrational excitations. The formalism of the collective model parallels very closely the linear-response theory of macroscopic transport coefficients.

To introduce the concepts of the linear-response theory, we begin by studying the interaction of a system with a weak, externally applied field. We have chosen to illustrate this with a simple example from heavy-ion reactions, the Coulomb excitation of nuclear states. We then discuss the computation of the microscopic system's response to an externally-applied force in a simple model for the intrinsic degrees of freedom, the independent-particle model. With this background, we proceed to the actual formulation of collective dynamics, in which the external force on the microscopic or intrinsic motion is governed by the collective motion. In turn, the intrinsic degrees of freedom exert forces on the collective variables, and the self-consistent treatment of this interaction leads to the random-phase approximation (RPA) for nuclear vibrations, as well as the cranking model of slow collective motion. In later chapters, we will apply these concepts to rotations (Chapter 9), and to the case of aperiodic motion, by means of the cranking model, using the notion of a relaxation time for the microscopic system's memory of the applied forces. With it, we shall be equipped with most of the conceptual and formal framework for the discussions of low-energy nuclear collisions (Chapter 11).

8.1   POLARIZATION AND RESPONSE

Consider a system whose state is specified by a set of intrinsic coordinates x_i, and whose motion is governed by the intrinsic Hamiltonian $H_0(x_i, p_i)$ where the p_i are the momenta associated with the coordinates x_i. The example we have in mind is the nucleus, where the x_i's are the positions and spins of the nucleons, but the initial discussion is of very general validity (see e.g. Ashcroft and Mermin). In the presence of an external classical field A, the system's Hamiltonian is augmented by a perturbation δH, which may be taken to be proportional to A, if the external field is sufficiently weak. More generally, we can consider the influence of a set of external fields A_μ, which may be taken to be functions of some externally-determined, classical (C-number) coordinates Q_α. If all the fields are weak enough, their interaction with the intrinsic system will be characterized by the interaction Hamiltonian

$$\delta H = \sum_\mu F_\mu A_\mu(\{Q_\alpha\}) \tag{8.1.1}$$

where the coefficients of proportionality F_μ depend on the intrinsic coordinates, $F_\mu = F_\mu(\{\vec{r}_i\}, \{\vec{p}_i\})$. In nuclear physics, the F_μ are usually referred to as form factors. A familiar example of such a bilinear coupling is the interaction of a nucleus with an external electrostatic field, where the A_μ are the multipole components of the potential, and the F_μ are the multipole moments of the nuclear charge distribution. In that example, the multipole moments of the external field are in turn determined by the coordinates Q_α of the charges causing the field. Since δH is Hermitean, we may assume without loss of generality that A_μ is real and F_μ Hermitean.

In this section, we want to study the effect of the intrinsic system on the external, classical coordinates Q_α. The mean force $\delta F_\alpha(t)$ on the component Q_α due to the intrinsic system is given by

$$\delta F_\alpha(t) = -\left\langle \frac{\partial H}{\partial Q_\alpha} \right\rangle_t = -\sum_\mu \langle F_\mu \rangle_t \frac{\partial A_\mu}{\partial Q_\alpha} \tag{8.1.2}$$

where $\langle\ \rangle_t$ denotes the expectation value over the intrinsic coordinates, and $H = H_0 + \delta H$ is the total Hamiltonian. We see that the effect of the intrinsic motion on Q_α appears only as a force proportional to $\langle F_\mu \rangle_t$.

8.1 A. Forces Induced by the External Field

Since we have assumed δH is weak, we may compute $\langle \psi(t)|F_\mu|\psi(t)\rangle$ in first-order perturbation theory. The wave function $|\psi(t)\rangle$, initially in its (non-degenerate) ground state $|gs\rangle$ at time $t= -\infty$, is then given to first order in δH by

$$|\psi(t)\rangle = |gs\rangle - \frac{i}{\hbar} \sum_{n \neq gs} |n\rangle \int_{-\infty}^{t} dt' \langle n|\delta H(t')|gs\rangle\, e^{i(E_n-E_0)(t'-t)/\hbar} \tag{8.1.3}$$

where the states $|n\rangle$ are eigenstates of H_0,

$$H_0|n\rangle = E_n|n\rangle, \tag{8.1.4}$$

and E_0 is the ground state energy.

Since δH depends on time only because of the dependence of A_μ on $Q_\alpha(t)$, the matrix element of δH in (8.1.3) factorizes:

$$\langle n|\delta H(t)|gs\rangle = \sum_\mu \langle n|F_\mu|gs\rangle A_\mu(\{Q_\alpha(t)\}). \tag{8.1.5}$$

Using eqs. (8.1.3) and (8.1.5), we find, to first order in δH,

$$\langle \psi(t)|F_\mu|\psi(t)\rangle = \langle gs|F_\mu|gs\rangle - \sum_\nu \int_{-\infty}^{\infty} dt'\, \tilde{\chi}_{\mu\nu}(t-t')A_\nu(t') \tag{8.1.6}$$

where we have introduced the notation

$$\tilde{\chi}_{\mu\nu}(t) = \frac{i}{\hbar}\Theta(t>0) \sum_n \Big(\langle gs|F_\mu|n\rangle e^{-i(E_n-E_0)t/\hbar}\langle n|F_\nu|gs\rangle$$

$$- \langle gs|F_\nu|n\rangle e^{i(E_n-E_0)t/\hbar}\langle n|F_\mu|gs\rangle \Big) \tag{8.1.7}$$

which may be written more concisely in the form

$$\tilde{\chi}_{\mu\nu}(t) = \frac{i}{\hbar}\Theta(t > 0)\left\langle gs\left|\left[e^{iH_0t/\hbar}F_\mu e^{-iH_0t/\hbar}, F_\nu\right]\right|gs\right\rangle. \tag{8.1.8}$$

The function $\tilde{\chi}_{\mu\nu}$ is called the response function of the intrinsic system. From eq. (8.1.6), we see that $\tilde{\chi}_{\mu\nu}$ $(t-t')$ has a very simple interpretation: it is the change in the moment $\langle\psi|F_\mu|\psi\rangle$ of the intrinsic system at time t, due to the action of the field component A_ν at the earlier time t'. The Θ-function in the definition of $\tilde{\chi}_{\mu\nu}$, eq. (8.1.7), assures that causality is not violated. Note that $\tilde{\chi}_{\mu\nu}$ is real since the F's are Hermitean.

The interpretation of the response function $\tilde{\chi}$ may be further illuminated by considering the case when the external field has a harmonic time dependence, $A_\mu(t) = A_\mu^0\cos\omega t = \text{Re}A_\mu^0 e^{-i\omega t}$. In that case, we find from eq. (8.1.6) that

$$\langle\psi(t)|F_\mu|\psi(t)\rangle = \langle gs|F_\mu|gs\rangle - \text{Re}\sum_\nu \chi_{\mu\nu}(\omega)A_\nu^0 e^{i\omega t} \tag{8.1.9}$$

where we have introduced the Fourier transform of $\tilde{\chi}$

$$\chi_{\mu\nu}(\omega) = \int_{-\infty}^{\infty} dt\,\tilde{\chi}_{\mu\nu}(t)\,e^{i\omega t}. \tag{8.1.10}$$

$\chi(\omega)$ is known as the polarizability tensor, since we see from eq. (8.1.9) that it is just the constant of proportionality between the applied field and the induced moment (we note that the two need not be in phase, since $\chi(\omega)$ may be complex, see eqs. (8.1.18) through (8.1.22)). The most familiar example of the polarizability tensor is in the electrodynamics of media, where A_μ is a component of the electric field, and $\langle\psi|F_\mu|\psi\rangle$ is the additional average field due to the polarization of the medium.

We thus conclude that the interaction of the intrinsic system with a weak external field is characterized solely by two quantities, the static moment $\langle gs|F_\mu|gs\rangle$, and the response function $\tilde{\chi}_{\mu\nu}$, which contains all the dynamical information that matters to the external field. In particular, a knowledge of these quantities enables us to compute the rate at which the external fields supply energy to the intrinsic system. This may be obtained by summing, for each coordinate Q_α, its velocity times the force on the intrinsic system, which is given by the negative of eq. (8.1.2):

$$\frac{dE}{dt} = \sum_\alpha \frac{dQ_\alpha}{dt}\sum_\mu\langle\psi|F_\mu|\psi\rangle\frac{\partial A_\mu}{\partial Q_\alpha} = \sum_\mu \frac{dA_\mu}{dt}\langle\psi(t)|F_\mu|\psi(t)\rangle \tag{8.1.11}$$

The rate of energy transfer may be either positive or negative, as the external field can either receive energy from, or deposit it in, the intrinsic system. Indeed, if the time variation of the external field is slow enough, we expect the interaction to proceed adiabatically and therefore reversibly, so that all of the energy lost to the internal system will be recovered by the external field, provided that A_μ returns to its initial value of 0 as $t \to +\infty$. The forces associated with such a reversible transfer of energy are conservative. However, if the time dependence of the external fields is more rapid, they may give rise to the irreversible excitation of the intrinsic system. Thus the force on the external field will, in general, be partly dissipative and partly conservative.

To separate the dissipative part of the force from its conservative part, we find the total energy supplied by the external field in the case where it returns to its initial value $A_\mu(-\infty) = A_\mu(+\infty)$: from eqs. (8.1.6) and (8.1.11),

$$\Delta E = \int_{-\infty}^{\infty} dt \frac{dE}{dt} = \int_{-\infty}^{\infty} dt \sum_\mu \frac{dA_\mu}{dt} \langle gs|F_\mu|gs\rangle$$

$$- \sum_{\mu\nu} \int_{-\infty}^{\infty} dt \int_{-\infty}^{\infty} dt' \frac{dA_\mu(t)}{dt} A_\nu(t') \tilde{\chi}_{\mu\nu}(t-t') \qquad (8.1.12)$$

The first term which arises from the static moment $\langle gs|F_\mu|gs\rangle$ is zero; obviously a constant force does not contribute to the dissipation. The second term may be integrated by parts to give

$$\Delta E = \int_{-\infty}^{\infty} dt \int_{-\infty}^{\infty} dt' \sum_{\mu\nu} A_\mu(t) A_\nu(t') \frac{d}{dt} \tilde{\chi}_{\mu\nu}(t-t') \qquad (8.1.13)$$

Thus we conclude that the dissipative forces arise only from the part of the response function which is antisymmetric(see eq. 8.1.1b).

8.1 B. Description of the Response and Polarizability Functions

Without loss of generality, the response function may be separated into a symmetric part $\tilde{\chi}'$ and an antisymmetric part $\hat{\chi}''$,

$$\tilde{\chi}_{\mu\nu} = \tilde{\chi}'_{\mu\nu} + i\tilde{\chi}''_{\mu\nu} \qquad (8.1.14)$$

where $\tilde{\chi}'$ and $\tilde{\chi}''$ are taken to have the symmetry relations

$$\tilde{\chi}'_{\mu\nu}(-t) = \tilde{\chi}'_{\nu\mu}(t) \qquad (8.1.15)$$

$$\tilde{\chi}''_{\mu\nu}(-t) = -\tilde{\chi}''_{\nu\mu}(t). \qquad (8.1.16)$$

According to eq. (8.1.13) the dissipative forces arise from $\tilde{\chi}''$, while $\tilde{\chi}'$ only gives rise to conservative forces.

From the symmetry relations (8.1.15) and (8.1.16), together with the observation that $\tilde{\chi}(t)$ vanishes for negative times, it is easy to show that

$$\tilde{\chi}_{\mu\nu}(t) = 2i\Theta(t > 0)\tilde{\chi}''_{\mu\nu}(t) = 2\Theta(t > 0)\tilde{\chi}'_{\mu\nu}(t). \tag{8.1.17}$$

At first sight, this separation into symmetric and anti-symmetric parts may appear pedantic, since they are evidently proportional (see fig. 8.1). The significance of eqs. (8.1.15) and (8.1.16) may perhaps be more easily appreciated by considering their consequences for the Fourier transforms of $\tilde{\chi}'$ and $\tilde{\chi}''$:

$$\chi_{\mu\nu}(\omega) = \chi'_{\mu\nu}(\omega) + i\chi''_{\mu\nu}(\omega) \tag{8.1.18}$$

$$\chi'^*_{\mu\nu}(\omega) = \chi'_{\mu\nu}(\omega) \tag{8.1.19}$$

$$\chi''^*_{\mu\nu}(\omega) = \chi''_{\mu\nu}(\omega) \tag{8.1.20}$$

$$\chi'_{\mu\nu}(-\omega) = \chi'_{\nu\mu}(\omega) \tag{8.1.21}$$

$$\chi''_{\mu\nu}(-\omega) = -\chi''_{\nu\mu}(\omega) \tag{8.1.22}$$

Considering for simplicity the diagonal part of the polarization $\mu = \nu$, we see by comparing eqs. (8.1.19) and (8.1.20) with eq. (8.1.9) that the conservative part of the response function $\tilde{\chi}'$ gives rise to a polarization of the intrinsic system which is in phase with the applied field, while the dissipative part $\tilde{\chi}''$ induces a polarization which is out of phase with the external field. This result is familiar in the case of the electrodynamics of an isotropic medium, where χ' and χ'' are the real and imaginary parts of the dielectric polarizability. Note that these relations (8.1.18) - (8.1.22) all arise from the causality property of the response function.

Another insight into the separation of the conservative and dissipative parts of the response function comes when we look at them in terms of the eigenstates $|n\rangle$

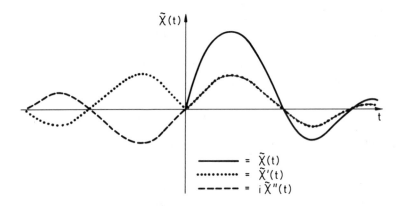

Figure 8.1   Relation between the response function $\tilde{\chi}(t)$ (full line), its symmetric part $\tilde{\chi}'(t)$ (dotted line) and antisymmetric part $\tilde{\chi}''(t)$ (dashed line).

of the intrinsic system. From eqs. (8.1.7), (8.1.16) and (8.1.17), we find

$$\tilde{\chi}''_{\mu\nu}(t) = \frac{1}{2\hbar} \sum_n \left(\langle gs|F_\mu|n\rangle\langle n|F_\nu|gs\rangle e^{-i\Omega_{no}t} - c.c. \right) \tag{8.1.23}$$

$$\chi''_{\mu\nu}(\omega) = \frac{\pi}{\hbar} \sum_n (\langle gs|F_\mu|n\rangle\langle n|F_\nu|gs\rangle \delta(\omega - \Omega_{no})$$

$$- \langle gs|F_\nu|n\rangle\langle n|F_\mu|gs\rangle \delta(\omega + \Omega_{no})) \tag{8.1.24}$$

where

$$\hbar\Omega_{no} = E_n - E_0. \tag{8.1.25}$$

To compute χ', we use the Kramers-Kronig relation

$$\chi'_{\mu\nu}(\omega) = P \int \frac{d\omega'}{\pi} \frac{\chi''_{\mu\nu}(\omega')}{\omega' - \omega} \tag{8.1.26}$$

which is a consequence of eqs. (8.1.15), (8.1.16) and (8.1.17) together with the relation $\int dt\Theta(t>0)\, e^{i\omega t} = P(\omega^{-1}) + i\pi\delta(\omega)$. Then $\chi'(\omega)$ is easily found from eq. (8.1.24):

$$\chi'_{\mu\nu}(\omega) = \frac{1}{\hbar} \sum_n \left(\langle gs|F_\mu|n\rangle\langle n|F_\nu|gs\rangle \frac{P}{\Omega_{no} - \omega} \right.$$

$$\left. + \langle gs|F_\nu|n\rangle\langle n|F_\mu|gs\rangle \frac{P}{\Omega_{no} + \omega} \right). \tag{8.1.27}$$

Comparing the expressions (8.1.24) and (8.1.27), we see that the difference between χ'' and χ' is that the former has δ-functions where the latter has principal values of pole terms. In the language of time-dependent perturbation theory, this means that χ'' corresponds to "real" transitions of the intrinsic system, and χ' to "virtual" transitions. Thus the energy lost from the macroscopic motion via χ'' reappears as excitations of the intrinsic system.

8.1 C.   Example: Coulomb Excitation

As a simple example of the application of linear response theory, consider the excitation of a heavy, spin-zero nucleus by the Coulomb field of a passing point charge (see fig. 8.2). H_0 is the nuclear Hamiltonian, and $\vec{Q}(t)$ is the coordinate of the projectile relative to the nucleus. The interaction Hamiltonian is conveniently written in its multipole expansion

$$\delta H = \sum_{\ell=0}^{\infty} \sum_{m=0}^{\ell} A_{\ell m}(\vec{Q}) F^E_{\ell m}(\{\vec{r}_i\}). \tag{8.1.28}$$

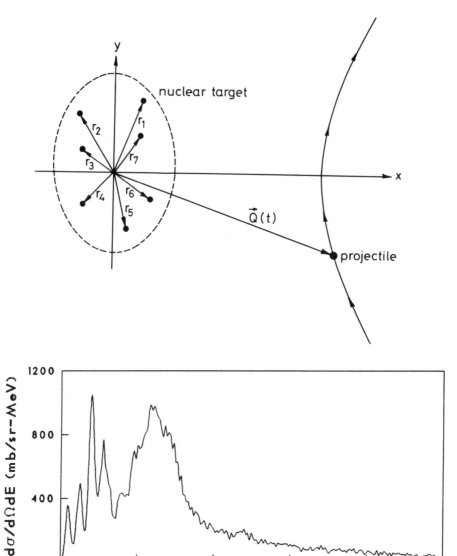

Figure 8.2   Coulomb excitation of a nucleus by a passing projectile. Above:Schematic picture. $\vec{Q}(t)$ is the projectile's trajectory in the target's coordinate system; $\vec{r}_1, \vec{r}_2$, etc. are the coordinates of nucleons in the target. Below: Measured differential cross section $d^2\sigma/d\Omega\,dE$ for 2.6° inelastic scattering of 84 MeV per nucleon ^{17}O by ^{208}Pb. [F. E. Bertrand, J. R. Beene, D. J. Horen, R. L. Auble, B. L. Burks, J. Gomez del Campo, M. L. Halbert, Y. Schutz, N. Alamanos, J. Barrette, F. Auger, B. Fernandez, A. Gillibert, W. Mittig, B. Haas, J. P. Vivien, data from GANIL experiment #83b.].

(Note that the two indices ℓ and m together constitute the label μ of eq. (8.1.1)). The intrinsic form factors $F_{\ell m}^E(x_i)$ are real combinations of the multipole operators on the proton coordinates x_i of the target:

$$F_{\ell m}^E(\{\vec{r}_i\}) = \frac{e}{2}\sum_{i=1}^{z}|\vec{r}_i|^\ell \left[Y_\ell^m(\theta_i, \phi_i) + (-1)^m Y_\ell^{-m}(\theta_i, \phi_i)\right] \qquad (8.1.29)$$

while the coupling coefficients $A_{\ell m}$ are similarly constructed from the multipole components of the electrostatic potential of the projectile's charge z acting on the nucleus

$$A_{\ell m}(\vec{Q}) = \frac{4\pi z e}{\left|\vec{Q}\right|^{\ell+1}}\frac{(-1)^\ell}{2(2\ell+1)}\left[Y_\ell^m(\theta_Q, \phi_Q) + (-1)^m Y_\ell^{-m}(\theta_Q, \phi_Q)\right]. \qquad (8.1.30)$$

The nucleus is initially in its ground state $|gs\rangle$. Since the ground state has total angular momentum 0, inspection of eqs. (8.1.23), (8.1.24) and (8.1.27) shows that, since all the intrinsic states $|n\rangle$ are eigenfunctions of the angular momentum, the response function $\chi_{\ell m, \ell' m'}$ must be diagonal in the angular momentum, i.e.

$$\chi_{\ell m, \ell' m'} = \chi_{\ell m}\delta_{\ell, \ell'}\delta_{m, m'}. \qquad (8.1.31)$$

To calculate the mean excitation energy, we use eq. (8.1.12): the static moments give 0, so

$$\Delta E = -\sum_{\ell m}\int_{-\infty}^{\infty}dt\int_{-\infty}^{\infty}dt'\,\frac{dA_{\ell m}(t)}{dt}A_{\ell m}(t')\tilde{\chi}_{\ell m}(t-t') \qquad (8.1.32)$$

which may be Fourier transformed to give

$$\Delta E = -i\sum_{\ell m}\int_{-\infty}^{\infty}\frac{d\omega}{2\pi}\overline{A}_{\ell m}(-\omega)\overline{A}_{\ell m}(\omega)\chi_{\ell m}(\omega)\omega$$

where

$$\overline{A}_{\ell m}(\omega) = \int_{-\infty}^{\infty}dt\,A_{\ell m}\left(\vec{Q}(t)\right)e^{i\omega t}. \qquad (8.1.33)$$

Inserting the spectral representations of χ' and χ'', eqs. (8.1.24) and (8.1.27), we find that the mean energy loss is a sum of terms, each arising from the excitation of a different intrinsic state $|n\rangle$:

$$\Delta E = \sum_n (E_n - E_0)P_n \qquad (8.1.34)$$

where

$$P_n = \sum_{\ell=0}^{\infty} \sum_{m=0}^{\ell} |\langle gs|F_{\ell m}^E|n\rangle|^2 \cdot |\overline{A}_{\ell m}(\Omega_{n0})|^2/\hbar^2. \tag{8.1.35}$$

It is natural to interpret P_n as the transition probability to the state $|n\rangle$.

From the definition of $F_{\ell m}^E$, eq. (8.1.29), we can see that there is only one non-zero term in the sum over ℓ in (8.1.35), namely when $\ell = \ell_n$, the angular momentum of the state $|n\rangle$. Furthermore, m must be equal to $|m_n|$, the absolute value of the state's angular momentum projection. Each level n must be part of a degenerate set of $2\ell + 1$ states with $m_n = -\ell$ to $m_n = \ell$. We can apply the Wigner-Eckart theorem to the non-zero matrix elements of $F_{\ell m}^E$ to show that they are all the same, except for $m = 0$ which is twice as big as the others because of the coherence of the two terms in (8.1.29). The total probability of exciting the states with energy E_n is therefore

$$P_n' = \sum_{m_n=-\ell_n}^{\ell_n} P_n = \frac{B(E\ell_n, 0 \to n)}{2(2\ell_n + 1)} \sum_{m=0}^{\ell_n} (1 + \delta_{m0}) |\overline{A}_{\ell_n m}(\Omega_{n0})|^2 \tag{8.1.36}$$

where we have introduced the electric multipole transition strength from the ground state to the level n,

$$B(E\ell_n, 0 \to n) = e^2 \sum_{m=-\ell_n}^{\ell_n} \sum_{m_n=-\ell_n}^{\ell_n} \left| \left\langle n\ell_n m_n \left| \sum_{i=1}^{Z} r_i^{\ell_n} Y_{\ell_n}^m(\theta_i, \phi_i) \right| gs \right\rangle \right|^2. \tag{8.1.37}$$

The multipole transition strength is one of the important characteristics of the structure of the states $|n\rangle$, see sect. (10.3.B).

P_n' is the product of two factors. The first is the multipole transition strength. The second factor is determined by the particle's trajectory $\vec{Q}(t)$, and is essentially the square of the Fourier amplitude of the multipole field of the projectile acting on the target, evaluated at the quantum-mechanical Bohr frequency of the excitation. The computation of this factor is a straightforward problem in classical mechanics. If we neglect the influence of the loss of angular momentum and energy on the projectile's trajectory, the cross section for excitation of the states $|n\rangle$ is given by

$$\frac{d\sigma}{d\Omega} = \left(\frac{d\sigma}{d\Omega}\right)_{\text{Rutherford}} \cdot P_n'. \tag{8.1.38}$$

Thus measurements of the cross section permit one to deduce P_n'. Since $\overline{A}_{\ell m}(\omega)$ is straightforward to compute [see for example Alder and Winther], we can use the measured transition probabilitites to determine $B(E\ell_n, 0 \to n)$, which is the only way that information about the structure of the states $|n\rangle$ appears in lowest-order perturbation theory. It is just the strength of the poles and δ-functions at $\omega = E_n$, appearing in the polarizability tensor $\chi(\omega)$. We observe from fig. 8.2 that a few states are much more strongly excited than the others. To wee what is special about these strongly-excited states, we need to understand more about the nucleus' response function.

8.2   NUCLEAR RESPONSE IN THE INDEPENDENT-PARTICLE MODEL

To provide a first concrete example of the computation of response and polarization functions, we choose the simple example of the independent-particle model (ipm). The ipm forms the starting point for many theories of nuclear structure, so it is an instructive first case to consider before discussing the effects of correlations. In fact, the linear response function of the ipm is the starting point for the simplest approaches to collective correlations: the collective model, and the random-phase approximation.

In the ipm, the nuclear interactions are approximated by a one-body potential, which determines a nucleon's motion in the field of the other nucleons. In addition to its dependence on the nucleon's coordinates and momenta, the mean field may also depend on other collective variables $Q = \{Q_\alpha\}$ such as the shape and orientation of the nucleus, or the external electric fields in which it finds itself. Thus the nucleonic coordinates $(\vec{r}_i, \vec{p}_i)$ appear in the Hamiltonian (4.2.1) in the form

$$H_{\text{ipm}} = \sum_i H^{\text{ipm}}(\vec{r}_i, \vec{p}_i, \{Q_\alpha\}). \tag{8.2.1}$$

To use the linear response theory, we must separate the ipm Hamiltonian into an intrinsic H_0 and a separable perturbation δH (cf. eq. (8.1.1)). The choice of this separation will depend on the application, but usually H_0 will be given by eq. (8.2.1) with $\{Q_\alpha\}$ chosen to have some suitable values $\{Q_\alpha^0\}$. For example, in Coulomb excitation, H_0 would be chosen as the mean field with the projectile at infinity, $Q^0 = \infty$. Thus we write

$$H_{\text{ipm}} = H_0 + \delta H \tag{8.2.2}$$

where

$$H_0 = \sum_i H^{\text{ipm}}(\vec{r}_i, \vec{p}_i, \{Q_\alpha^0\}). \tag{8.2.3}$$

Similarly, the perturbation field is also of the one-body type

$$\delta H = \sum_\mu A_\mu(\{Q_\alpha\}, \{Q_\alpha^0\}) \sum_i F_\mu^{\text{ipm}}(\vec{r}_i, \vec{p}_i, \{Q_\alpha^0\}). \tag{8.2.4}$$

We see that eq. (8.2.4) is of the form of eq. (8.1.1) with

$$A_\mu = A_\mu(\{Q_\alpha\}, \{Q_\alpha^0\}) \tag{8.2.5}$$

$$F_\mu = \sum_i F_\mu^{ipm}(\vec{r}_i, \vec{p}_i, \{Q_\alpha^0\}). \tag{8.2.6}$$

Note that both H_0 and F_μ depend on Q^0; thus χ and $\tilde{\chi}$ will also depend on $\{Q_\alpha^0\}$. For notational simplicity, we shall often drop the label Q^0 but it should be remembered that the implicit dependence is always there.

The eigenstates of H_0 are conveniently expressed as

$$|n\rangle = \prod_k (a_k^\dagger)^{n_k(n)} |vac\rangle \tag{8.2.7}$$

where $a_k^\dagger$ is a creation operator for a particle in the single-particle state $|k\rangle$ with single-particle energy $\hat{\varepsilon}_k$, and the occupation number $n_k(n)$ for the single-particle state $|k\rangle$ in the many-body state $|n\rangle$ can have the values 0 or 1. (Thus, the sum of the $n_k(n)$ for any many-body state $|n\rangle$ must be equal to the number of fermions.) The form factors F_μ entering into eqs. (8.1.23), (8.1.24) and (8.1.27) can then be written

$$F_\mu = \sum_{ik} \langle i|F_\mu^{ipm}|k\rangle a_i^\dagger a_k. \tag{8.2.8}$$

Inserting these expressions into eqs. (8.1.23) - (8.1.27), we find the response functions of the independent-particle model:

$$\tilde{\chi}_{\mu\nu}^{ipm\,\prime\prime}(t) = \frac{1}{2\hbar} \sum_{ik} N_{ik}[\langle i|F_\mu^{ipm}|k\rangle \langle k|F_\nu^{ipm}|i\rangle e^{i\omega_{ki}t} - c.c.] \tag{8.2.9}$$

$$\chi_{\mu\nu}^{ipm\,\prime\prime}(\omega) = \frac{\pi}{\hbar} \sum_{ik} N_{ik}[\langle i|F_\mu^{ipm}|k\rangle \langle k|F_\nu^{ipm}|i\rangle \delta(\omega - \omega_{ki})$$
$$- \langle i|F_\mu^{ipm}|k\rangle \langle k|F_\nu^{ipm}|i\rangle \delta(\omega + \omega_{ki})] \tag{8.2.10}$$

$$\chi_{\mu\nu}^{ipm\,\prime}(\omega) = \frac{1}{\hbar} \sum_{ik} N_{ik}[\langle i|F_\mu^{ipm}|k\rangle \langle k|F_\nu^{ipm}|i\rangle \frac{P}{\omega_{ki} - \omega}$$
$$+ \langle i|F_\mu^{ipm}|k\rangle \langle k|F_\nu^{ipm}|i\rangle \frac{P}{\omega_{ki} + \omega}] \tag{8.2.11}$$

where

$$\hbar\omega_{ki} = \hat{\varepsilon}_k - \hat{\varepsilon}_i \tag{8.2.12}$$

and

$$N_{ik} = n_i(gs)[1 - n_k(gs)]. \tag{8.2.13}$$

N_{ik} is the probability that the s.p. state $|i\rangle$ is occupied, and state $|k\rangle$ unoccupied. In the nuclear ground state, the factor N_{ik} merely restricts the summations to state i below and k above the Fermi surface.

For small-amplitude variations of the one-body field H^{ipm}, it is always possible to approximate the coupling in the bilinear form (8.2.4), by simply expanding H^{ipm} in powers of $Q_\alpha - Q_\alpha^0$. Keeping only the leading term, we find

$$A_\mu = Q_\mu - Q_\mu^0 \tag{8.2.14}$$

$$F_\mu^{ipm}(\vec{r}_i, \vec{p}_i, \{Q_\alpha^0\}) = \frac{\partial}{\partial Q_\mu} H^{ipm}(\vec{r}_i, \vec{p}_i, \{Q_\alpha\}) \Bigg|_{\{Q_\alpha\}=\{Q_\alpha^0\}}. \tag{8.2.15}$$

8.3   VIBRATIONS OF THE NUCLEAR SHAPE IN THE INDEPENDENT-PARTICLE MODEL

A famous example of this coupling is applied in Bohr and Mottelson's treatment of vibrations of spherical nuclei. They consider a small perturbation of the nuclear shape about its spherical equilibrium, expanded in multipole coefficients $\alpha_{\ell m}$ as in eq. (5.3.1):

$$H^{ipm}(\vec{r}, \vec{p}, \{Q_\alpha\}) = \frac{\vec{p}^2}{2m_N} + \hat{U}\left(|\vec{r}| - R(A)\sum \alpha_{\ell m} Y_\ell^m(\theta, \phi)\right). \tag{8.3.1}$$

The collective coordinates are taken to be the multipole coefficients $\alpha_{\ell m}$. Then, using eqs. (8.2.15) and (8.3.1) with $Q^0 = 0$,

$$F_{\ell m}^{ipm}(\vec{r}) = -R(A)\frac{\partial \hat{U}}{\partial r}(|\vec{r}|)\, Y_\ell^m(\theta, \phi). \tag{8.3.2}$$

For $m \neq 0$, this definition of F is only Hermitean if the $\alpha_{\ell m}$ are constrained by the relation $\alpha_{\ell m}^* = (-1)^m \alpha_{\ell,-m}$. We ought thus to consider real combinations of Y_ℓ^m and Y_ℓ^{-m} as in eq. (8.1.29). In that case, χ will turn out to be diagonal, for the reasons discussed in connection with eq. (8.1.29).

The particular cases $\ell = 0$ and 1, the isoscalar monopole and dipole modes, are badly described in the independent-particle model. The monopole excitation involves a change in the nuclear volume, and must depend on the compressibility of nuclear matter which is determined by the short-range correlations. The isoscalar dipole mode corresponds to translation of the nucleus, while the ipm artifically restricts the motion of the nucleus' center of mass, as described in Sect. 4.5. Both these types of nuclear motion can be treated by linear-response theory, but not within the ipm. We will return to consideration of these modes in Sect. 8.5.

Another, particularly instructive example of the coupling (8.2.15), also treated by Bohr and Mottelson, is the case when H^{ipm} is a deformed harmonic oscillator:

$$H^{ipm}(\vec{r}, \vec{p}, \{\Omega_\mu\}) = \frac{\vec{p}^2}{2m_N} + \frac{m_N}{2}\sum_{\mu=x,y,z} r_\mu^2 \Omega_\mu^2 \tag{8.3.3}$$

where $\{r_\mu\}$ are the components of $\vec{r}$ along the cartesian axes, and $\{\Omega_\mu\}$ are the "spring constants", which determine the shape of the nuclear single-particle field. In order to use the ipm, it is necessary to restrict consideration to changes of shape which maintain the volume of the nucleus, a requirement which may be approximately formulated as (see eq. (5.3.4))

$$\Omega_x \Omega_y \Omega_z = \text{constant}. \tag{8.3.4}$$

Thus only two of the Ω_μ are independent, with the third determined by eq. (8.3.4). Choosing Ω_x and Ω_y as the independent coordinates Q_μ, we find from eqs. (8.2.15), (8.3.3) and (8.3.4).

$$F_\mu^{\text{ipm}} = m_N r_\mu^2 \, \Omega_\mu^0 + m_N z^2 \, \Omega_z^0 \left. \frac{\partial \Omega_z}{\partial \Omega_\mu} \right|_{\Omega_\mu = \Omega_\mu^0} \tag{8.3.5}$$

where $\{\Omega_\mu^0\}$ are the equilibrium values, determined in Chapter 5.

The matrix elements of F_μ^{ipm} are easily evaluated. Representing the eigenstates of H^{ipm} by

$$|i\rangle = |n_x(i)\, n_y(i)\, n_z(i)\rangle \tag{8.3.6}$$

with the eigenvalue

$$\hat{\varepsilon}_i = \hbar \sum_\mu \left(n_\mu(i) + \tfrac{1}{2} \right) \Omega_\mu^0 \tag{8.3.7}$$

we find for the response functions $\tilde{\chi}_{\mu\nu}^{\text{ipm}}(\mu, \nu = x, y)$

$$\tilde{\chi}_{\mu\nu}^{\text{ipm}\,\prime\prime}(t) = \sigma_\mu \delta_{\mu\nu} \sin\left(2\Omega_\mu^0 t\right) + \sigma_z \frac{\left(\Omega_z^0\right)^2}{\Omega_\mu^0 \Omega_\nu^0} \sin\left(2\Omega_z^0 t\right) \tag{8.3.8a}$$

$$\chi_{\mu\nu}^{\text{ipm}\,\prime}(\omega) = \sigma_\mu \delta_{\mu\nu} \frac{2\hbar\Omega_\mu^0 P}{\left(2\Omega_\mu^0\right)^2 - \omega^2} + \sigma_z \frac{\left(\Omega_z^0\right)^2}{\Omega_\mu^0 \Omega_\nu^0} \frac{2\hbar\Omega_z^0 P}{\left(2\Omega_z^0\right)^2 - \omega^2} \tag{8.3.8b}$$

$$\chi_{\mu\nu}^{\text{ipm}\,\prime\prime}(\omega) = \pi \sigma_\mu \delta_{\mu\nu} \left[\delta\left(\omega - 2\Omega_\mu^0\right) - \delta\left(\omega + 2\Omega_\mu^0\right) \right] \hbar$$

$$+ \pi \sigma_z \frac{\left(\Omega_z^0\right)^2}{\Omega_\mu^0 \Omega_\nu^0} \left[\delta\left(\omega - 2\Omega_z^0\right) - \delta\left(\omega + 2\Omega_z^0\right) \right] \hbar \tag{8.3.8c}$$

where

$$\sigma_\mu = 2 \sum_i^< \left(n_\mu(i) + 1 \right)\left(n_\mu(i) + 2 \right) \sum_j^> \delta_{n_\mu(j), n_\mu(i)+2} \tag{8.3.8d}$$

and the sums over i and j are over occupied and unoccupied s.p. orbitals, respectively. We see that the response consists of components of two frequencies. The first terms in eqs. (8.3.8a-c), with frequency $2\Omega_\mu$, correspond to excitations in which the particles are excited to states two shells above the ground-state wave functions (single-shell excitations are not excited because of parity).

Inspecting eq. (8.3.8d), we see that only single-particle states near the Fermi surface are excited. This excitation has a simple physical interpretation as a density wave, excited by squeezing the nucleus along the μ-axis, which propagates across the nuclear diameter and bounces back at a time $2\pi/2\Omega_\mu$ later. This time is just the time it takes a classical particle in the well to cross the nucleus once. Since, in the harmonic oscillator, this time is the same for all particles traveling in the μ-direction, they all move in phase and bump up against the opposite side at the same time. There, they bounce and continue to slosh back and forth indefinitely.

The remaining terms in eqs. (8.3.8a-c), with frequency $2\Omega_z$, correspond to similar excitations along the z-axis, induced by the volume-conservation condition: if one squeezes the nucleus along the x-axis, it bulges along the z-axis. This effect of the incompressibility of the nuclear matter also produces a force along the z-axis, if the nucleus is squeezed along the x-axis: note the term for $\mu \neq \nu$ in eqs. (8.3.8a-c).

We can look for a set of collective coordinates in which χ is diagonal. Since all nuclear ground states are axially symmetric, we need only consider the case $\Omega_x^0 = \Omega_y^0 = \Omega_\perp^0$; then the appropriate linear combinations are given by the following unitary transformation U:

$$\begin{pmatrix} Q_\beta \\ Q_\gamma \end{pmatrix} = \frac{1}{\sqrt{2}} \begin{pmatrix} 1 & 1 \\ 1 & -1 \end{pmatrix} \begin{pmatrix} \Omega_x - \Omega_\perp^0 \\ \Omega_y - \Omega_\perp^0 \end{pmatrix} \equiv U \begin{pmatrix} \Omega_x - \Omega_\perp^0 \\ \Omega_y - \Omega_\perp^0 \end{pmatrix} \tag{8.3.9}$$

The corresponding response and polarizability tensors are given by the transformation $\chi \rightarrow U\chi U^\dagger$, so we get e.g.

$$\chi_{\beta\beta}^{ipm}{}'(\omega) = \sigma_\perp \frac{2\hbar\Omega_\perp^0 P}{(2\Omega_\perp^0)^2 - \omega^2} + 2\sigma_z \left(\frac{\Omega_z^0}{\Omega_\perp^0}\right)^2 \frac{2\hbar\Omega_z^0 P}{(2\Omega_z^0)^2 - \omega^2} \tag{8.3.10a}$$

$$\chi_{\gamma\gamma}^{ipm}{}'(\omega) = \sigma_\perp \frac{2\hbar\Omega_\perp^0 P}{(2\Omega_\perp^0)^2 - \omega^2} \tag{8.3.10b}$$

with similar results for $\chi^{ipm}{}''(\omega)$ and $\tilde{\chi}^{ipm}{}''(t)$. The corresponding form factors are

$$F_\beta = \frac{m_N}{\sqrt{2}\,\Omega_\perp^0} \left[(x^2 + y^2)\left(\Omega_\perp^0\right)^2 - 2z^2 \left(\Omega_z^0\right)^2\right] \tag{8.3.11a}$$

$$F_\gamma = \frac{m_N}{\sqrt{2}\,\Omega_\perp^0} (x^2 - y^2)\left(\Omega_\perp^0\right)^2 . \tag{8.3.11b}$$

Q_β and F_β correspond to a change of shape in which the nucleus becomes shorter and thicker, or longer and thinner, retaining axial symmetry. Q_γ and F_γ correspond to a departure from axial symmetry in which the equator of the nucleus deforms into an ellipse in the xy plane, with the length along the z-axis unchanged. The lowest-frequency modes corresponding to these shape changes are known as β and γ vibrations, respectively.

The high degree of coherence in the ipm response to F_β and F_γ is a special feature of the oscillator potential. For a more realistic potential with spin-orbit terms and a sharper surface, the excitations would be spread over a band of frequencies $\Delta\Omega$ around the mean shell spacing, and would go out of phase after a time

$$\tau_L \approx (\Delta\Omega)^{-1} \tag{8.3.12}$$

known as the collective coherence time. Thus $\tilde{\chi}^{ipm}(t)$ would not oscillate indefinitely, as it does in eq. (8.3.8a), but would instead die away, after a time τ_L. In addition, overtones appear due to excitation of states across several major shells. Fig. 8.3 shows an example of the response function for a heavy (prolate) nucleus in the ipm. The form factor corresponds to a change in the length of the nucleus. The first few oscillations show the "beats" characteristic of the superposition of oscillations with the two shell frequencies; these beats become weaker for long times as the incoherence of the ipm response within a major shell becomes important. Residual interactions neglected in the ipm make the response functions fall off for large times.

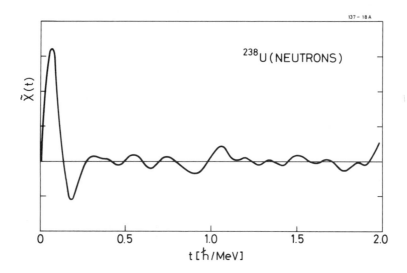

Figure 8.3   Response function $\tilde{\chi}(t)$ of neutrons in a nucleus of ^{238}U in the independent-particle model. The form factor is $e^{z/a}$ where a is the surface thickness of the ipm, and z is the axis of symmetry of the prolate ground state. [*from P. Johansen, H. Hofmann, A. Jensen and P. Siemens, Nucl. Phys.* **A288** *(1977) 52.*]

The decay of $\tilde{\chi}^{ipm}(t)$ for large t has a very important consequence: it becomes possible to study slow, large-amplitude collective motion within the linear-response theory, as we shall show in Sect. 8.6. Before considering large-amplitude motion, however, we will discuss the simpler case of small-amplitude harmonic vibrations.

8.4   HARMONIC VIBRATIONS IN THE RANDOM PHASE APPROXIMATION

In this section we want to study the small-amplitude vibrations in the neighborhood of Q^0, which are implied by the ipm Hamiltonian (8.2.1). We will follow the approach of the cranking model in which the parameters Q_α which characterize the one-body field are allowed to be functions of time, $Q_\alpha = Q_\alpha(t)$. We then use linear response theory to look for the equations of motion of $Q_\alpha(t)$. The result is identical with the Random Phase Approximation.

In the model described in this chapter, Q_α is the main dynamical variable. The influence of Q_α's time dependence on the motion of the intrinsic coordinates is treated perturbatively, then averaged over to obtain an equation of motion for $Q_\alpha(t)$ by itself. However, the effect of the intrinsic motion on Q_α is not a perturbation. This "favoritism", in which Q_α is treated better than the $\vec{r}_i$ and $\vec{p}_i$, is gained at the expense of a classical treatment of Q_α, in contrast to the quantum motion of the intrinsic degrees of freedom.

8.4 A.   Constructing the Single-particle Hamiltonian

The starting point for the cranking model is the assumption that we know a one-body operator $H_{sp}(\{\vec{r}_i\}, \{\vec{p}_i\}, \{Q_\alpha(t)\})$, which is the Hamiltonian operator corresponding to the true energy of the system. This assumption is in the spirit of the mean-field or a Hartree-Fock-Brueckner model. Indeed, we might be tempted to think of H_{sp} as a parametrization of the family of mean fields U found in these models, where the dependence on $\{Q_\alpha\}$ could arise through the introduction of a suitable constraining external field. However, we must recall (eqs. (4.2.9), (4.2.12)) that, even in a simple Hartree-Fock theory, the energy is not given by the expectation value of the mean field, which over-counts the 2-body interaction:

$$\sum_{j=1}^{A} \varepsilon_j = \langle \psi_{HF} | \sum_{i=1}^{A} \frac{p_i^2}{2m_N} | \psi_{HF} \rangle + \langle \psi_{HF} | \sum_{ij} V(\vec{r}_i - \vec{r}_j) | \psi_{HF} \rangle \tag{8.4.1}$$

while

$$\langle \psi_{HF} | H | \psi_{HF} \rangle = \langle \psi_{HF} | \sum_{i=1}^{A} \frac{p_i^2}{2m_n} | \psi_{HF} \rangle$$
$$+ \frac{1}{2} \langle \psi_{HF} | \sum_{ij} V(\vec{r}_i - \vec{r}_j) | \psi_{HF} \rangle. \tag{8.4.2}$$

Thus the mean fields cannot be used as a Hamiltonian.

Fortunately, Strutinsky showed that the overcounted potential energy is a smooth function of the collective coordinates $\{Q_\alpha\}$. Using the results of Sect. 5.6, we can find the correction to the independent-particle-model energy

$$\hat{\mathcal{E}} = \sum_{i=1}^{A} \hat{\varepsilon}_i. \tag{5.6.7}$$

Using eqs. (5.6.12) - (5.6.14) we have

$$\langle \psi_{\mathrm{HF}} \left| \mathrm{H} \right| \psi_{\mathrm{HF}} \rangle = \hat{\mathcal{E}} + \delta\widetilde{\mathcal{E}} \tag{8.4.3}$$

where

$$\delta\widetilde{\mathcal{E}}(\{Q_\alpha\}) = \widetilde{\mathrm{E}} - \widetilde{\mathcal{E}} = -\mathrm{B}_{\mathrm{LD}} - \widetilde{\mathcal{E}} \tag{8.4.4}$$

is the difference between the liquid-drop energy and the smoothed single-particle energy $\widetilde{\mathcal{E}}$. The same correction $\delta\widetilde{\mathcal{E}}$ could have been obtained if we had started with a Slater determinant corresponding to an excited state of $\mathrm{H}_{\mathrm{ipm}}$.

We thus construct our model Hamiltonian

$$\mathrm{H}_{\mathrm{sp}} = \sum_{i=1}^{A} \mathrm{H}^{\mathrm{sp}} \left(\vec{r}_i, \vec{p}_i, \{Q_\alpha\} \right) \tag{8.4.5}$$

where

$$\mathrm{H}^{\mathrm{sp}} \left(\vec{r}, \vec{p}, \{Q_\alpha\} \right) = \mathrm{H}^{\mathrm{ipm}} \left(\vec{r}, \vec{p}, \{Q_\alpha\} \right) + \frac{\delta\widetilde{\mathcal{E}}(\{Q_\alpha\})}{A} \tag{8.4.6}$$

differs from the phenomenological independent-particle model only by the function $\delta\widetilde{\mathcal{E}}(\{Q_\alpha\})/A$ which does not depend on the intrinsic coordinates $\vec{r}$ and $\vec{p}$. Inspecting eqs. (8.2.3) and (8.2.15), we see that this change will add a numerical constant to H_0 and F_μ:

$$\mathrm{H}_0^{\mathrm{sp}} = \mathrm{H}^{\mathrm{ipm}}(\vec{r}, \vec{p}, \{Q_\alpha^0\}) + \delta\widetilde{\mathcal{E}}(\{Q_\alpha^0\})/A \tag{8.4.7}$$

$$\mathrm{F}_\mu^{\mathrm{sp}} = \frac{\partial \mathrm{H}^{\mathrm{sp}}}{\partial Q_\mu} \big|_{\{Q_\alpha = Q_\alpha^0\}} = \mathrm{F}_\mu^{\mathrm{ipm}} + \frac{\partial}{\partial Q_\mu^0} \frac{\delta\widetilde{\mathcal{E}}(\{Q_\alpha^0\})}{A}. \tag{8.4.8}$$

This will change $\langle \mathrm{gs}|\mathrm{F}_\mu|\mathrm{gs}\rangle$ and will shift the single-particle energies $\hat{\varepsilon}_j$ by a uniform amount, but will leave the response function unchanged from the ipm because the change involves only diagonal matrix elements, which do not appear in the response function.

8.4 B.   Constructing the Equation of Motion

We have found an $\mathrm{H}_{\mathrm{sp}} \left(\{\vec{r}_i\}, \{\vec{p}_i\}, \{Q_\alpha\} \right)$ which adequately represents the energy of a nucleus, as well as its spectrum of elementary (single-particle) excitations. The classical equations of motion for $\{Q_\alpha(t)\}$ can now be obtained very simply: we just have to require that $\{Q_\alpha(t)\}$ be such that $\langle\psi|H_{sp}|\psi\rangle$ is conserved. That requirement follows from the Ehrenfest equation of motion for $\langle H_{sp}\rangle$:

$$
\begin{aligned}
0 &= \frac{d}{dt}\langle\psi(t)|H_{sp}|\psi(t)\rangle = \left\langle\psi(t)\left|\frac{\partial}{\partial t}H_{sp}(\{Q_\alpha(t)\})\right|\psi(t)\right\rangle \\
&= \sum_\mu \frac{dQ_\mu}{dt}\left\langle\psi(t)\left|\frac{\partial H_{sp}}{\partial Q_\mu}\right|\psi(t)\right\rangle .
\end{aligned}
\tag{8.4.9}
$$

The energy should be conserved along any physically-realizable trajectory in the $\{Q_\alpha\}$ space. Then, except for the trivial solution $Q_\mu = \text{constant}$, eq. (8.4.9) implies

$$
\left\langle\psi(t)\left|\frac{\partial H_{sp}}{\partial Q_\mu}\right|\psi(t)\right\rangle = 0. \tag{8.4.10}
$$

To allow for oscillatory motion, we have to expand the single-particle Hamiltonian (compare eqs. 8.2.2–4 and 8.2.14– 8.1.15) to second order in $Q_\mu - Q_\mu^0$:

$$
\begin{aligned}
H_{sp} &\approx \sum_i H_0^{sp}(\vec{r}_i,\vec{p}_i) + \sum_\mu (Q_\mu - Q_\mu^0)\sum_i F_\mu^{sp}(\vec{r}_i,\vec{p}_i,\{Q_\alpha^0\}) \\
&+ \frac{1}{2}\sum_{\mu,\nu}(Q_\mu - Q_\mu^0)(Q_\nu - Q_\nu^0)\sum_i \frac{\partial^2 H^{sp}}{\partial Q_\mu \partial Q_\nu}(\vec{r}_i,\vec{p}_i,\{Q_\alpha^0\}) .
\end{aligned}
\tag{8.4.11}
$$

The force on the collective coordinate Q_μ is now given by

$$
\begin{aligned}
-\left\langle\psi(t)\left|\frac{\partial H_{sp}}{\partial Q_\mu}\right|\psi(t)\right\rangle &= 0 \approx -\left\langle\psi(t)\left|\sum_i F_\mu^{sp}(\vec{r}_i,\vec{p}_i,\{Q_\alpha^0\})\right|\psi(t)\right\rangle \\
&- \sum_\nu (Q_\nu(t) - Q_\nu^0)\left\langle\psi(t)\left|\sum_i \frac{\partial^2 H^{sp}}{\partial Q_\mu \partial Q_\nu}(\vec{r}_i,\vec{p}_i,\{Q_\alpha^0\})\right|\psi(t)\right\rangle .
\end{aligned}
\tag{8.4.12}
$$

Corresponding to the evaluation of $\langle\psi|H_{sp}|\psi\rangle$ to second order in $Q_\mu - Q_\mu^0$, we must evaluate the expectation values in (8.4.12) to first order in $Q_\mu - Q_\mu^0$. Since the second term already has a factor $Q_\mu - Q_\mu^0$, the expectation value can be approximated by its value in the ground state

$$
\kappa_{\mu\nu} \equiv -\left\langle gs\left|\sum_i \frac{\partial^2 H^{sp}}{\partial Q_\mu \partial Q_\nu}(\vec{r}_i,\vec{p}_i,\{Q_\alpha^0\})\right|gs\right\rangle . \tag{8.4.13}
$$

Using the linear-response eq. (8.1.6) to evaluate the first term, we find the equation of motion

$$
\begin{aligned}
0 &= \int_{-\infty}^{\infty} dt' \sum_\nu \tilde{\chi}_{\mu\nu}^{ipm}(t-t')(Q_\nu(t') - Q_\nu^0) - \left\langle gs\left|\sum_i F_\mu^{sp}\right|gs\right\rangle \\
&+ \sum_\nu \kappa_{\mu\nu}(Q_\nu(t) - Q_\nu^0) .
\end{aligned}
\tag{8.4.14}
$$

8.4 C.   Small-amplitude Vibrational Solutions

The simplest solutions to the equations of motion are found by assuming that $Q_\mu - Q_\mu^0$ remains small for all times. For studying small vibrations, it is convenient to choose the equilibrium values of $\{Q_\alpha^0\}$ given by

$$\left\langle \text{gs} \left| \sum_i F_\mu^{\text{sp}} (\{Q_\alpha^0\}) \right| \text{gs} \right\rangle = 0.$$

We will adopt this convention for the rest of this section, returning to the more general case in Sect. 8.6. Under this assumption, we can look for solutions of the form $Q_\nu(t) = Q_\nu^0 + A_\nu^n e^{-i\omega_n t + \phi}$, which should be found for every value of the arbitrary phase ϕ. Then we find, analogous to eq. (8.1.9), an eigenvalue equation for A_ν^n and ω_n:

$$\sum_\nu \left[\kappa_{\mu\nu} + \chi_{\mu\nu}^{\text{ipm}} (\omega_n) \right] A_\nu^n = 0. \tag{8.4.15}$$

The homogeneous eq. (8.4.15) only has solutions when

$$\det \left| \kappa_{\mu\nu} + \chi_{\mu\nu}^{\text{ipm}} (\omega_n) \right| = 0. \tag{8.4.16}$$

This is an equation for the eigenfrequencies ω_n, known as the characteristic equation of the RPA (Random Phase Approximation).

To understand the RPA eigenvalue eq. (8.4.16), we can consider the simple case when $\kappa_{\mu\nu}$ and $\chi_{\mu\nu}^{\text{ipm}}$ are diagonal. Then eq. (8.4.16) has roots whenever

$$\chi_\mu^{\text{ipm}} (\omega_n) = -\kappa_\mu. \tag{8.4.17}$$

In the ipm, we have from eqs. (8.2.10), (8.2.11), and (8.1.18)

$$\chi_\mu^{\text{ipm}} (\omega) = \sum_{jk} \frac{N_{jk}}{\hbar} \left| \langle j \left| F_\mu^{\text{ipm}} \right| k \rangle \right|^2$$

$$\left[\frac{2\omega_{kj} P}{\omega_{kj}^2 - \omega^2} + i\pi \left[\delta (\omega - \omega_{kj}) - \delta (\omega + \omega_{kj}) \right] \right]. \tag{8.4.18}$$

We see that eq. (8.4.17) has solutions for real values of ω_n, when

$$\chi_\mu^{\text{ipm}\,\prime} (\omega_n) = \sum_{jk} \frac{N_{jk}}{\hbar} \left| \langle j \left| F_\mu^{\text{ipm}} \right| k \rangle \right|^2 \left[\frac{2\omega_{kj}}{\omega_{kj}^2 - \omega_n^2} \right] = -\kappa_\mu. \tag{8.4.19}$$

The graphical solution of eq. (8.4.19) is illustrated schematically in fig. 8.4.

For oscillations of the nuclear shape, κ_μ is negative. For example, choosing $Q_\mu = \alpha_{\ell 0}$ in eq. (8.3.1), we see that eqs. (5.3.2) and (5.3.3) imply that κ is diagonal in the angular momentum, ℓ, with

$$-\kappa_\ell = \langle \mathrm{gs} | \frac{\partial^2 \mathrm{H}_{\mathrm{sp}}}{\partial Q_\mu^2} | \mathrm{gs} \rangle \qquad (8.4.20)$$

$$\approx \sum_i \langle \mathrm{gs} | [R(A)^2 \frac{\partial^2 \hat{U}(r_i)}{\partial r_i^2} Y_\ell^0(\theta_i, \phi_i)^2 + 2R(A) \frac{\partial \hat{U}(r_i)}{\partial r_i} Y_0^0(\theta_i, \phi_i)^2] | \mathrm{gs} \rangle.$$

Since the largest term, $R^2 \partial^2 \hat{U} / \partial r^2$, is positive inside the nucleus, the mean value is positive, so κ_ℓ is negative. (The second derivative of $\delta \widetilde{\mathcal{E}}/A$ doesn't change this result since $\delta \widetilde{\mathcal{E}}$ is very flat for small deformations.) This situation corresponds to the upper broken line in fig. 8.4. We see that it is possible for eq. (8.4.19) to have a low-frequency solution for ω_1 less than any of the single-particle excitation frequencies ω_{kj}. The remaining eigenfrequencies are sandwiched in between the ω_{kj}.

We shall see in Sect. 8.5 that the nucleus also has vibrational modes for which κ_μ is positive. In this case there are no low-frequency solutions to eq. (8.4.19). Instead,

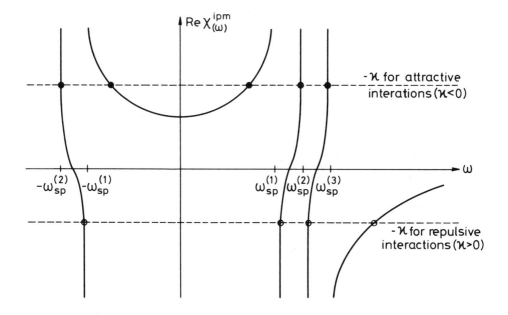

Figure 8.4   Graphical solution of the RPA dispersion relation, eq. (8.4.19). Solid lines: $\mathrm{Re}\chi^{\mathrm{ipm}}(\omega)$. The $\omega_{\mathrm{sp}}^{(i)}$ are various Bohr transition frequencies ω_{kj} of the independent-particle model. Dashed lines: κ for attractive, repulsive interactions. Solid dots: eigenvalues for attractive interaction. Open circles: eigenvalues for repulsive interaction.

a high-frequency solution appears with an energy greater than the single-particle excitation energies of the ipm. This situation corresponds to the lower broken line in fig. 8.4.

8.4 D.   RPA Response to an External Field

In order to interpret more clearly these solutions to the collective equation of motion, it is useful to return to the problem of the response of the nucleus to external fields $Q_\mu^{ext}(t)$:

$$H = \sum_i H^{sp}\left(\vec{r}_i, \vec{p}_i, \{Q_\alpha(t)\}\right) + H^{ext}\left(\vec{r}_i, \vec{p}_i, \{Q_\mu^{ext}(t)\}\right). \tag{8.4.21}$$

For a weak external field, we may suppose H^{ext} to be linear in Q_μ^{ext}:

$$H^{ext} = \sum_\mu F_\mu^{ext}\left(\vec{r}_i, \vec{p}_i\right) Q_\mu^{ext}(t). \tag{8.4.22}$$

Let us further suppose that we can contrive to make the external field mimic the effect of a change in the shape of the nuclear mean field

$$F_\mu^{ext}\left(\vec{r}_i, \vec{p}_i\right) \approx F_\mu^{ipm}\left(\vec{r}_i, \vec{p}_i\right). \tag{8.4.23}$$

In the case of an external electromagnetic field, eq. (8.4.23) can be approximately satisfied because both form-factors, r^ℓ and $\partial\hat{U}/\partial r$, are largest in the nuclear surface. The interaction Hamiltonian perturbing the nucleonic degrees of freedom is now, analogous to eq. (8.1.1),

$$\delta H = \sum_\mu \left(Q_\mu - Q_\mu^0\right) \sum_i F_\mu^{sp}\left(\vec{r}_i, \vec{p}_i\right) + \sum_\mu Q_\mu^{ext} \sum_i F_\mu^{ipm}\left(\vec{r}_i, \vec{p}_i\right). \tag{8.4.24}$$

The force on the external field Q_μ^{ext} is given by

$$-\frac{\partial}{\partial Q_\mu^{ext}} \langle \psi(t)|\delta H|\psi(t)\rangle = -\left\langle \psi(t)\left|\sum_i F_\mu^{ipm}\left(\vec{r}_i, \vec{p}_i\right)\right|\psi(t)\right\rangle. \tag{8.4.25}$$

This force may then be evaluated in linear-response theory, analogous to eq. (8.1.6);

$$\left\langle \psi(t)\left|\sum_i F_\mu^{ipm}\left(\vec{r}_i, \vec{p}_i\right)\right|\psi(t)\right\rangle = \left\langle gs\left|\sum_i F_\mu^{ipm}\left(\vec{r}_i, \vec{p}_i\right)\right|gs\right\rangle$$

$$-\sum_\nu \int_{-\infty}^{\infty} dt'\, \tilde{\chi}_{\mu\nu}^{ipm}(t-t') \left[Q_\nu(t') - Q_\nu^0 + Q_\nu^{ext}(t')\right] \tag{8.4.26}$$

since the nucleons are affected by the self-consistent field as well as the external field. The presence of the term proportional to Q_ν^{ext} in eq. (8.4.26) is necessary because any response of the nucleons to the external field will modify the mean field due to the nucleons' interactions.

The rate of change of energy of the nucleus is now, according to eq. (8.1.11),

$$\sum_\mu \frac{dQ_\mu^{ext}}{dt} \left\langle \psi(t) \left| \sum_i F_\mu^{ipm}(\vec{r}_i, \vec{p}_i) \right| \psi(t) \right\rangle$$

$$= \frac{dE}{dt} \equiv \frac{d}{dt} \left\langle \psi(t) \left| H(t) \right| \psi(t) \right\rangle. \tag{8.4.27}$$

Using the Ehrenfest Theorem, we find

$$\frac{d}{dt} \left\langle \psi(t) \left| H(t) \right| \psi(t) \right\rangle = \left\langle \psi(t) \left| \frac{\partial H}{\partial t} \right| \psi(t) \right\rangle$$

$$= \sum_\mu \frac{dQ_\mu^{ext}}{dt} \left\langle \psi(t) \left| \sum_i F_\mu^{ipm}(\vec{r}_i, \vec{p}_i) \right| \psi(t) \right\rangle \tag{8.4.28}$$

$$+ \sum_\mu \frac{dQ_\mu}{dt} \left\langle \psi(t) \left| \frac{\partial H_{sp}}{\partial Q_\mu} \right| \psi(t) \right\rangle.$$

Comparing (8.4.27) and (8.4.28), we see that (8.4.9) is still valid even though there is an external field. Thus we can use our earlier results (8.4.10), (8.4.12), (8.4.13), and (8.4.26) to eliminate $Q_\mu(t)$ by Fourier transforms, obtaining after some computation

$$\left\langle \psi(t) \left| \sum_i F_\mu^{ipm}(\vec{r}_i, \vec{p}_i) \right| \psi(t) \right\rangle = \left\langle gs \left| \sum_i F_\mu^{ipm}(\vec{r}_i, \vec{p}_i) \right| gs \right\rangle$$

$$- \sum_\nu \int_{-\infty}^\infty dt' \, \tilde{\chi}_{\mu\nu}^{RPA}(t - t')(Q_\nu^{ext}(t') + \sum_\sigma (\kappa^{-1})_{\nu\sigma} \langle gs | F_\sigma | gs \rangle. \tag{8.4.29}$$

where we have introduced the RPA response tensor

$$\tilde{\chi}_{\mu\nu}^{RPA}(t) = \frac{1}{2\pi} \int d\omega \, e^{-i\omega t} \chi_{\mu\nu}^{RPA}(\omega) \tag{8.4.30}$$

$$\chi_{\mu\nu}^{RPA}(\omega) = \sum_{\rho\lambda} \kappa_{\mu\lambda} \left[\kappa + \chi^{ipm}(\omega) \right]_{\lambda\rho}^{-1} \chi_{\rho\nu}^{ipm}(\omega). \tag{8.4.31}$$

The RPA response tensor describes the total force on the external field due to the nucleons' rearrangement, including the effect of their rearrangement on the self-consistent field. This is analogous to the dielectric theory of electromagnetism. Q^{ext} is like the D-field, which is created by the external sources; $Q - Q^0$ is the polarization field (the P-field), and $Q + Q^{ext}$ is the actual field (the E-field) in the medium.

The RPA polarization tensor $\chi_{\mu\nu}^{\text{RPA}}$ is proportional to the inverse of the matrix in eq. (8.4.15) whose determinant vanishes at the RPA eigenfrequencies ω_n. This implies that $\chi_{\mu\nu}^{\text{RPA}}$ has poles at the ω_n, and can be written

$$\chi_{\mu\nu}^{\text{RPA}}(\omega) = \sum_n D_{\mu\nu}^n \left[\frac{P}{\omega_n - \omega} + i\pi\delta(\omega_n - \omega) \right] \tag{8.4.32}$$

where

$$D_{\mu\nu}^n = -\lim_{\epsilon \to 0} \epsilon \sum_{\rho\lambda} \kappa_{\mu\lambda} \left[\kappa + \chi^{\text{ipm}}(\omega_n + \epsilon) \right]_{\lambda\rho}^{-1} \chi_{\rho\nu}^{\text{ipm}}(\omega_n). \tag{8.4.33}$$

At the eigenfrequencies ω_n, the polarizability develops an imaginary part, corresponding to the nucleus absorbing energy from the external field. To inspect the nature of these singularities, we return to the simple case of one collective variable. Then the pole strength is given by

$$D^n = \frac{\kappa^2}{\partial\chi^{\text{ipm}'}/\partial\omega}\bigg|_{\omega_n}. \tag{8.4.34}$$

D^n is a measure of how easy it is to excite the nucleus by an external field of frequency ω_n (see sect. 8.1). The most easily excited transitions are those for which $\partial\chi^{\text{ipm}'}/\partial\omega$ is smallest. Inspecting fig. 8.4, we see that these are just the eigenfrequencies which lie farthest from the ipm excitation energies. The corresponding states are called "collective", because they are most easily excited by an external field F^{ext} which acts equally–collectively–on all the particles. In a nuclear model like the spherical harmonic oscillator where all the s.p. levels within a shell are degenerate, D^n is non-zero only for the collective excitations.

The linear response theory is only one of many ways to derive the RPA, using similar assumptions about the physical nature of the nuclear system (saturating, short-range interactions, etc.). The linear response theory is one of the simplest ways to understand the RPA, but has the apparent formal disadvantage of introducing extra coordinates $\{Q_\alpha\}$ in addition to the "natural" nucleon variables $\vec{r}_i, \vec{p}_i$. As we have seen, a good guess at the nature of the collective motion permits us to use only a few collective variables, but in its most general form the RPA requires a complete set of collective coordinates, often chosen as the values of the field at each point $\vec{\mu}$, so that $F_{\vec{\mu}}(\vec{r}) = \delta(\vec{r} - \vec{\mu})$. Thus we would in principle need an infinity of collective coordinates. Among methods for avoiding the collective coordinates, the mean-field (Hartree-Fock) theory is probably the closest to a microscopic many-body theory. By letting the mean field depend on time, similar equations of motion are obtained for small-amplitude vibrations [see Thouless]. Another, related way to avoid introducing collective coordinates is to note that the mean force on the nucleons would be the same if we had considered a perturbation Hamiltonian

$$\delta H = \frac{1}{2} \sum_{ij} \sum_{\mu\nu} F_\mu^{\text{sp}}\left(\vec{r}_i, \vec{p}_i, \{Q_\alpha^0\}\right) F_\nu^{\text{sp}}\left(\vec{r}_j, \vec{p}_j, \{Q_\alpha^0\}\right) (\kappa^{-1})_{\mu\nu} \tag{8.4.35}$$

·and then used the mean-field approximation. Thus eq. (8.4.35) is often taken as the starting point for a discussion of collective motion. This and other methods are clearly described in Brown's book.

8.4E. Vibrational Parameters: the Coupling Constant and the Sum Rule

The most important step in looking for nuclear collective vibrational modes is to determine the collective degree of freedom Q and its associated form factor,

$$F_\mu^{sp} = \frac{\partial H^{sp}}{\partial Q_\mu}|_{\{Q_\alpha = Q_\alpha^0\}} \tag{8.4.8}$$

where Q_α^0 represents the equilibrium (ground state) configuration. We suppose that we know the single-particle field of the ground state, $H_0^{sp} \equiv H^{sp}(\{Q_\alpha^0\})$, and its eigenstates $|k\rangle$ and eigenvalues $\hat{\epsilon}_k$:

$$H_0^{sp}|k\rangle = \hat{\epsilon}_k|k\rangle. \tag{8.4.36}$$

The most obvious way to calculate κ is from its definition

$$\kappa_{\mu\nu} \equiv -\langle gs|\partial^2 H_{sp}/\partial Q_\mu \partial Q_\nu|_{\{Q_\alpha = Q_\alpha^0\}}|gs\rangle. \tag{8.4.13}$$

For example, Bohr and Mottelson consider vibrations about a spherical equilibrium using the single-particle potential defined in eq. (8.3.1) and obtain the estimate of the coupling constant, from eq. (8.4.20) with (5.3.2):

$$
\begin{aligned}
\kappa_\ell &= -R(A)^2 \int_0^\infty dr\ r^2\hat{n}(r) \left(\frac{\partial^2 \hat{U}}{\partial r^2} + \frac{2}{R(A)}\frac{\partial \hat{U}}{\partial r}\right) \\
&\approx -R(A)^2 \int_0^\infty dr\ r^2\hat{n}(r) \left(\frac{\partial^2 \hat{U}}{\partial r^2} + \frac{2}{r}\frac{\partial \hat{U}}{\partial r}\right) \\
&= \frac{R(A)^2}{4\pi} \int d^3\vec{r}\,\vec{\nabla}\hat{n}(\vec{r}) \cdot \vec{\nabla}\hat{U}(\vec{r})
\end{aligned}
\tag{8.4.37}
$$

This result is independent of multipole order when the liquid-drop corrections to $\hat{U}$ (see eq. 8.4.6) are neglected which is a good approximation for isoscalar shape vibrations (except the monopole and dipole). However, this method for calculating κ is quite sensitive to the choice of $H_{sp}(Q)$, since it involves second-order derivatives. Thus, for example, the volume conservation term, eq. (5.3.3), is needed to obtain the result above. For other vibrational degrees of freedom, an additional complication arises in relating the single-particle potential to the liquid-drop energy, which is most naturally estimated from the nucleonic density rather than the potential. The connection between the single-particle potential and the density has to be estimated very accurately, to second order in the departure from equilibrium. Thus, apparently, $\kappa_{\mu\nu}$ seems much harder to estimate than F_μ^{sp}, which only needs to be found to first order in the departure from equilibrium.

Luckily, there is an alternative way of estimating $\kappa_{\mu\nu}$ which depends only on quantities which may be calculated to first order in the departure from equilibrium. This method depends on a knowledge of the nuclear force to infer, as in Sect. 5.3, the shape of the potential from the shape of the density, since the density causes the potential, eq. (4.2.26). Therefore, for any given density distribution which the nucleus acquired during its vibrational motion, we can evaluate the mean value of the formfactor F_μ^{sp} and then find the corresponding potential parameters $Q_\mu - Q_\mu^0$ by the self-consistency requirement that the potential is determined by the density via the nuclear force. Once we know the expectation value of F_μ^{sp} corresponding to a given potential shape, we can use eq. (8.4.12) to find, to first order in the deviation from equilibrium,

$$\kappa_{\mu\nu} = \frac{\partial}{\partial Q_\nu} \langle \{Q_\alpha\}| \sum_i F_\mu^{sp} |\{Q_\alpha\}\rangle|_{\{Q_\alpha^0\}}. \qquad (8.4.38)$$

where $|\{Q_\alpha\}\rangle\langle\{Q_\alpha\}|$ is the density matrix corresponding to the parameters $\{Q_\alpha\}$ via self-consistency.

This method of evaluating the coupling constants $\kappa_{\mu\nu}$ was invented and widely applied by Bohr and Mottelson [see vol. 2]; we should remark that their form factor F is sometimes, but not always, defined to include a factor κ^{-1}. They exploit, in addition, the short range of the nuclear force to simplify the connection between the density and the potential. If the range of the potential is very small, as in the Skyrme parametrization of the nuclear force, then the shapes of the potential and density distributions must be the same, as is evident from inspecting eq. (4.2.33). We used this simplified self-consistency condition for static shapes in Sect. 5.3; Bohr and Mottelson showed how to apply it to time-dependent shapes, too, leading to very simple explanations of many features of nuclear vibrations. For example, we can easily see from eq. (8.4.38) that κ is negative if a change in the density leads to a deeper potential, and positive if it makes the potential less attractive. Applying the method to the case of a deformed Woods-Saxon potential, we merely use eq. 5.3.1 to describe the deformation both of the potential $\hat{U}(\vec{r})$ and the density $\hat{n}(\vec{r})$. Since (see eq. 8.3.2) $F_{\ell 0}^{ipm}$ is proportional to $Y_\ell^0(\theta, \phi)$, (we choose m=0 because $F_{\mu\nu}$ is assumed real), the only part of the density that contributes to the expectation value in eq. (8.4.38) is the part with the same angular dependence,

$$\delta n_{\ell 0}(\vec{r}) = -\alpha_{\ell 0} Y_\ell^0(\theta, \phi) R(A) \frac{\partial}{\partial r} \hat{n}(\vec{r}) \qquad (8.4.39)$$

Using eqs. (8.3.2) and (8.4.39), we find

$$\kappa_\ell = \frac{\partial}{\partial \alpha_{\ell 0}} \int_0^\infty r^2 dr \frac{\partial \hat{U}}{\partial r} \frac{\partial \hat{n}}{\partial r} \cdot \int d\Omega Y_\ell^0(\theta, \phi)^2 \alpha_{\ell 0} R(A)^2$$

which is exactly the same as eq. (8.4.37).

Before turning to specific applications we want to define an important quantity called the energy-weighted sum rule,

$$S_{\mu\nu} = \frac{1}{2}\langle gs|[\sum_i F_\mu^{sp}, [H_{sp}(\{Q_\alpha^0\}), \sum_i F_\nu^{sp}]]|gs\rangle, \qquad (8.4.40)$$

which, by inserting the complete set of eigenstates $|n\rangle$ of the Hamiltonian $H_{sp}(\{Q_\alpha^0\})$ and their eigenvalues E_n, can be rewritten as

$$S_{\mu\nu} = \sum_n (E_n - E_0)\langle gs|\sum_i F_\mu^{sp}|n\rangle\langle n|\sum_i F_\nu^{sp}|gs\rangle. \qquad (8.4.41)$$

Comparison with eq. (8.1.24) then gives immediately

$$S_{\mu\nu} = \frac{\hbar^2}{2\pi} \int_{-\infty}^{\infty} d\omega \chi_{\mu\nu}^{ipm\prime\prime}(\omega) \cdot \omega. \qquad (8.4.42)$$

In the RPA, the Hamiltonian we solved was given by eq. (8.4.11) with the replacement, eq. (8.4.13), of the second-derivative term by its mean value,

$$H_{sp}(\{Q_\alpha\}) \approx \sum_i H_0^{sp}(\vec{r}_i, \vec{p}_i) + \sum_\mu (Q_\mu - Q_\mu^0) \sum_{\mu\nu} F_\mu^{sp}(\vec{r}_i, \vec{p}_i, \{Q_\alpha^0\})$$
$$-\frac{1}{2}\sum_{\mu\nu} \kappa_{\mu\nu}(Q_\mu - Q_\mu^0)(Q_\nu - Q_\nu^0). \qquad (8.4.43)$$

Even though we didn't explicitly construct its solutions, we were able to find its eigenvalues and response function.

The commutator in eq. (8.4.40) would be the same if we had used $H_{sp}(\{Q_\alpha\})$ instead of $H_{sp}(\{Q_\alpha^0\})$; thus eqs. (8.4.41) and 8.4.42) remain true, with the same value of $S_{\mu\nu}$, if the eigenstates, eigenvalues and response functions of the RPA are inserted:

$$S_{\mu\nu} = \frac{\hbar^2}{2\pi} \int_{-\infty}^{\infty} d\omega \chi_{\mu\nu}^{RPA\prime\prime}(\omega) \cdot \omega \qquad (8.4.44)$$

or, with eq. (8.4.32)

$$S_{\mu\nu} = \hbar^2 \sum_{\omega_n > 0} D_{\mu\nu}^n \cdot \omega_n \qquad (8.4.45)$$

where $D_{\mu\nu}^n$ is the squared transition matrix element for the RPA state $|n\rangle$, compare eq. (8.4.33).

The interest in eq. (8.4.44) lies in the fact that S only depends on the nuclear ground state and the operators F_μ^{sp} and H_{sp}, whereas the right-hand side is the sum of contributions from the individual RPA vibrational solutions. When only some of these are known one can deduce, from knowledge of S, whether the remaining solutions contribute (are "collective") or stay as incoherent nucleon excitations with small D (see eq. (8.4.34)).

From the inverse of the Fourier transformation defining $\chi_{\mu\nu}(\omega)$, eq. (8.4.30), we find another expression for the sum rule:

$$S_{\mu\nu} = \hbar^2 \frac{d}{dt} \tilde{\chi}_{\mu\nu}^{RPA}(t)|_{t=0} = \hbar^2 \frac{d}{dt} \tilde{\chi}_{\mu\nu}^{ipm}(t)|_{t=0}. \qquad (8.4.46)$$

This gives us some insight both into the meaning of $S_{\mu\nu}$ and into the RPA: we see that $S_{\mu\nu}$ represents the short-time behavior of the response function $\tilde{\chi}_{\mu\nu}$, and that this is the same in the RPA and the ipm. The interactions of the self-consistent field affect only the long-time behavior of the response.

8.5   PHENOMENOLOGY OF NUCLEAR VIBRATIONS

The general theory developed in the previous subsections is directly applicable to a variety of collective nuclear vibrations. We will now discuss a series of these vibrations by estimating their characteristics in simple models and comparing to experimental information when it is available.

8.5 A.   Isoscalar Shape Vibrations

The most straightforward treatment of isoscalar shape vibrations is to parametrize them by a deformed Woods-Saxon potential, as in eq. (5.3.1) and (8.3.1), leading to expressions (8.3.2) for F and (8.4.37) for κ. Except for the monopole state, where κ has to include liquid-drop corrections (see Sect. 8.5.D below), this procedure gives a reasonable description of the isoscalar shape vibrations. However, χ^{ipm} has to be computed numerically. We can learn a lot about the vibrations by considering a harmonic-oscillator approximation instead, since then we can calculate most quantities analytically.

For comparison with experiments on Coulomb excitation, it is useful to use the multipole-operator form factors $F_{\ell m}^E$ defined by eq. (8.1.29). However, these act only on the protons, while the nuclear mean field acts on both protons and neutrons. For isoscalar vibrations, the neutrons and protons move together, so the transition rates calculated by using the isoscalar multipole form factors,

$$F_{\ell m}^0(\{\vec{r}_i\}) = \frac{1}{2}\sqrt{\frac{2}{1+\delta_{m0}}} \sum_{i=1}^{A} r_i^\ell \left(Y_\ell^m(\theta_i, \phi_i) + (-1)^m Y_\ell^{-m}(\theta_i, \phi_i) \right) \qquad (8.5.1)$$

need merely to be multiplied by a factor $(eZ/A)^2$ to get the rates corresponding to $F_{\ell m}^E$.

We will explicitly work through examples in the harmonic oscillator with $\ell=0$, 1, 2. For these values of ℓ, the potential obtained by adding the perturbing field $Q_{\ell m}F^0_{\ell m}$ to the original harmonic-oscillator Hamiltonian merely produces a new oscillator Hamiltonian, which simplifies the computations and arguments. The application of the harmonic-oscillator approximation to higher values of ℓ requires some subtlety, but can be done quite simply as demonstrated by Bohr and Mottelson at many places in Volume 2 of their textbook.

As we have mentioned before, the monopole ($\ell=0$) and dipole ($\ell=1$) form factors correspond to compressional and translational modes respectively, which need special arguments. The isoscalar vibrations for $\ell \geq 2$ are the most straightforward, so we will discuss them first.

8.5 B. Quadrupole and Higher Multipole Isoscalar Vibrations

We begin by studying the isoscalar quadrupole vibration in the harmonic-oscillator approximation. Adding the terms $Q_{2m} F^0_{2m}$ to a spherically-symmetric oscillator (the generalization to deformed ground states is quite straightforward) we obtain the approximate single-particle potential

$$\hat{U}^{\mathrm{ipm}}_2(r, \theta, \phi) = \frac{1}{2}m_N\Omega^2_0 r^2[1 - \sum_{m=0}^{2} \alpha_{2m}\sqrt{\frac{2}{1 + \delta_{m0}}}\left(Y^m_\ell + (-1)^m Y^{-m}_\ell\right)$$

$$+ \frac{5}{4\pi}\sum_{m=0}^{2} \alpha^2_{2m}] \tag{8.5.2}$$

where we have introduced the notation

$$\alpha_{2m} \equiv -(m_N\Omega^2_0)^{-1}Q_{2m} \tag{8.5.3}$$

by analogy with eq. (5.3.1). We have added the quadratic term as well to ensure a deformation-independent volume inside any equipotential surface, to second order in α_{2m}, as in eq. (5.3.2). This volume-conserving addition would make the Strutinsky and liquid-drop corrections to κ_2 vanish if we compute it using $\partial^2 H^{\mathrm{sp}}/\partial Q^2$, eq. (8.4.13), but doesn't come in when we use the self-consistency condition, eq. (8.4.38).

On the face of it, the potential $\hat{U}^{\mathrm{ipm}}_2$ of eq. (8.5.2) appears to have three degrees of freedom, i.e. m=0,1,2. In deformed nuclei, however, only two of these correspond to vibrations, m=0 and 2, since the third, m=1, corresponds to a change in the orientation of the nuclear field rather than a change in its shape. Thus the m=1 degree of freedom is in fact of rotational nature and will be discussed in Chapter 9. The remaining two degrees of freedom in deformed nuclei correspond to the beta and gamma vibrations of Sect. 8.3. For spherical nuclei, any real linear combination of the Q_{2m} corresponds to the nuclear field becoming longer or shorter along some axis of symmetry; thus it is sufficient to consider only one of the three degrees of freedom.

We simplify our task by considering a spherical nucleus, and choose the form factor F_{20}^0. By inserting the expressions for the spherical harmonics in eq. (8.5.1), we find

$$F_{20}^0(\{\vec{r}_i\}) = \sqrt{\frac{5}{4\pi}} \cdot \frac{1}{2} \sum_{i=1}^{A}(2z_i^2 - x_i^2 - y_i^2). \tag{8.5.4}$$

Comparing to eq. (8.3.11), and realizing that to give the same Hamiltonian we must have $Q_\beta F_\beta = Q_{20}F_{20}^0$, we conclude that

$$Q_{20} = -m_N\Omega_0\sqrt{\frac{8\pi}{5}}Q_\beta. \tag{8.5.5a}$$

Since we have already found $\chi_{\beta\beta}^{\text{ipm}}$, eq. (8.3.10), we can use that result. Noting that $\chi_{\mu\nu}^{\text{ipm}}$ is proportional to $F_\mu F_\nu$, we see that

$$\chi_2^{\text{ipm}\prime}(\omega) = \frac{5\hbar}{4\pi m_N^2\Omega_0} \cdot \frac{\sigma}{(2\Omega_0)^2 - \omega^2} \tag{8.5.6}$$

with corresponding expressions for the imaginary part. Here we have introduced the number $\sigma = \sigma_x + \sigma_y + \sigma_z$, in the notation of Sect. 8.3, to denote the sum over single-particle quantum numbers corresponding to transitions across the Fermi level, eq. (8.3.8d).

To estimate κ_2 from the consistency of the potential with the density, eq. (8.4.38), we can extend the arguments of Sect. 5.3. To apply these arguments we notice that when $Q_{21}=Q_{22}=0$, the potential $\hat{U}_2^{\text{ipm}}$, eq. (8.5.2), has the form of a cartesian deformed oscillator with (to first order in Q_{20})

$$\Omega_x^2 = \Omega_0^2[1 - \sqrt{\frac{5}{4\pi}}(m_N\Omega_0^2)^{-1}Q_{20}] \tag{8.5.7a}$$

$$\Omega_y^2 = \Omega_0^2[1 - \sqrt{\frac{5}{4\pi}}(m_N\Omega_0^2)^{-1}Q_{20}] \tag{8.5.7b}$$

$$\Omega_z^2 = \Omega_0^2[1 + \sqrt{\frac{5}{4\pi}}(m_N\Omega_0^2)^{-1}2Q_{20}]. \tag{8.5.7c}$$

The ellipsoidal equipotential surfaces have axes a_x, a_y, a_z which are inversely proportional to Ω_x, Ω_y, and Ω_z, see eq. (5.3.5). The ratios of these axes ought to be proportional to the moments of the density distributions,

$$\langle x^2\rangle/a_x^2 = \langle y^2\rangle/a_y^2 = \langle z^2\rangle/a_z^2 \tag{8.5.8}$$

which implies

$$\langle x^2\rangle\Omega_x^2 = \langle y^2\rangle\Omega_y^2 = \langle z^2\rangle\Omega_z^2 = \frac{1}{3}\langle r^2\rangle\Omega_0^2. \tag{8.5.9}$$

We cannot use eq. (5.3.6) to get the density moments, because away from equilibrium the wave functions are not stationary states as assumed there. But $\langle F_{20}^0 \rangle$ depends directly on those moments, see eq. (8.5.4). Inserting eq. (8.5.9) in the expectation value of eq. (8.5.4) we find using eq. (8.5.7)

$$\langle Q_{20}|F_{20}^0|Q_{20}\rangle = -\frac{5}{4\pi}(m_N\Omega_0^2)^{-1}A\langle r^2\rangle Q_{20} + \mathcal{O}(Q^2) \qquad (8.5.10)$$

so that κ_2 has the value from eq. (8.4.38)

$$\kappa_2 = -\frac{5}{4\pi}A\frac{\langle r^2\rangle}{m_N\Omega_0^2}. \qquad (8.5.11)$$

With this estimate of κ_2, we can find the RPA eigenvalues from eq. (8.4.17), together with our expression for $\chi_2^{\mathrm{ipm}\prime}(\omega)$, eq. (8.5.6). There is only one eigenvalue, given by

$$\omega_2 = 2\Omega_0\sqrt{1 + \chi_2^{\mathrm{ipm}}(0)/\kappa_2}. \qquad (8.5.12)$$

The quantities $\langle r^2\rangle$ and σ appearing in κ_2 and $\chi_2^{\mathrm{ipm}}(0)$ can be evaluated, as in chapter 5, in terms of the principal quantum number N_F of the last occupied level. Expressing N_F in terms of the nucleon number A, eq. (5.2.6), we obtain the results

$$\sigma = (\frac{3}{2}A)^{4/3} \qquad (8.5.13a)$$

$$\chi_2^{\mathrm{ipm}}(0)/\kappa_2 = -\frac{1}{2} + \mathcal{O}(A^{-1/3}) \qquad (8.5.13b)$$

which leads to the RPA energy

$$\hbar\omega_2 \equiv \sqrt{2}\hbar\Omega_0 \approx 58\mathrm{MeV}\ A^{-1/3}. \qquad (8.5.14)$$

This can be compared to the measured values for many nuclei throughout the periodic table, fig. 8.5, which fall within a narrow band

$$E^{\mathrm{exp}}(\ell = 2) = (64 \pm 4)\mathrm{MeV}/A^{1/3}. \qquad (8.5.15)$$

The agreement is more than satisfactory considering that eq. (8.5.14) was obtained as the leading term in an oversimplified independent particle model. The result gives one excitation energy where the measurements find a distribution of excited states of various transition strengths, distributed over an energy interval whose width is also shown in fig. 8.5.

To interpret the transition probabilities we can estimate the energy-weighted sum rule, eq. (8.4.40). The commutators are easy to evaluate, since the only part of H_{sp} that doesn't commute with F^{sp} is the kinetic energy. The result is independent of m for $\ell = 2$:

$$S_2 = \frac{5}{4\pi}\frac{\hbar^2}{m_N}A\langle r^2\rangle. \qquad (8.5.16)$$

On the other hand, we can evaluate the transition strength D_n for the RPA eigenstate by using eqs. (8.4.34), (8.5.6), (8.5.11), (8.5.13) and (8.5.14). Applying the result in eq. (8.4.45) it is not surprising to find that this single state gives a contribution to S_2 equal to its total value, eq. (8.5.16). We say that the state "exhausts the sum rule." For that reason it is called a giant quadrupole vibration. In more realistic cases the strength is distributed over more than one RPA solution and less than 100% of S in the known vibrations indicates the existence of other interesting states.

The observed strength of the vibration in eq. (8.5.15) varies between 40% and 100%. The dependence on A is not smooth, indicating that the strength of the

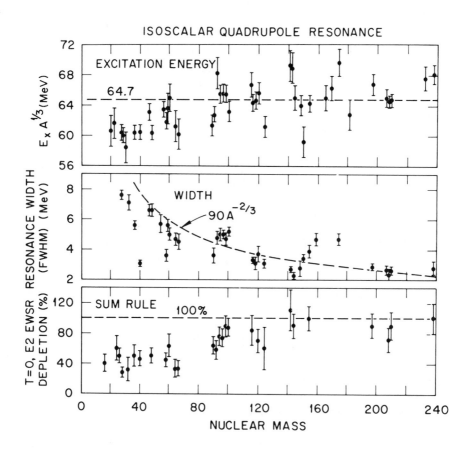

Figure 8.5   Systematics for the excitation energy, width, and percentage of sum rule of the isoscalar giant quadrupole resonance [*from F. Bertrand, Nucl. Phys.* **A354** *(1981) 129c*].

quadrupole vibration, unlike its excitation energy, depends considerably on the nuclear shell structure (see fig. 8.5).

The level schemes for real nuclei are different from the rather schematic harmonic-oscillator example considered above. In particular, one major shell is not completely degenerate, as seen already from the presence of the spin-orbit interaction. The form factors, $F_{2m}^0(m = 0, 2)$, therefore contribute non-vanishing matrix elements between states within the same major shell, but on opposite sides of the Fermi level. This leads to RPA solutions of energy below the major shell splitting (see fig. 8.4). The actual vibrational energies depend on the details of the shell structure, i.e. the number of neutrons and protons, the equilibrium deformation and residual interactions like the pairing force. For example, in deformed nuclei both the low-frequency and giant quadrupole vibrations split into two modes each, corresponding to the β and γ degrees of freedom of Sect. 8.3.

The spin-orbit splitting, eq. (5.2.10), is $20 \text{ MeV} \times (\ell + 1/2)/A^{2/3}$. For the highest ℓ values the splitting varies between 3 and 4 MeV. The smallest two-quasiparticle excitation energy $2\Delta = 24\text{MeV}/\sqrt{A}$, eq. (6.6.2), varies between 3.5 and 1.5 MeV for $50 < A < 256$. The RPA solution must be smaller than these energies and can therefore be expected to be of the order of 1 MeV with a tendency to decrease with A.

The experimental results can be seen in fig. 8.6, where the regions I, II and III correspond to rotational states (see Chapter 9). For the present purpose they should therefore be ignored. The rather strong shell-structure dependence is apparent. In fig. 8.7 we show the fraction of the energy-weighted sum rule exhausted by these low-energy vibrations. The values, varying between 5 and 10% are in general much smaller than those of the giant quadrupole vibrations.

Higher multipoles can in principle be treated analogously. However the estimates are much more difficult to carry out. For example the octupole vibrations ($\ell = 3$) have form factors which are third-degree polynomials of negative parity. They therefore connect states differing both by one and three quanta. The strength can then be expected to be distributed over two types of solutions, i.e. the highest energy somewhat below $3\hbar\Omega_0$ and the low-energy solution below $\hbar\Omega_0$.

For realistic estimates the spin-orbit interaction must be included as for the low-energy quadrupole vibrations. The shell structure is expected to be important for the low-energy solution, although less so than for the $\ell = 2$ case, since odd-parity states have a tendency (opposed by the spin-orbit force) to be $\hbar\Omega_0$ apart.

The experimental results are shown in fig. 8.8. They scatter around the curve given by $30 \text{ MeV}/A^{1/3}$ and carry between about 10 to 20% of the strength. There is much less known about the high-energy solutions. Indications are energies about $120 \text{ MeV}/A^{1/3}$.

For hexadecapole vibrations the form factors are fourth-degree polynomials of even parity. They connect states differing by 0, 2 and 4 oscillator quanta and consequently three types of solutions can be expected. No indication of the highest has been observed. The second-highest, somewhat below $2\hbar\Omega_0$, coincide in energy with the giant quadrupole. A few analyses including such a state claim evidence for

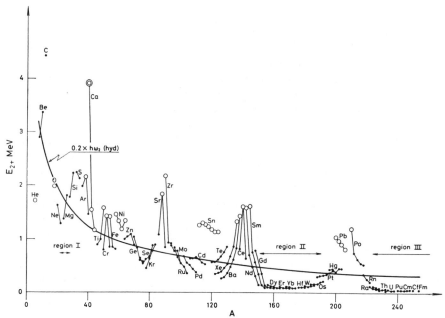

Figure 8.6   The energy of the first excited state of $J^{\pi} = 2^+$ in even-even nuclei. [*from O. Nathan and S. Nilsson, in "Alpha-, Beta-, and Gamma-Ray Spectroscopy," K. Siegbahn, ed., North-Holland, Amsterdam, 1965*]

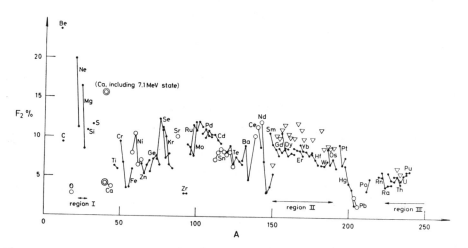

Figure 8.7   The fraction F_2 of the energy weighted sum rule for $T = 0$ excitations exhausted by the lowest $J^{\pi} = 2^+$ excitations in even-even nuclei. [*from O. Nathan and S. Nilsson, loc. cit.*]

20-40% of the hexadecapole strength at this energy, but separation from the giant quadrupole transitions makes analysis of measurements difficult. Low-energy collective states with $\ell=4$ are common in heavier nuclei, with energies usually somewhat above the low-lying octupole states.

8.5 C.   Isoscalar Monopole Vibrations

The isoscalar monopole vibration, also known as the "breathing mode," consists of the nucleus expanding and contracting rhythmically. If we begin with a form factor like eq. (8.5.1) we get nowhere because for $\ell = 0$ it is just a constant. In principle, the monopole form factor would have two components describing changes of the nuclear interior density and of its surface thickness. Since it has not yet been possible experimentally to determine the combination of these form factors, we can simplify our consideration by taking only one, the scaling transformation

$$\hat{U}^{\mathrm{ipm}}(\vec{r}, \alpha_0) = \hat{U}^{\mathrm{ipm}}(\alpha_0\vec{r}, 1). \tag{8.5.17}$$

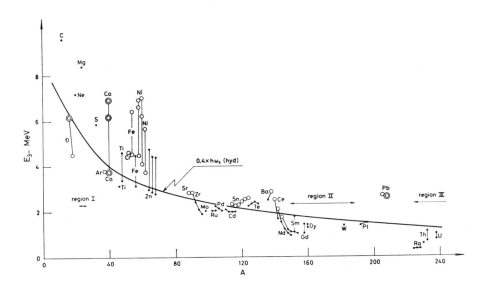

Figure 8.8   The energies of the isoscalar $J^\pi = 3^-$ collective vibrational states in even-even nuclei. (The open circles represent closed shells.) The collective character of these states has in most cases been inferred from the large excitation cross-sections found in inelastic scattering processes. [*from O. Nathan and S. Nilsson, loc. cit.*]

Unlike the case of isoscalar surface vibrations, we cannot neglect the smooth energy correction $\delta\tilde{\mathcal{E}}(\{Q_\alpha\})$ of eq. (8.4.4).

For the case of a harmonic-oscillator potential it is as convenient to find the coupling constant κ_0 from its definition, eq. (8.4.13), as from the consistency condition, eq. (8.4.38). This is because Strutinsky's smoothed single-particle energy $\tilde{\mathcal{E}}$ is proportional to the oscillator parameter $\hbar\Omega_0$. Since

$$\hat{U}^{sp}(\vec{r},\alpha_0) = \frac{1}{2}m_N\Omega_0^2\alpha_0^2 r^2 - (\tilde{\mathcal{E}}(\alpha_0) + B_{LD}(\alpha_0))/A \qquad (8.5.18)$$

the single-particle wavefunctions and eigenvalues will be those of an oscillator with frequency $\Omega_0\alpha_0$. Thus

$$\tilde{\mathcal{E}}(\alpha_0) = \alpha_0\tilde{\mathcal{E}}(\alpha_0 = 1) \qquad (8.5.19)$$

and its second derivative vanishes, so we may ignore it in estimating κ_0 from eq. (8.4.13). The contribution from $\partial^2\hat{U}^{ipm}/\partial\alpha_0^2$ is easy to estimate, since it is just twice the potential energy in equilibrium. Finally, we need the dependence of the liquid-drop binding energy on the nuclear density. Parametrizing the nuclear-matter energy per nucleon by its linear compression modulus K_∞,

$$E_{NM}(n)/A = -b_V + \frac{1}{18}K_\infty(1 - n/n_0)^2 \qquad (8.5.20)$$

where n_0 is the equilibrium density, we find

$$B_{LD}(\alpha_0)/A = b_V - \frac{1}{18}K_\infty(1 - \alpha_0^3)^2 - b_c Z^2 A^{-4/3}\alpha_0 + \mathcal{O}(A^{-1/3}). \qquad (8.5.21)$$

Altogether, we obtain

$$\kappa_0 = -\langle gs|\partial^2\hat{U}^{sp}/\partial\alpha_0^2|gs\rangle_{\alpha_0=1}$$

$$= -A(m_N\Omega_0^2\langle r^2\rangle + K_\infty) \qquad (8.5.22a)$$

$$\approx -A(35\,\text{MeV} + K_\infty) \qquad (8.5.22b)$$

where we have used eqs. (5.2.8) and (5.2.9).

Since F_0^0 is proportional to $\vec{r}^2 = x^2 + y^2 + z^2$, χ_0^{ipm} involves the same quantity $\sigma = \sigma_x + \sigma_y + \sigma_z$ as appeared in the giant quadrupole vibration (see eqs. 8.3.8d and 8.5.6):

$$\chi_0^{ipm\prime}(\omega) = \sigma\frac{2\hbar\Omega_0^3}{(2\Omega_0)^2 - \omega^2}. \qquad (8.5.23)$$

Using the estimates of eq. (8.5.13a) for σ and (5.2.8) for $\langle r^2\rangle$ we find, analogous to eq. (8.5.12),

$$\omega_0 = 2\Omega_0[1 + \chi_0^{ipm}(0)/\kappa_0]^{1/2} \qquad (8.5.24a)$$

$$= 2\Omega_0[1 + \frac{3}{4}(\frac{3}{2}A)^{1/3}\hbar\Omega_0/K_\infty]^{-1/2} \qquad (8.5.24b)$$

or, using eq. (5.2.9) for Ω_0,

$$\hbar\omega_0 = \frac{82\text{MeV}}{A^{1/3}[1 + 35\text{MeV}/K_\infty]^{1/2}}.$$
$$\approx 76\text{MeV}/A^{1/3}$$

(8.5.25)

where the last estimate was obtained from $K_\infty \approx 200$ MeV.

The most decisive factor leading to eq. (8.5.25) is the r^2 dependence of the form factor resulting in the basic particle-hole excitations of $2\hbar\Omega_0$. This assumption of scaling the density distribution is clearly not unique; for example, the surface could vibrate leaving the interior at rest. Eq. (8.5.25) is fairly close to the experimental values for A>120 and gives an increasing overestimate for decreasing A. This is perhaps indicative of a change in the importance of the surface versus volume degrees of freedom (see fig. 8.9).

The energy weighted sum rule for the scaling mode can be calculated directly (see also problem 8.4) to give

$$S_0 = 2m_N \hbar^2 \Omega_0^4 A \langle r^2 \rangle$$

(8.5.26)

which equals $2(\hbar\Omega_0)^2\sigma$, i.e. the right-hand side of eq. (8.4.45). The sum rule is exhausted as it should be from our construction of one vibrational energy. Experimentally, however the strength increases from about 20% at $A \approx 60$ to about 100% for $A \geq 100$. This is perhaps another indication of the interplay between surface and volume.

8.5 D.   Isoscalar Dipole Vibrations and the Spurious State

As we have explained before in Sect. 5.3, a dipole change of the nuclear shape corresponds to a translation of the nucleus as a whole. We saw in Sect. 4.5 that the mean-field picture treats the translational motion incorrectly, with a degenerate set of localized solutions corresponding to different positions of the center of the potential. We will now show that this degeneracy appears in the RPA as a zero-frequency vibrational mode.

We choose to consider translational motion along the z direction, corresponding to the Y_1^0 multipole. It is convenient to choose the collective coordinate as Z, the center of the single-particle potential

$$H^{sp}(\vec{r}_i, \vec{p}_i, Z) = H^{sp}(\vec{r}_i - Z\hat{z}, \vec{p}_i, 0).$$

(8.5.27)

Since the Strutinsky correction does not depend on Z, the form factor is

$$F_Z^{sp} = F_Z^{ipm}(\vec{r}_i) = \partial H^{sp}/\partial Z$$
$$= -\partial H^{sp}(\vec{r}_i)/\partial z_i = -i[p_{zi}, H^{sp}]/\hbar$$

(8.5.28)

so that

$$F_Z = \sum_i F_Z^{sp}(\vec{r}_i) = -i[P_z, H_{sp}]/\hbar \qquad (8.5.29)$$

where P_z is the total momentum in the z direction. The static response is (see eq. 8.1.27)

$$\chi_Z^{ipm}(\omega = 0) = 2 \sum_{n \neq gs} (E_n - E_0)^{-1} \langle gs|F_Z|n\rangle\langle n|F_Z|gs\rangle \qquad (8.5.30)$$

where we can evaluate the matrix elements using (8.5.29):

$$\langle gs|F_Z|n\rangle = -i(E_n - E_0)\langle gs|P_z|n\rangle/\hbar. \qquad (8.5.31)$$

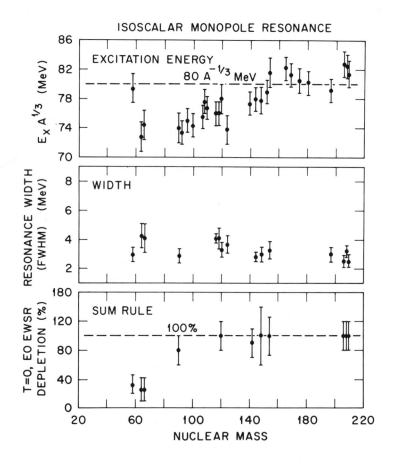

Figure 8.9   Systematics for the excitation energy, width, and percentage of sum rule of the isoscalar giant monopole resonance. [*from F. Bertrand, loc. cit.*]

Combining the last two equations we have

$$\chi_Z^{\text{ipm}}(\omega = 0) = -\frac{i}{\hbar}\sum_n (\langle gs|P_z|n\rangle\langle n|F_z|gs\rangle - \langle gs|F_z|n\rangle\langle n|P_z|gs\rangle)$$

$$= -\frac{i}{\hbar}\langle gs|[P_z, F_z]|gs\rangle \tag{8.5.32}$$

$$= \langle gs|\partial^2 H_{sp}/\partial Z^2|gs\rangle = -\kappa_Z.$$

Comparing with the RPA eigenvalue eq. (8.4.19), we see that $\omega_1 = 0$ is an RPA eigenfrequency for the collective translational motion along the z-axis.

This zero-frequency collective mode is known as a **spurious state** because it is an artifact of the mean-field approximation. As we saw in Chapter 4, the correct description of collective translations is somewhat different: while the uniform translational motion is indeed very slow, it corresponds to continued motion in the same direction instead of an oscillatory vibration.

The other RPA eigenfrequencies for the dipole mode correspond to nuclear excited states. Since they are orthogonal to the spurious mode, they do not have any center-of-mass motion mixed into them, unlike the dipole states of the independent-particle model. However, they have little dipole transition strength: in the harmonic oscillator, the entire dipole sum rule is exhausted by the spurious state.

8.5 E.   Isovector Vibrations

When neutrons and protons oscillate against each other in opposite phase, the normal modes are called isovector vibrations. The isovector monopole and dipole modes are not particularly different from the other isovector multipoles.

In the Woods-Saxon independent-particle model, the isovector form factors and coupling constants may be obtained by extending the arguments of Sect. 7.4 for the time-independent mean field. We assume a zero-range isovector two-body interaction, and find the self-consistent potential arising from opposing changes in the neutron and proton densities. For simplicity, we take the case m=0, for which the density of nucleons with isospin projection t_3 may be parametrized by

$$\hat{n}(\vec{r}, t_3, \alpha_{\ell 0}^t) = n_0 f\left(|\vec{r}| + 2t_3\alpha_{\ell 0}^t R(A) Y_\ell^0(\theta, \phi)\right)\frac{N(t_3)}{A} \tag{8.5.33}$$

where $f(r)$ is the Woods-Saxon distribution, and $N(t_3) = N$ for neutrons and Z for protons. Assuming a spherical ground state, the first-order change in the above density is

$$\delta\hat{n}_{\ell 0}^t(\vec{r}, t_3) = 2t_3\alpha_{\ell 0}^t Y_\ell^0(\theta, \phi) R(A)\frac{\partial}{\partial r}\hat{n}(\vec{r})\frac{N(t_3)}{A}. \tag{8.5.34}$$

where $\hat{n}$ is the total density in equilibrium. The corresponding change in the single-particle potential may be evaluated from (7.4.3) as

$$\alpha_{\ell 0}^t F_{\ell 0}^t(\vec{r}, t_3) = \frac{\Im^t}{2}\frac{A}{2N(t_3)}\delta\hat{n}_{\ell 0}^t(\vec{r}, t_3) \tag{8.5.35}$$

which gives with eq. (8.5.34)

$$F_{\ell 0}^t(\vec{r}, t_3) = \frac{t_3}{2} Y_\ell^0(\theta, \phi) R(A) \frac{\partial \hat{n}}{\partial r} \Im^t. \tag{8.5.36}$$

The coupling constant is then obtained from eq. (8.4.38) with (8.5.35):

$$\kappa_\ell^t = \frac{4}{\Im^t} \int d^3\vec{r} |F_{\ell 0}^t(\vec{r}, t_3)|^2 \tag{8.5.37}$$

which gives with eq. (8.5.36)

$$\kappa_\ell^t = \frac{1}{4} \int_0^\infty r^2 dr \left(\frac{\partial \hat{n}}{\partial r}\right)^2 R(A)^2 \Im^t. \tag{8.5.38}$$

We see that $\kappa_\ell^t > 0$, since $\Im^t > 0$ (see eq. 7.4.4). We realize that the RPA frequencies will be larger than the ipm excitation energies: the attractive force between neutrons and protons makes them harder to separate than they would be in the independent-particle model.

For isovector monopole vibrations, the Woods-Saxon estimate of the density variation, eq. (8.5.34), is not reasonable because it does not conserve the numbers of neutrons and protons. Instead, we can use a scaling picture similar to that for the isoscalar vibrations. We take

$$\hat{n}(\vec{r}, t_3, \alpha_0^t) = (1 + 2t_3\alpha_0^t)^3 \hat{n} \left((1 + 2t_3\alpha_0^t)\vec{r}, t_3, 0\right) \tag{8.5.39}$$

so that the first-order change in the density is

$$\delta \hat{n}_0^t(\vec{r}, t_3) = 2t_3\alpha_0^t \frac{N(t_3)}{A} \left(3 + r\frac{\partial}{\partial r}\right) \hat{n}(\vec{r}). \tag{8.5.40}$$

The form factor becomes, using eq. (8.5.35),

$$F_0^t(\vec{r}, t_3) = \frac{t_3}{2} \Im^t (3 + r\frac{\partial}{\partial r}) \hat{n}(\vec{r}), \tag{8.5.41}$$

and the corresponding coupling constant is obtained from eq. (8.5.37),

$$\kappa_0^t = \pi \Im^t \int_0^\infty r^2 dr (3\hat{n} + r\frac{\partial \hat{n}}{\partial r})^2. \tag{8.5.42}$$

For simple, quantitative estimates, we once again employ the harmonic-oscillator approximation to the independent-particle model for $\ell = 1$ and 2. We choose the density variation

$$\delta \hat{n}_{\ell m}^t(\vec{r}, t_3) = \alpha_{\ell m}^t m_N \Omega_0^2(2t_3) \frac{2N(t_3)}{A} n_0 \left(\frac{2}{1 + \delta_{m0}}\right)^{1/2}$$

$$[Y_\ell^m(\theta, \phi) + (-1)^m Y_\ell^{-m}(\theta, \phi)] r^\ell \Theta(r < R(A)). \tag{8.5.43}$$

We can use eq. (8.5.35) to find the corresponding form factor, which leads to the coupling constant from eqs. (8.5.37) and (7.4.4)

$$\kappa_\ell^t = (m_N\Omega_0^2)^2 U_0^t A^2 \langle r^{2\ell} \rangle / \pi. \tag{8.5.44}$$

The response functions for the form factors (8.5.35) are straightforward to calculate if the Θ function in (8.5.43) is ignored. The quadrupole computation is just like the one described in Sect. 8.5.B, and the dipole is easier. Assuming that the nucleus has closed major shells for both neutrons and protons, the results are

$$\chi_{1t}^{ipm\prime}(\omega) = \frac{3A}{4\pi} m_N \Omega_0^4 \frac{1}{\Omega_0^2 - \omega^2} (AU_0^t)^2 \tag{8.5.45a}$$

$$\chi_{2t}^{ipm\prime}(\omega) = \frac{5}{4\pi} \hbar\Omega_0^3 \left(\frac{3}{2}A\right)^{4/3} \frac{1}{(2\Omega_0)^2 - \omega^2} (AU_0^t)^2. \tag{8.5.45b}$$

In each case, the RPA equation has only one solution, with eigenvalues

$$\hbar\omega_{1t} = \hbar\Omega_0 \sqrt{1 + \tfrac{3}{4} U_0^t A / (m_N \Omega_0^2 \langle r^2 \rangle)} \tag{8.5.46a}$$

$$\hbar\omega_{2t} = 2\hbar\Omega_0 \sqrt{1 + U_0^t \left(\tfrac{3}{2}A\right)^{4/3} 5\hbar / (16 m_N^2 \Omega_0^3 \langle r^4 \rangle)}. \tag{8.5.46b}$$

Using numerical values from eqs. (7.4.4) and (5.2.9), with a sphere of radius $R(A) = 1.2$ fm $A^{1/3}$ to estimate the moments as in eq. (5.2.4), we obtain

$$\hbar\omega_{1t} = 72\text{MeV } A^{-1/3} \tag{8.5.47a}$$

$$\hbar\omega_{2t} = 128\text{MeV } A^{-1/3}. \tag{8.5.47b}$$

The isovector dipole resonance has been observed in nuclei throughout the periodic table. Its excitation energy, width, and sum-rule fraction are shown in fig. 8.10. We see that for heavy nuclei, the giant dipole exhausts the sum rule and has an energy in fair agreement with the estimate of eq. (8.5.47a). In deformed nuclei, the giant dipole vibration splits into two components representing vibrations along and perpendicular to the symmery axis. The corresponding frequencies are obtained from eq. (8.5.46a) by replacing the $\hbar\Omega_0$ in front of the square root with $\hbar\Omega_z$ and $\hbar\Omega_\perp$ respectively (the directional variation of the factors corresponding to Ω_0^2 and $\langle r^2 \rangle$ under the square root cancel each other). Since each component has a width similar to those in spherical nuclei, the two components overlap, leading to a very broad distribution of transition strength. This distribution is skewed toward the high-frequency side, because the higher-frequency vibration associated with perpendicular motion is doubly degenerate with modes corresponding to both x and y axes. The giant dipole resonance has recently been observed in photon-emission spectra from very highly excited nuclei. When the angular momentum of these nuclei is large enough, the dipole strength function's asymmetry shifts to lower energy, because the rapidly-rotating nuclei have an oblate shape with $\Omega_\perp < \Omega_z$, as we shall see in Chapter 9.

The isovector giant quadrupole resonance has also been observed in a number of nuclei. Its energy, shown in fig. 8.11, lies in the region 120–140 MeV/$A^{1/3}$, in good agreement with the estimate of eq. (8.5.47b).

To apply the oscillator model to the isovector monopole vibration, we choose the scaling approximation analogous to eq. (8.5.39):

$$\hat{U}^{\text{ipm}}(\vec{r}, t_3, \alpha_0^t) = \hat{U}^{\text{ipm}}\left((1 + 2t_3\alpha_0^t)\vec{r}, t_3, 0\right). \tag{8.5.48}$$

Then the form factor is just ± 1 for neutrons and protons, respectively, times the form factor for the isoscalar monopole,

$$\begin{aligned}
F_0^t(\vec{r}, t_3) &= 2t_3 m_N \Omega_0^2 r^2 \Theta\left(r < R(A)\right) \\
&= 2t_3 F_0^0(\vec{r}).
\end{aligned} \tag{8.5.49}$$

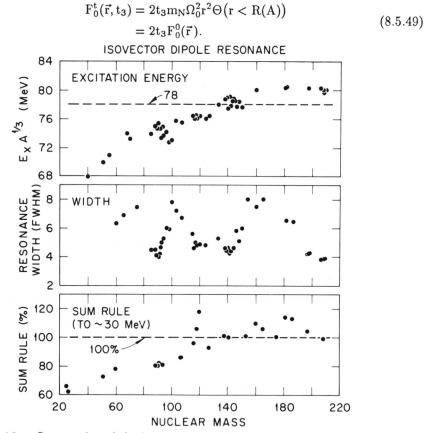

Figure 8.10   Systematics of the isovector giant dipole excitation energy, width, and sum rule depletion [*from F. Bertrand, loc. cit.*]

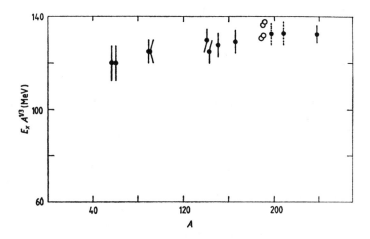

Figure 8.11 Excitation energy $\hbar\omega_{2t}A^{1/3}$ plotted against A for the isovector giant quadrupole resonance. [*from J. Speth and A. van der Woude, Rep. Prog. Phys.* **44** *(1981) 46*]

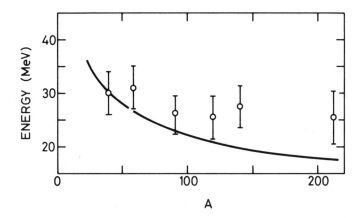

Figure 8.12 Energies of isovector monopole vibrations in various nuclei. The energies are estimated as the mean of their isobaric-analog states in neighboring nuclei with charges Z±1. Vertical bars indicate experimental uncertainties. The smooth curve is 103 MeV/$A^{1/3}$, eq. (8.5.52). [*Data from J. D. Bowman, in "Nuclear Structure 1985," R. Broglia, G. Hagemann, and B. Herskind, eds., North-Holland, Amsterdam, 1985, p. 549*]

Consequently the response function is the same as for the isoscalar monopole given in eq. (8.5.23),

$$\chi_{0t}^{\text{ipm}\prime}(\omega) = \chi_0^{\text{ipm}\prime}(\omega) = \left(\tfrac{3}{2}A\right)^{4/3} \cdot \frac{2\hbar\Omega_0^3}{(2\Omega_0)^2 - \omega^2}. \tag{8.5.50}$$

The coupling constant may be estimated from eqs. (8.5.37), (8.5.49), and (7.4.4). The result is the same as for the isovector quadrupole except for a factor $4\pi/(AU_0^t)^2$ which enters because F_0^t differs from F_{20}^t by a factor $Y_2^0 n_0 \Im^t$,

$$\kappa_0^t = 4\pi\kappa_2^t / \left(AU_0^t\right)^2 = 4(m_N\Omega_0^2)^2 \langle r^4 \rangle / U_0^t. \tag{8.5.51}$$

The RPA eigenvalue from eq. (8.4.17) is then

$$\hbar\omega_{0t} = 2\hbar\Omega_0\sqrt{1 + (\tfrac{3}{2}A)^{4/3} U_0^t \hbar / 8m_N^2\Omega_0^3 \langle r^4 \rangle}$$

$$\approx 2\hbar\Omega_0\sqrt{1 + U_0^t / 166\text{MeV}} \tag{8.5.52}$$

$$\approx 103\text{MeV}/A^{1/3}.$$

The sum rule estimate is identical to that of the isoscalar monopole scaling vibration. The isovector monopole has been identified experimentally in pion charge-exchange scattering. Its energies for various nuclei are shown in fig. 8.12. The observed energies are in fair agreement with the above estimate.

8.5 F.   Spin Vibrations

Isovector vibrations were characterized as neutrons and protons moving against each other. In complete analogy, spin-up nucleons can move against spin-down nucleons. The operators are just like the isovector operators, but with s_z in place of t_3. The spin dependence of the two-body effective interaction may be characterized by a zero-range term

$$\Im_s(\vec{r}_{12}, \vec{s}_1, \vec{s}_2) = \vec{s}_1 \cdot \vec{s}_2 \Im^s \delta(\vec{r}_{12}) \tag{8.5.53}$$

analogous to the isospin interaction of eq. (7.4.1). Then all the expressions for isovector vibrations may be adapted directly to spin vibrations merely by substituting $\Im^s$ for $\Im^t$ and $U_0^s = n_0 \Im^s / A$ for U_0^t. Unfortunately we cannot estimate U_0^s from the spin dependence of the phenomenological optical potential $\hat{U}^s$, which would by analogy to eq. (7.4.3) be proportional to the difference of spin-up and spin-down nucleon densities which is very small in nuclear ground states. Collective isoscalar spin oscillations have not yet been observed, since it is difficult to find a probe that excites them strongly.

The charge-exchange reactions (p, n) and (n, p), on the other hand, can excite isovector vibrations very conveniently. For nucleon energies below about 100

MeV, they excite mainly the isobaric analogs of the target's ground state. How-ever for higher energies the nucleon-nucleon charge-exchange effective interaction is predominantly of the spin-flip type,

$$\Im_{st}(\vec{r}_{12}, \vec{t}_1, \vec{t}_2, \vec{s}_1, \vec{s}_2) = (\vec{s}_1 \cdot \vec{s}_2)(\vec{t}_1 \cdot \vec{t}_2)\Im^{st}(\vec{r}_{12}). \tag{8.5.54}$$

We saw in Chapter 2 that the exchange of pions gives rise to such an effective force, to which exchange of ρ mesons also contributes. These vibrations are known as Gamow-Teller resonances, by analogy with the Gamow-Teller beta-decay interaction (Sect. 10.4) which contributes (very slightly) to the $(t_{1\pm}t_{2\mp})(s_{1\pm}s_{2\mp})$ part of $\Im_{st}$.

The observed isovector spin vibrations are of two types, with monopole and dipole spatial factors corresponding to $\ell=0$ and $\ell=1$. They are seen in medium to heavy nuclei, and their energies and strengths depend significantly on the neutron excess (see fig. 8.13). The $\ell=0$, S=1, T=1 vibrations seen in the (p, n) reaction lie about one to three MeV above the isobaric analog of the target's ground state (a handy reference point to account for the Coulomb energy shift of changing a neutron to a proton). Like the low-lying isoscalar vibrations, they are due to the spin-orbit force bringing an opposite-parity single-particle state into the closed oscillator shell, and cannot be described by our crude oscillator estimates. Detailed RPA computations with a realistic independent-particle model show a substantial repulsion for $\Im^{st}$. The $\ell=1$, S=1, T=1 vibrations are seen in the (p, n) reaction to lie near or above the shell spacing $\hbar\Omega_0$, and also depend strongly on the neutron excess which induces an (N-Z) dependence in the ipm response function. Fig. 8.14 shows that their energies are well fit by

$$E^{st}_{\ell=1}(\text{fit}) = 69\text{MeV}\,A^{-1/3} - 27.6\text{MeV}\frac{N-Z}{A}. \tag{8.5.55}$$

The coefficient of (N-Z)/A is smaller than the value of 48 MeV that would have been expected from a simple comparison of states with successive values of $T = \frac{1}{2}(N-Z)$ and $T = \frac{1}{2}(N-Z)-1$ in the estimate, eq. (7.4.11), of the interaction energy of the spin-independent part of the isovector effective interaction.

8.5 G.   Beyond the RPA

We have seen that simple oscillator-model estimates of the RPA vibrational fre-quencies give a fairly good description of giant resonances. Detailed calculations with realistic mean-field potentials, and with effective interactions of the Skyrme type, have been used to give a much more detailed description of both energies and transition strengths of low-lying as well as giant collective oscillations [Gogny]. The differences we have encountered between our simple estimates and the observed gi-ant resonance energies are not only due to the oversimplification of the oscillator model, but also to the limitations of the mean field approximation which lies behind the RPA.

We should recall that the mean field gives an incomplete account of the single-particle energies of nuclei, which lie closer together near the Fermi surface than

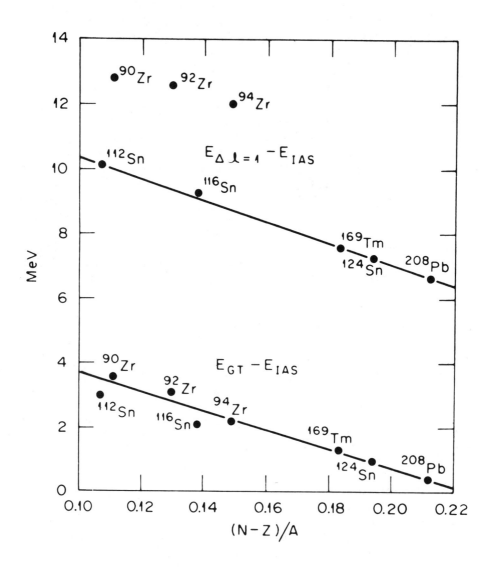

Figure 8.13   Energies of spin-flip isovector states with $\ell=0$ (GT) and $\ell=1$ excited by the (p, n) reaction at $\varepsilon_p=200$ MeV. The masses and atomic designations are those of the target nucleus before the reaction. [*from D. Horen, C. Goodman, D. Bainum, C. Foster, C. Gaarde, C. Goulding, M. Greenfield, J. Rapaport, T. Taddeucci, E. Sugarbaker, T. Masterson, S. Austin, A. Galonsky and W. Sterrenburg, Phys. Letters* **99 B** *(1981) 383*]

the mean field predicts (Sect. 4.4.B), and which have widths associated with the single-particle lifetimes due to the nucleons' finite mean free path (Sect. 4.4.A). The RPA model can be adapted to take these effects into account approximately, by adjusting the mean field to correspond to the observed single-particle properties. This ad hoc procedure, however, only partially compensates for the inadequacy of the mean-field approximation. The effective interaction also has to be modified due to many-body effects [Nakayama, Krewald and Speth]. Thus, for example, the observed energies of isovector modes show that the effective interaction $\Im_t$ is different in these excited states than the effective interaction that produces the symmetry energy of the ground state.

The study of collective nuclear vibrations uncovers fascinating many-body physics which is the subject of continuing theoretical and experimental investigations. For example, recent advances in inelastic electron scattering make it possible to display the detailed radial dependence of the density fluctuations associated with collective motion (fig. 8.15). These studies exploit the unique experimental accessibility of the nucleus's microscopic structure to teach us the detailed interplay between single- particle and collective motion in a many-body system.

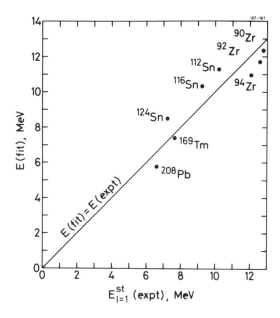

Figure 8.14   Comparison of observed energies for $\ell=1$ isovector spin-flip vibrations with fit by formula (8.5.55). [*Data from fig. 8.13*]

8.6   SLOW, LARGE-AMPLITUDE COLLECTIVE MOTION

The RPA allows a theoretical understanding of a very broad class of nuclear vibrational motion. Any periodic excitation where the dynamics are dominated by the

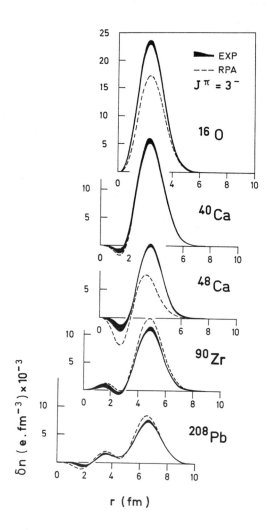

Figure 8.15   Transition charge densities $e\delta n(r, t_3 = \frac{1}{2})$ for the lowest collective isoscalar
octupole vibrations in various nuclei. The broad curve is the experimental result; its width
indicates the uncertainty. The dashed curves are results of RPA calculations with density-
dependent effective interactions. [*from B. Frois and C. N. Papanicolas, Annual Review
of Nuclear and Particle Science, Palo Alto, to be published*]

mean field can be treated, including isospin vibrations and excitations of the pairing field, as well as shape vibrations. The equation of motion (8.4.14) for the collective variable is, however, derived in perturbation theory, and therefore only valid when the influence of the collective variables on the nucleons' motion is weak at every time. This condition can be satisfied in two very different ways, depending on the physics of the nucleonic system and on the type of collective motion. The first and most obvious way is to consider only small-amplitude collective motion, such as small vibrations about an equilibrium Q^0. Such vibrations are always possible in classical physics, where the amplitude may be arbitrarily small. In quantum mechanics, the condition of a small perturbation must hold for collective amplitudes larger than the zero-point amplitude of the oscillation. This condition places no limitation on the frequency ω_n of the oscillation, and the RPA loses its usefulness only for frequencies so high that the time for the effective forces to act becomes significant.

The other, less obvious case in which the equation of motion (8.4.14) retains validity occurs when the response function $\tilde{\chi}_{\mu\nu}^{ipm}(t)$ vanishes for times larger than some relaxation time τ. In this case, slow collective motion can be described by a simple generalization of eq. (8.4.14). To see this, we need only use eq. (8.1.10) with $\omega = 0$ to write eq. (8.4.14) as

$$\mathcal{K}_\mu(\{Q_\alpha(t)\}) + \sum_\nu \int_{-\infty}^{\infty} dt' \tilde{\chi}_{\mu\nu}^{ipm}(t - t')(Q_\nu(t') - Q_\nu(t)) = 0 \qquad (8.6.1)$$

where

$$\mathcal{K}_\mu(\{Q_\alpha\}) = \sum_\nu (Q_\nu - Q_\nu^0)\left(\kappa_{\mu\nu} + \chi_{\mu\nu}^{ipm\prime}(\omega = 0)\right) - \langle gs| \sum_i F_\mu^{sp} |gs\rangle \qquad (8.6.2)$$

is the static force. We see from eq. (8.6.1) that $\{Q_\alpha(t)\}$ is determined by the values of $\{Q_\alpha\}$ at previous times t' for which $\tilde{\chi}_{\mu\nu}^{ipm}(t - t')$ is non-vanishing. When H^{sp} is such that $\tilde{\chi}_{\mu\nu}^{ipm}(t - t')$ becomes small for large times $t - t' \gg \tau$, the earlier history of the collective motion is forgotten and does not influence the motion at time t, to leading order in the perturbation on the intrinsic system.

We conclude that we have found an equation of motion, eq. (8.6.1), valid in a time interval during which $\{Q_\alpha(t)\}$ does not differ too much from $\{Q_\alpha^0\}$. For this equation of motion to be useful, it is necessary that the time interval during which the perturbation condition is fulfilled be at least as large as the relaxation time of the response function. This condition is significantly less restrictive than the usual perturbation condition: by choosing different values of Q_α^0 appropriate to different time intervals, it may be possible to satisfy the perturbation condition sequentially for a much longer time than if one were restricted to a single choice of Q_α^0.

If this weaker perturbation condition is fulfilled, we are free to choose $\{Q_\alpha^0\}$ in accordance with the actual orbit $\{Q_\alpha(t)\}$. A natural choice would be $Q_\alpha^0 = Q_\alpha(t)$. However, it is more convenient to make the choice in a way which is symmetric in

t' and t, for example, $Q_\alpha^0 = \frac{1}{2}(Q_\alpha(t) + Q_\alpha(t'))$ or $Q_\alpha^0 = Q_\alpha(\frac{t+t'}{2})$. Such a symmetric prescription will preserve the association of χ' and χ'' with conservative and dissipative processes, as discussed in connection with eq. (8.1.13). The differences among various prescriptions for choosing $\{Q_\alpha^0\}$ are of higher order in $\{(Q_\alpha - Q_\alpha^0)\}$ and thus have to be small if the perturbative assumption is valid. The requirement that these higher-order terms be small may be estimated by comparing the equation of motion that would be found if we expand about $\{Q_\alpha^{(1)}\} = \{Q_\alpha(t)\}$ with the equation obtained by expanding about $\{Q_\alpha^{(2)}\} = \{Q_\alpha(t - \tau)\}$. Considering the second term in eq. (8.6.1), we see that the two perturbative expansions will lead to the same value of the integral provided that

$$\tau \sum_\lambda \frac{dQ_\lambda^0}{dt} \frac{\partial \tilde\chi_{\mu\nu}^{\text{ipm}}(t, \{Q_\alpha^0\})}{\partial Q_\lambda^0} \ll \tilde\chi_{\mu\nu}^{\text{ipm}}(t', \{Q_\alpha^0\}) \qquad (8.6.3)$$

for all t', $t < \tau$. This condition clearly requires that the collective motion be slow on the scale of the relaxation time, but the severity of this requirement depends on how sensitive $\tilde\chi^{\text{ipm}}$ is to $\{Q_\alpha^0\}$.

The condition (8.6.3) is clearly sufficient to ensure that the second term in eq. (8.6.1) is insensitive to the choice of $\{Q_\alpha^0\}$. If it is fulfilled, then the first term $\mathcal{K}_\mu$ is also insensitive to $\{Q_\alpha^0\}$. To see this we note that

$$\mathcal{K}_\mu = -\frac{\partial}{\partial Q_\mu} \langle H_{\text{sp}}(Q) \rangle_Q \qquad (8.6.4)$$

where $\langle H_{\text{sp}}(Q) \rangle_Q$ is the static energy of the nucleon system, correct to second order in $\{Q_\alpha - Q_\alpha^0\}$. This follows from computing the force as we did in eqs. (8.4.11–14), but with $Q_\nu(t)$ taken to be independent of time; the expression for the force which emerges is then just eq. (8.6.2).

The equation of motion (8.6.1) has a simple interpretation: the static force at each time must just balance the dynamic force, which remembers how different the nucleus' configuration was at earlier times. The response function describes this memory, in a linearized approximation which is valid when the response function itself doesn't change very much during its memory time, the relaxation time. The equation of motion (8.6.1) therefore is valid both for small-amplitude vibrations, or for slow collective motion, provided that we take account of the dependence on Q_ν of both $\mathcal{K}_\mu$ and $\tilde\chi_{\mu\nu}^{\text{ipm}}$.

If the collective motion is very slow on the scale of the relaxation time, we can further simplify eq. (8.6.1). Then, we can expand $Q_\nu(t')$ in a Taylor series around $t' = t$:

$$Q_\nu(t') = Q_\nu(t) + (t' - t)\dot Q_\nu(t) + \frac{1}{2}(t' - t)^2 \ddot Q_\nu(t) + \dots \qquad (8.6.5)$$

where the dots above Q_ν indicate time derivatives. Using the expansion (8.6.5) in eq. (8.6.1) we get

$$\mathcal{K}_\mu(\{Q_\alpha(t)\}) - \sum_\nu \gamma_{\mu\nu} \dot Q_\nu(t) = \sum_\nu M_{\mu\nu} \ddot Q_\nu(t), \qquad (8.6.6)$$

where the integrals over t' now appear in the coefficients

$$\gamma_{\mu\nu} = -\int_{-\infty}^{\infty} dt'\, \tilde{\chi}_{\mu\nu}^{\rm ipm}(-t')t', \tag{8.6.7a}$$

$$M_{\mu\nu} = -\frac{1}{2}\int_{-\infty}^{\infty} dt'\, \tilde{\chi}_{\mu\nu}^{\rm ipm}(-t')t'^2. \tag{8.6.7b}$$

We recognize the equation of motion (8.6.6) as Newton's second law. On the left, we have the static force plus a force proportional to the velocity $\dot{Q}_{\nu}$. On the right, we see an inertial term proportional to the acceleration $\ddot{Q}_{\nu}$, with the inertial coefficients $M_{\mu\nu}$ playing the role of the mass.

We have thus obtained a familiar result, Newton's second law of motion, as a limiting case of our collective equation of motion when the collective coordinates change slowly compared to the time over which the individual nucleons move. The force proportional to the velocity, with coefficients $\gamma_{\mu\nu}$, is perhaps a bit surprising but not unwelcome: we are familiar with such forces in macroscopic physics. For example, a particle moving in a magnetic field experiences the Lorentz force which is proportional to its velocity but perpendicular to it, which is a conservative force because it doesn't change the particle's kinetic energy. As another example, an object moving slowly through a gas or liquid experiences a viscous force proportional to the velocity but opposed to it, which is a dissipative force because it reduces the kinetic energy of the object by heating up the fluid. Thus the velocity-dependent force, which originated in the memory term of eq. (8.6.1), may contain both conservative and dissipative forces, as we discussed in sect. 8.1.

Our expressions for the velocity-dependent forces and inertia, eq. (8.6.7), are most conveniently expressed in terms of the frequency-dependent polarizability $\chi_{\mu\nu}^{\rm ipm}(\omega)$. Using eq. (8.1.10), it is easy to show that

$$\gamma_{\mu\nu} = \frac{1}{i}\frac{\partial}{\partial\omega}\chi_{\mu\nu}^{\rm ipm}(\omega)|_{\omega=0}, \tag{8.6.8a}$$

$$M_{\mu\nu} = \frac{1}{2}\frac{\partial^2}{\partial\omega^2}\chi_{\mu\nu}^{\rm ipm}(\omega)|_{\omega=0}. \tag{8.6.8b}$$

The derivatives are easily evaluated from eqs. (8.2.10-11). We note that, because of the limit $\omega = 0$ in eqs. (8.6.8), the dissipative forces will vanish in this slow-motion limit unless the independent particle model includes occupied states in the continuum, since otherwise the δ-functions in eq. (8.2.10) would be zero for small frequencies. Thus slow collective motion built on the nuclear ground state is not damped. This is the case for the low-frequency shape vibrations described in the preceding section, and for the even slower rotational motion discussed in Chapter 9 below. The higher-frequency vibrational modes, on the other hand, are damped; this is why the high-frequency shape vibrations, the giant isoscalar and isovector dipole resonances, appear as broad peaks in the excitation spectra, with widths inversely proportional to their damping times. Since the vibrational periods of these collective modes are less than the relaxation time of $\tilde{\chi}_{\mu\nu}^{ipm}$ (see eq. (8.3.12)), they cannot be described by the slow-motion Newton eq. (8.6.6). A quantitative account of their damping is not currently available.

In addition to rotation, another slow, large-amplitude collective motion is found in fission. We will discuss fission in Sect. 10.2.E, and find that both the inertial and damping forces are important. Another important application of Newton's law for collective motion, eq. (8.6.6), has been the description of the collisions of heavy nuclei at small relative velocities. We will discuss these collisions in Chapter 11. In these collisions, the translational motion of the nuclei is very important. Fortunately, it is easy to see that the inertial parameter (8.6.8b) for translational motion has exactly the correct value Am_N (compare Sect. 4.5 and problem 8.7). This result is a good check on the validity of the RPA approach to collective motion.

PROBLEMS

8.1 Show that the volume inside a given equipotential surface of the potential in eq. (8.5.2) is independent of the deformation parameter to second order.

8.2 Calculate the leading term of the smooth energy $\tilde{\mathcal{E}}$ for an anisotropic harmonic oscillator. Then calculate the coupling constant κ_2 corresponding to α_{20} in eq. (8.5.2) by using eqs. (8.4.13), (8.4.6) and (8.4.4). Compare to eq. (8.5.11).

8.3 Calculate $\chi_2^{ipm}(0)$ of eq. (8.5.12) to leading order in nucleon number A in the case of doubly-closed-shell nuclei. Use this to derive eq. (8.5.13b).

8.4 Derive the values of the sum rules S_0 and S_2, eqs. (8.5.26) and (8.5.16). Then compute the contribution to each sum rule from the RPA modes of eqs. (8.5.25) and (8.5.14) respectively.

8.5 Justify eq. (8.5.21).

8.6 Find expressions for the RPA energies of the giant quadrupole vibrations corresponding to the beta and gamma degrees of freedom, assuming κ_2 is the same as for spherical nuclei. What are the energies for ^{20}Ne?

8.7 Show that the inertial parameter (8.6.8b) for translational motion is exactly the correct result Am_N. Hint: use eqs. (8.5.27–29, 31) and eq. (8.1.27), and obtain another equation from evaluating $[H_{sp}, Z]$.

8.8 Assume the nuclear gound state is the BCS wavefunction in eq. (6.4.21) and the excited states are n-quasiparticle states of form $\alpha_{k_1}^+ \cdot \alpha_{k_2}^+ \ldots \alpha_{k_n}^+ |gs\rangle$. Derive the corresponding generalizations of eqs. (8.2.9-11). Compute the mass parameters for very slow collective motion.

8.9 Assume the time dependent ipm response function is given by

$$\tilde{\chi}^{ipm}(t) = C \sin(\lambda t)e^{-\Gamma t}\Theta(t > 0)$$

where C, λ and Γ are positive constants. Calculate the related response and polarizability functions $\tilde{\chi}'(t)$, $\tilde{\chi}''(t)$, $\chi'(\omega)$ and $\chi''(\omega)$. Solve the RPA eigenvalue equation, eq. (8.4.17) and give the RPA response functions as a function of the coupling constant κ. What is the physical interpretation of the complex eigenvalues? Derive the friction coefficient γ and mass parameter M, and discuss the small-amplitude collective motion when it is very slow. Compare the eigenfrequencies of Newton's equation with the RPA in the case $\Gamma \ll \lambda$.

8.10 Repeat the various steps in problem 8.9 for

$$\tilde{\chi}^{ipm}(t) = C \sinh(\lambda t)e^{-\Gamma t}\Theta(t > 0)$$

where C, λ and Γ are positive constants and $\Gamma > \lambda$.

9

Rotational Motion

9.1 CLASSICAL PICTURE OF ROTATIONAL MOTION FOR DEFORMED NUCLEI

The lowest excited states of nuclei with spherical ground states are vibrational collective states at excitation energies of a few MeV, described by the RPA picture as coherent oscillations of the nuclear shape around its equilibrium (spherical) shape. Deformed nuclei also have vibrational states, with energies comparable to those of spherical nuclei with nearby masses. But the vibrational states of deformed nuclei are never their lowest excited states.

Instead, each deformed nucleus has a set of excited states with energies much less than 1 MeV. These low-lying states appear to be collective: like the vibrations, they are strongly excited by the Coulomb fields of passing projectiles. Their energies, too, vary smoothly with the masses of the deformed nuclei; in fact, they are even more regular than the vibrational energies, with groups of states of various angular momenta occuring in typical patterns of excitation energies. For example, a deformed nucleus with a spin-zero ground state always has as its lowest excited states a series of states with even angular momentum $\hbar J = 2\hbar$, $4\hbar$ etc. whose excitation energies $E_J^* = E_J - E_0$ are well approximated by

$$E_J^* = BJ(J + 1). \tag{9.1.1}$$

Such a set of states is called a rotational band. For some examples, see figs. 9.1 and 9.2.

The form of the observed spectra (9.1.1) suggests an immediate analogy with the spectrum of the lowest excited states of molecules. These states are rotational states, corresponding to the molecule's rotating with angular momentum $\hbar J$. It is natural to picture these nuclear states as the deformed nucleus rotating about an axis. As in the case of molecules, we may guess that the Hamiltonian describing the motion might be

$$H_{\text{rot}} = \frac{\vec{J}^2}{2\mathcal{I}} \tag{9.1.2}$$

us of eq. (9.1.1): we relate the constant B to the moment of inertia

$$B = \frac{\hbar^2}{2\mathcal{I}}.$$

(9.1.3)

The simplest way to estimate the moment of inertia is to assume that the whole nucleus revolves like a rigid body, as in the case of most molecules. Then an axially-symmetric nucleus would have two different moments of inertia, one about its symmetry axis

$$\mathcal{I}_z = m_N \langle gs|x^2 + y^2|gs\rangle$$

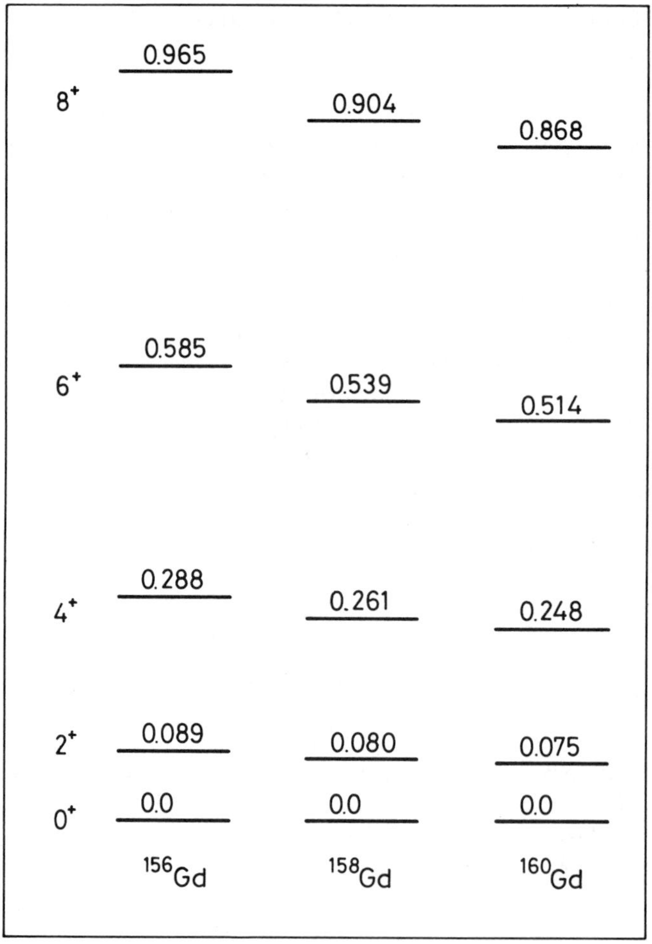

Figure 9.1 The lowest five states of three isotopes of gadolinium (Z=64). Their excitation energies are shown in MeV, with J^π on the left.

and one about the perpendicular axes,

$$\mathcal{I}_{\rm rig} = m_{\rm N} \langle {\rm gs}|x^2 + z^2|{\rm gs} \rangle \approx \frac{2}{5} m_{\rm N} r_0^2 A^{5/3}. \tag{9.1.4}$$

The rigid-body inertia gives the order of magnitude of the observed moments of inertia for even-even deformed nuclei, giving typically a factor of two too large. But the spectrum (9.1.1) comes from quantizing an isotropic rotor, not an anisotropic one; there is no sign of the rotations about the symmetry axis. Besides, the rigid-body picture gives no clue why only deformed nuclei have low-lying rotational bands.

9.2   CRANKING PICTURE OF ROTATIONAL MOTION FOR DEFORMED NUCLEI

To get a simple quantum-mechanical picture of rotations, we can apply the results of linear response theory. The collective coordinate that describes a rotation of the

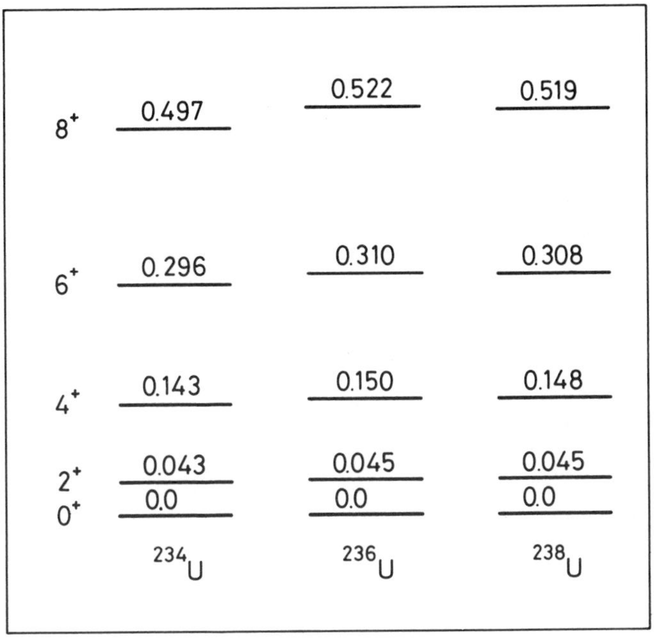

Figure 9.2   The lowest five states of three isotopes of uranium (Z=92). Their excitation energies are shown in MeV, with J^π on the left.

To get a simple quantum-mechanical picture of rotations, we can apply the results of linear response theory. The collective coordinate that describes a rotation of the nucleus' mean field is the orientation of its axis of symmetry: for example, if the axis of symmetry $\hat{n}$ is rotating in the x-z plane, then the potential's orientation in this plane is described by the symmetry axis' angle about the y-axis θ_y:

$$\hat{n} \cdot \hat{z} = \cos \theta_y$$

$$\hat{n} \cdot \hat{x} = \sin \theta_y .$$

9.2 A. The Cranking Inertia

Since the frequency of the collective motion is observed to be very slow, we can use the adiabatic approximation of Sect. 8.6. The restoring force $\partial \langle H \rangle / \partial \theta_y$ is zero, since the energy of the deformed nucleus cannot depend on its orientation. The inertia $M_{\theta\theta} = \mathcal{I}_{cr}$ is given by using eq. (8.1.27) in eq. (8.6.8b):

$$\mathcal{I}_{cr} = 2\hbar^2 \sum_{n \neq gs} \frac{|\langle gs \, |\partial H_{sp}/\partial \theta_y| \, n \rangle|^2}{(E_n - E_0)^3} . \tag{9.2.1}$$

Using the identity

$$i\hbar \left\langle gs \left| \frac{\partial H_{sp}}{\partial \theta_y} \right| n \right\rangle = \langle gs \, |[H_{sp}, J_y]| \, n \rangle = (E_0 - E_n) \langle gs \, |J_y| \, n \rangle \tag{9.2.2}$$

we obtain Inglis' formula for the moment of inertia

$$\mathcal{I}_{cr} = 2 \sum_{n \neq gs} \frac{|\langle n \, |J_y| \, gs \rangle|^2}{(E_n - E_0)} . \tag{9.2.3}$$

This is known as the **cranking inertia**, because it is based on the nucleus' response to an effort to turn it about its y-axis.

From the cranking model, we can immediately see why there are no rotational states from rotations about the mean field's symmetry axis $\hat{z}$: $\partial H / \partial \phi_z = 0$, so that $\mathcal{I}_{cr} = 0$ also. This explanation applies **a fortiori** to spherical nuclei. It also gives very good values for the moments of inertia, provided pairing correlations are taken into account in computing $\mathcal{I}_{cr}$.

9.2 B. Cranking the Harmonic Oscillator

As a simple example of the cranking inertia, we turn once again to the deformed, axially symmetric harmonic oscillator. We computed ground-state deformations in this model in Sect. 5.4, and studied its vibrations in Sect. 8.3.

First, we rewrite eq. (9.2.3) in terms of single-particle eigenstates $|k\rangle$,

$$\mathcal{I}_{\mathrm{cr}} = 2 \sum_{jk} N_{kj} \frac{|\langle j \, |j_y| \, k \rangle|^2}{\varepsilon_k - \varepsilon_j}. \tag{9.2.4}$$

The single-particle eigenstates are characterized by their quantum numbers n_x, n_y and n_z. Since $j_y = z p_x - x p_z$, we can find its matrix elements

$$\langle n'_x n'_y n'_z \, |j_y| \, n_x n_y n_z \rangle = \frac{i\frac{1}{2}\hbar\delta_{n_y,n'_y}}{\sqrt{\Omega_\perp \Omega_z}}$$

$$\left[(\Omega_z - \Omega_\perp) \left(\sqrt{n_x n_z} \, \delta_{n'_x, n_x-1} \delta_{n'_z, n_z-1} - \sqrt{n'_x n'_z} \, \delta_{n'_x, n_x+1} \delta_{n'_z, n_z+1} \right) \right.$$

$$\left. + (\Omega_z + \Omega_\perp) \left(\sqrt{n'_x n_z} \, \delta_{n'_x, n_x+1} \delta_{n'_z, n_z-1} - \sqrt{n_x n'_z} \, \delta_{n'_x, n_x-1} \delta_{n'_z, n_z+1} \right) \right]. \tag{9.2.5}$$

We see from (9.2.5) that J_y connects states whose energies differ either by $(\hbar\Omega_z - \hbar\Omega_\perp)$ or $(\hbar\Omega_z + \hbar\Omega_\perp)$. Applying (9.2.5) to (9.2.4) one obtains a result which depends on the same quantities as appeared in the deformation energy (eq. (5.3.6)):

$$\mathcal{I}_{\mathrm{cr}}^{\mathrm{HO}} = \frac{\hbar}{2\Omega_\perp \Omega_z} \left[\frac{(\Omega_\perp + \Omega_z)^2}{(\Omega_\perp - \Omega_z)} (\Sigma_z - \Sigma_x) + \frac{(\Omega_\perp - \Omega_z)^2}{(\Omega_\perp + \Omega_z)} (\Sigma_x + \Sigma_z) \right]. \tag{9.2.6}$$

Using the self-consistency condition (5.3.7) together with the definition of $\vec{\Sigma}$, eq. (5.3.6), we find the surprising result

$$\mathcal{I}_{\mathrm{cr}}^{\mathrm{HO}} = m_N A \left(\langle z^2 \rangle + \langle x^2 \rangle \right) = \mathcal{I}_{\mathrm{rig}}^{\mathrm{HO}}. \tag{9.2.7}$$

Calculations in the independent-particle model with Woods-Saxon single-particle potentials, including spin-orbit terms, also give values for the cranking inertia very close to the rigid-body value, provided a self-consistent shape is chosen. We must remember that, in the mean-field picture, the spherical shape is only self-consistent for a closed-shell nucleus. If we tried to compute the cranking inertia for a spherical open-shell nucleus, we would get an indefinite answer: the numerator would vanish because $\partial H^{\mathrm{sp}}/\partial\theta_y$ is zero, but there would be a vanishing denominator too because the ground state is degenerate.

9.2 C. Success and Failure of the Cranking Picture

To obtain a significant deviation from the rigid moment of inertia, one has to go beyond the independent-particle model. The BCS model gives smaller inertias than the ipm. If we use the states of Sect. (6.4.B) in eq. (9.2.3), we get

$$\mathcal{I}_{\mathrm{cr}}^{\mathrm{BCS}} = 2 \sum_{k_1 k_2} \frac{|\langle k_1 \, |j_y| \, k_2 \rangle|^2}{\epsilon_{k_1} + \epsilon_{k_2}} \left(u_{k_1} v_{k_2} - v_{k_1} u_{k_2} \right)^2. \tag{9.2.8}$$

The pairing correlations reduce the moment of inertia in two ways. First, the energy gap makes the denominators larger. Second, the numerators get occupation factors which reduce the matrix elements. Together these account for the factor-of-two difference between $\mathcal{I}_{rig}$ and experiment.

Unfortunately, the cranking model does not give a fully satisfactory account of the observed rotational spectra. Although the moments of inertia inferred from the observed spectra via eqs. (9.1.1 - 9.1.3) are within 20% of the values predicted by formula (9.2.8), the model cannot explain the absence of odd values of J for spin-zero (even-even) nuclei. Nor does it give any clue to understanding the more complicated rotational spectra of odd nuclei, whose ground-state spins are different from zero.

The failure of the cranking model is due to its excessive reliance on the mean-field description, in which the nuclear ground state breaks the rotational symmetry of the Hamiltonian. We recall the failings of the mean-field picture in the case of translational symmetry: it had a set of degenerate solutions, with various centers of mass, instead of the correct non-degenerate ground state of unbroken symmetry and the correct spectrum of excited states corresponding to uniform translational motion. The mean-field picture also has a set of degenerate, broken-rotational-symmetry ground states, with different orientations of the remaining symmetry axis; it is not too surprising that these aspects of the mean-field picture are incorrect. Indeed, we notice that the mean-field ground state cannot be a state of zero angular momentum, since it is not rotationally symmetric; yet the observed ground states of even-even nuclei have J=0. To understand ground-state angular momenta, as well as rotational motion, we have to go beyond the mean-field picture.

9.3 GENERATOR-COORDINATE PICTURE OF DEFORMED SPIN-ZERO NUCLEI

We proceed by analogy with the case of translational symmetry, which is also broken by the mean-field approximation. In that case, we were able to form eigenstates of momentum $\vec{P}$ by making linear combinations of the degenerate set of wave functions generated by potentials displaced by $\vec{R}$, with a weighting function $e^{i\vec{P}\cdot\vec{R}/\hbar}$. Peierls and Yoccoz also applied this method to the case of broken rotational symmetry.

9.3 A. The Generator-coordinate Wave Function

In the case of broken rotational symmetry, the set of degenerate broken-symmetry wave functions in the mean-field picture corresponds to choosing various orientations of the mean field. Thus, if we start with a wave function with axial symmetry about the z-axis $\psi_0(\vec{r}_1, \ldots, \vec{r}_A)$, we can find another wave function $\psi_{\vec{\Omega}}$ symmetric about an axis whose polar angles are $\vec{\Omega} = (\theta, \phi)$:

$$\psi_{\vec{\Omega}}(\vec{r}_1, \ldots, \vec{r}_A) = \psi_0(\Re_{\vec{\Omega}}\vec{r}_1, \ldots, \Re_{\vec{\Omega}}\vec{r}_A) \tag{9.3.1}$$

where $\Re_{\vec{\Omega}}$ is the operator which rotates the z axis into the direction $\vec{\Omega}$: $\Re_{\vec{\Omega}}\hat{z} = \vec{\Omega}$, so that $\Re_{\vec{\Omega}}\vec{r}$ is rotated from $\vec{r}$ by the same rotation $\vec{\Omega}$. The angle ϕ specifies the direction in the x-y plane of the axis around which we rotate; the angle θ specifies the amount of the rotation (see fig. 9.3). $\psi_{\vec{\Omega}}$ is a Slater determinant of eigenfunctions of the rotated mean-field potentials

$$U_{\vec{\Omega}}^D(\vec{r}) = U_0^D(\Re_{\vec{\Omega}}\vec{r}) \tag{9.3.2a}$$

$$U_{\vec{\Omega}}^X(\vec{r}, \vec{r}') = U_0^X(\Re_{\vec{\Omega}}\vec{r}, \Re_{\vec{\Omega}}\vec{r}'). \tag{9.3.2b}$$

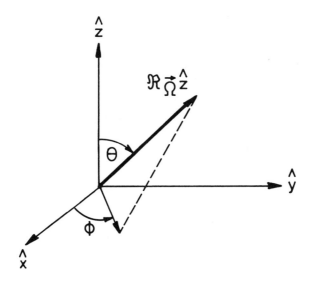

Figure 9.3   Coordinate systems related by the rotation $R_{\vec{\Omega}}$.

It is obvious that the energies of all the wave functions $\psi_{\vec{\Omega}}$ are the same. We say that the angular coordinates $\vec{\Omega} = (\theta, \phi)$ generate the set of wave functions $\psi_{\vec{\Omega}}$.

Just as the breaking of translational symmetry was shown by the fact that the mean-field wave functions $\psi_{\vec{k}}$ were not eigenfunctions of the momentum, the breaking of rotational symmetry appears in the failure of the $\psi_{\vec{\Omega}}$ to be eigenstates of angular momentum. The linear superposition of the $\psi_{\vec{\Omega}}$ which has total angular momentum $\vec{J}^2 = J(J+1)\hbar^2$ and z-component $J_z = \hbar M$ is

$$\psi_{JM}(\vec{r}_1, \ldots, \vec{r}_A) = \int d\Omega Y_J^M(\Omega)\psi_{\vec{\Omega}}(\vec{r}_1, \ldots, \vec{r}_A)$$

$$= \int_0^{2\pi} d\phi \int_0^{\pi} d\theta \sin\theta Y_J^M(\theta, \phi)\psi_0(\Re_{\vec{\Omega}}\vec{r}_1, \ldots, \Re_{\vec{\Omega}}\vec{r}_A). \tag{9.3.3}$$

It is not hard to show that ψ_{JM} is an eigenfunction of the total angular momentum and its z-component, with the desired eigenvalues (Problem 9.4).

We can already see the reason why the odd values of J are missing from the nuclear rotational spectra: if ψ_0 is a state of even parity, then

$$\psi_{JM} = 0 \qquad \text{for J odd.} \tag{9.3.4}$$

This is because the linear combination (9.3.3) contains equal but opposite amounts of $\psi_{\theta,\phi}$ and $\psi_{\pi-\theta,\pi+\phi}$ due to the symmetry $Y_J^M(\theta,\phi) = (-1)^J \, Y_J^M(\pi-\theta, \pi+\phi)$. But $\psi_{\theta\phi}(\vec{r}_1,\ldots,\vec{r}_A) = \psi_{\pi-\theta,\pi+\phi}(-\vec{r}_1,\ldots,-\vec{r}_A)$, which are equal if ψ_0 has even parity; thus the contributions to ψ_{JM} for the two hemispheres of $\vec{\Omega}$ cancel out for J odd. Similarly, if ψ_0 has odd parity, then only odd values of J will occur.

We can also see why spherical nuclei do not have rotational spectra. Think first of the case of a spherical nucleus in a spherically-symmetrical state ψ_0 of zero angular momentum. Then $\psi_{\vec{\Omega}}$ is independent of $\vec{\Omega}$, i.e. $\psi_{\vec{\Omega}} = \psi_0$, and we see immediately that eq. (9.3.3) gives $\psi_{JM} = 0$ except for J=M=0. In fact, if ψ_0 is an eigenstate of $\vec{J}^2$ and J_z with eigenvalues $J_0(J_0+1)\hbar^2$ and $\hbar M_0$, then $\psi_{JM} = 0$ unless both J=J_0 and M=M_0, in which case $\psi_{JM} = \psi_0$. We see that the operation described by (9.3.3) projects the component with angular momentum JM out of the state ψ_0.

9.3 B.   Rotational Energy in the Generator-coordinate Method

The generator-coordinate method can also explain the form of the rotational spectrum of excitation energies (9.1.1). To see this we compute the expectation value of the energy, following Peierls and Yoccoz:

$$
\begin{aligned}
E_{JM} &= \frac{\langle \psi_{JM} | H | \psi_{JM} \rangle}{\langle \psi_{JM} | \psi_{JM} \rangle} \\[2mm]
&= \frac{\int d\Omega \, d\Omega' \, \langle \psi_0(\Re_{\vec{\Omega}}\vec{r}_1,\ldots,\Re_{\vec{\Omega}}\vec{r}_A) | H | \psi_0(\Re_{\vec{\Omega}'}\vec{r}_1,\ldots,\Re_{\vec{\Omega}'}\vec{r}_A) \rangle \, Y_J^M(\vec{\Omega})^* Y_J^M(\vec{\Omega}')}{\int d\Omega \, d\Omega' \, \langle \psi_0(\Re_{\vec{\Omega}}\vec{r}_1,\ldots,\Re_{\vec{\Omega}}\vec{r}_A) | \psi_0(\Re_{\vec{\Omega}'}\vec{r}_1,\ldots,\Re_{\vec{\Omega}'}\vec{r}_A) \rangle \, Y_J^M(\vec{\Omega})^* Y_J^M(\vec{\Omega}')}.
\end{aligned}
\tag{9.3.5}
$$

The matrix elements in the integrands depend only on the angle Θ between the directions $\vec{\Omega}$ and $\vec{\Omega}'$. We can thus expand the matrix elements in Legendre polynomials of $\cos\Theta$:

$$
\begin{aligned}
\mathcal{H}(\Theta) &\equiv \langle \psi_0(\Re_{\vec{\Omega}}\vec{r}_1,\ldots,\Re_{\vec{\Omega}}\vec{r}_A) | H | \psi_0(\Re_{\vec{\Omega}'}\vec{r}_1,\ldots,\Re_{\vec{\Omega}'}\vec{r}_A) \rangle \\
&= \sum_L \mathcal{H}_L P_L(\cos\Theta),
\end{aligned}
\tag{9.3.6a}
$$

$$
\begin{aligned}
\mathcal{N}(\Theta) &\equiv \langle \psi_0(\Re_{\vec{\Omega}}\vec{r}_1,\ldots,\Re_{\vec{\Omega}}\vec{r}_A) | \psi_0(\Re_{\vec{\Omega}'}\vec{r}_1,\ldots,\Re_{\vec{\Omega}'}\vec{r}_A) \rangle \\
&= \sum_L \mathcal{N}_L P_L(\cos\Theta).
\end{aligned}
\tag{9.3.6b}
$$

Using the addition theorem $P_L(\cos\Theta) = 4\pi\sum_m Y_L^m(\vec{\Omega})^* Y_L^m(\vec{\Omega}')/(2L+1)$ and the orthonormality relations for the Y_J^M, we can substitute eqs. (9.3.6) into eq. (9.3.5) to find

$$E_{JM} = \frac{\mathcal{H}_J}{\mathcal{N}_J} \tag{9.3.7}$$

independent of M. The expansion coefficents are given by

$$\mathcal{H}_L = (L + \tfrac{1}{2}) \int \mathcal{H}(\Theta) P_L(\cos\Theta)\, d\cos\Theta \tag{9.3.8a}$$

$$\mathcal{N}_L = (L + \tfrac{1}{2}) \int \mathcal{N}(\Theta) P_L(\cos\Theta)\, d\cos\Theta. \tag{9.3.8b}$$

If the mean fields U^D and U^X are very anisotropic, as is the case for strongly-deformed nuclei, then the overlap between wave functions of different orientation will be small unless their symmetry axes are nearly the same. Thus we expect $\mathcal{H}(\Theta)$ and $\mathcal{N}(\Theta)$ to be very small unless Θ is small, so that we may expand the Legendre polynomials in the expressions for the expansion coefficents (9.3.8):

$$\mathcal{H}_L = (L + \tfrac{1}{2}) \int \mathcal{H}(\Theta)(1 - \tfrac{1}{2}L(L+1)\Theta^2 + \ldots) \sin\Theta\, d\Theta$$
$$= (L + \tfrac{1}{2})(H_0 - \tfrac{1}{2}L(L+1)H_2 + \ldots) \tag{9.3.9a}$$

$$\mathcal{N}_L = (L + \tfrac{1}{2}) \int \mathcal{N}(\Theta)(1 - \tfrac{1}{2}L(L+1)\Theta^2 + \ldots) \sin\Theta\, d\Theta$$
$$= (L + \tfrac{1}{2})(N_0 - \tfrac{1}{2}L(L+1)N_2 + \ldots) \tag{9.3.9b}$$

where

$$H_\nu = \int \mathcal{H}(\Theta)\Theta^\nu \sin\Theta\, d\Theta \tag{9.3.10a}$$

$$N_\nu = \int \mathcal{N}(\Theta)\Theta^\nu \sin\Theta\, d\Theta \tag{9.3.10b}$$

are the moments of $\mathcal{H}(\Theta)$ and $\mathcal{N}(\Theta)$. If $\mathcal{H}(\Theta)$ and $\mathcal{N}(\Theta)$ are strongly peaked near $\Theta = 0$, then the coefficents H_ν and N_ν are rapidly decreasing functions of ν, so that (9.3.7) becomes

$$E_{JM} \approx \frac{H_0 - \tfrac{1}{2}J(J+1)H_2}{N_0 - \tfrac{1}{2}J(J+1)N_2} \approx \frac{H_0}{N_0} + J(J+1)\left[\frac{N_2 H_0 - H_2 N_0}{2N_0^2}\right]. \tag{9.3.11}$$

We see that the excitation spectrum is indeed of the form (9.1.1); the expression in square brackets in (9.3.11) gives the coefficent B, which is related to the moment of inertia $\mathcal{I}$ by eq. (9.1.3), $B = \hbar^2/2\mathcal{I}$.

The generator-coordinate method of Peierls and Yoccoz, which we have followed in this section, gives several important, qualitative improvements on the mean-field picture of axially-symmetric spin-zero nuclei:

1. The ground state is spherically symmetric;

2. There is a rotational series of excited states with angular momentum $\hbar J$ and excitation energies approximately proportional to $J(J+1)$, if the deformation is large;

3. J takes only even (odd) values if the mean-field state has even (odd) parity;

4. Spherical nuclei have no rotational bands.

Unfortunately, the quantitative prediction for the moment of inertia is not very good. This difficulty was traced by Peierls and Thouless to the fact that the trial wave functions are all generated by stationary average potentials. Just as the method of Peierls and Yoccoz gave the wrong inertial mass for translational motion, it gives a poor approximation to the inertia of rotational motion.

In the case of translational motion, the wave functions could be improved by boosting them to a moving frame of reference. It is also possible to find a better set of trial wave functions for rotations by including wave functions representing a nucleus rotating with a uniform angular velocity. Unlike the case of Galilean invariance in translational motion, this boost to a rotating reference does not represent an exact symmetry of the Hamiltonian; thus we cannot expect to find an exact, model-independent result for the moment of inertia, which we did have for the translational mass. Unfortunately, the operators which boost a wave function to a rotating frame of reference are rather complicated; the resulting computation of the rotational energy is too lengthy to reproduce here. However, Peierls and Thouless showed that the result for the inertia is very simple: it is just the cranking formula (9.2.3) which we obtained from linear-response theory.

We have thus arrived at the somewhat paradoxical conclusion that the mean-field picture is quantitatively useful but qualitatively wrong, while the simple generator coordinate picture of Peierls and Yoccoz is qualitatively right but quantitatively poor. The picture of Peierls and Thouless, which is both qualitatively and quantitatively good, can be used as a guide to choose the best points of the simpler pictures. In the following sections we will describe some important results, motivating them with the simple pictures of generator coordinates and the mean field. We emphasize, however, that these results can be obtained in more rigorous ways.

9.4   ROTATIONAL BANDS WITH INTRINSIC ANGULAR MOMENTUM

While the ground states of even-even nuclei in the mean-field picture are always axially symmetric, there are also excited states corresponding to an axially-symmetric mean field. The wave functions of these states are eigenstates of J_z with an eigenvalue conventionally denoted $\hbar K$; they are not, however, eigenstates of J^2 unless the mean field is also spherically symmetric.

9.4 A.   Representation of the Rotation Operator

While only two angles are necessary to specify the orientation of the mean-field's symmetry axis, a third angle ψ is necessary to specify the most general rotation of the wave function. For example, after performing the rotation of fig. 9.3, we could rotate by an angle ψ about the new z axis. For an axially-symmetric wave function with K=0, this last rotation does not change the wave function; if K$\neq$0, however, it causes the wave function's phase to change by a factor $e^{iK\psi}$. We can denote the operator associated with this most-general rotation by $\Re_{\vec{\omega}}$ where $\vec{\omega}$ is shorthand notation for the set of Euler angles ϕ, θ, ψ. The $\Re_{\vec{\Omega}}$ used in Sect. 9.3 is just $\Re_{\vec{\omega}}$ with $\psi = 0, \vec{\Omega} = (\theta, \phi)$. We can formally write the rotation operator in terms of the angular-momentum operator $\vec{J}$:

$$\Re_{\vec{\omega}} = e^{-i\phi J_z} e^{-i\theta J_y} e^{-i\psi J_z} \qquad (9.4.1)$$

Recalling that $e^{-i\psi J_z}$ makes a rotation by $-\psi$ about the z-axis, we can read eq. (9.4.1) as saying that $\Re_{\vec{\omega}}$ consists of a rotation by $-\theta$ about the **new** y-axis, followed by a rotation by $-\phi$ about the final z-axis. The minus signs are because rotating the coordinate system in one direction, as in fig. 9.3, is the same as rotating the coordinates in the opposite direction.

Clearly, the operator $\Re_{\vec{\omega}}$ conserves the total angular momentum $\vec{J}^2$, whose value is independent of the orientation of the coordinate system. Therefore, the representation of $\Re_{\vec{\omega}}$ is blockwise diagonal in the basis $|JM)$ of eigenstates of $\vec{J}^2$ and J_z; we define its representation $\mathcal{D}$ by the equation

$$\langle JM | \Re_{\vec{\omega}} | J'M' \rangle = \delta_{JJ'} \mathcal{D}^J_{MM'}(\vec{\omega})^*. \qquad (9.4.2)$$

We follow here the convention of Bohr and Mottelson; other notations which look the same but differ in meaning are used by Rose and by Edmonds. The $\mathcal{D}$ functions may be thought of as generalizations of the spherical harmonics: in particular, $\mathcal{D}^J_{M0}(\vec{\omega}) = \sqrt{4\pi/(2J+1)}\, Y^M_J(\theta, \phi)$. For integer values of J, their orthonormality relation is

$$\int d^3\vec{\omega}\, \mathcal{D}^J_{MK}(\vec{\omega})^* \mathcal{D}^{J'}_{M'K'}(\vec{\omega}) = \int_0^\pi \sin\theta d\theta \int_0^{2\pi} d\phi \int_0^{2\pi} d\psi\, \mathcal{D}^J_{MK}(\vec{\omega})^* \mathcal{D}^{J'}_{M'K'}(\vec{\omega})$$

$$= \frac{8\pi^2}{2J+1} \delta_{JJ'} \delta_{MM'} \delta_{KK'}. \qquad (9.4.3)$$

They also form a complete set: two successive rotations can be expressed as a superposition of single rotations, according to

$$\mathcal{D}^{J_1}_{M_1 K_1}(\vec{\omega}) \mathcal{D}^{J_2}_{M_2 K_2}(\vec{\omega}) = \sum_{J=|J_1-J_2|}^{J_1+J_2} \langle J_1 M_1 J_2 M_2 | J, M_1 + M_2 \rangle \times$$

$$\langle J_1 K_1 J_2 K_2 | J, K_1 + K_2 \rangle\, \mathcal{D}^J_{M_1+M_2, K_1+K_2}(\vec{\omega}).$$

$$(9.4.4)$$

A more complete discussion of the $\mathcal{D}$ functions may be found in the books by Bohr and Mottelson, Rose, or Edmonds.

9.4 B.   Rotational Eigenfunctions for Even-A Nuclei

The corresponding generalization of eq. (9.3.3) to construct eigenstates of $\vec{J}^2$ and J_z from a state ψ_K is

$$\psi_{JM}^K(\vec{r}_1, \ldots, \vec{r}_A) = \int d^3\vec{\omega} \, \mathcal{D}_{MK}^J(\vec{\omega}) \psi_K(\Re_{\vec{\omega}}\vec{r}_1, \ldots, \Re_{\vec{\omega}}\vec{r}_A) \qquad (9.4.5)$$

where

$$\vec{J}^2 \psi_{JM}^K(\vec{r}_1, \ldots, \vec{r}_A) = \hbar^2 J(J+1) \psi_{JM}^K(\vec{r}_1, \ldots, \vec{r}_A) \qquad (9.4.6a)$$

$$J_z \psi_{JM}^K(\vec{r}_1, \ldots, \vec{r}_A) = \hbar M \psi_{JM}^K(\vec{r}_1, \ldots, \vec{r}_A), \qquad (9.4.6b)$$

provided

$$J_z \psi_K(\vec{r}_1, \ldots, \vec{r}_A) = \hbar K \psi_K(\vec{r}_1, \ldots, \vec{r}_A). \qquad (9.4.6c)$$

The quantum number K is said to be the angular-momentum projection in the **intrinsic** or **body-fixed** coordinates. It is clear from the definition of $\mathcal{D}_{MK}^J$, eq. (9.4.2), that we must choose $J \geq |K|$.

Usually, the mean-field states ψ_K possess additional symmetries. However, the degeneracy associated with these symmetries may not appear in the spectrum of rotational states. For example, nuclear states are almost always eigenstates of parity (in principle, the mean field can also break parity, but only a few examples of this are known). Then there are two intrinsic states of the same energy, namely ψ_K and ψ_{-K}. Up to a phase factor, ψ_{-K} can be obtained from ψ_K by rotation through 180° about an axis perpendicular to $\hat{z}$. The same rotation turns $\mathcal{D}_{MK}^J$ into $\mathcal{D}_{M-K}^J$, except for a phase factor. Thus the linear superposition of all orientations, used to construct ψ_{JM}^K (eq. (9.4.5)), only differs by a phase factor from ψ_{JM}^{-K}. The two intrinsic states ψ_K and ψ_{-K} lead to only one set of rotational wave functions: the degeneracy associated with parity is swallowed up in the degeneracy associated with J_z. Conventionally, we consider only states with $K \geq 0$ and let M have its $2J+1$ values from $-J$ to $+J$.

Each deformed mean-field state of the even-even nucleus thus leads to a rotational band, similar to that of the ground state, but where the lowest value of J is the projection K of the mean-field state's angular momentum,

$$J = K, \ K+1, \ K+2, \ldots. \qquad (9.4.7)$$

The energies of these states have spacings given approximately by (9.1.1), except that all the excitation energies are additionally displaced from the ground state energies by the excitation energy of the mean-field excited state above the mean-field ground state. Some examples of rotational bands based on excited states of an even-even nucleus are shown in fig. 9.4. Many nuclei with spherical ground states also have rotational bands built on non-spherical excited states.

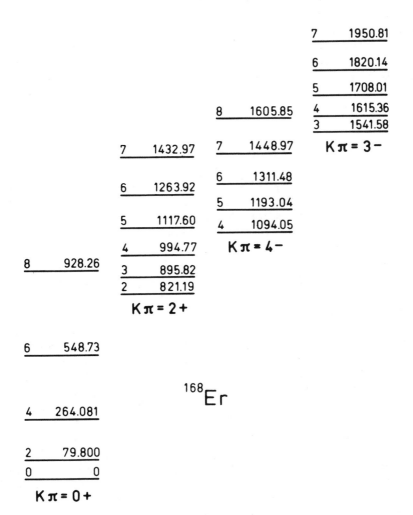

Figure 9.4   Rotational bands of ^{168}Er based on the ground state ($K\pi=0^{+}$) and on several other intrinsic states. Angular momenta given at left, energies in keV at right. [*from Bohr and Mottelson, vol. 2*]

9.4 C.   Rotational Eigenfunctions for Odd-A Nuclei

The preceeding discussion applies equally to states of odd nuclei, except that then K, J, and M take on half-integer values. The main modification is that, because the $\mathcal{D}$-functions with half-integer J change sign under rotation by 2π, the domain of Euler angles $d^3\vec{\omega}$ (defined in eq. (9.4.3)) must be doubled to retain the periodicity property characteristic of rotations. Thus, when K is a half-integer, the integrals $\int d^3\vec{\omega}$ must be understood to extend over a larger domain, for example by letting ψ range from 0 to 4π. Examples of rotational bands of an odd nucleus are shown in fig. 9.5.

9.5   DEPARTURES FROM THE IDEAL ROTOR

The spectra shown in the figures above show that, for many nuclei, the rotational states occur in bands with excitation energies approximately given by the ideal rotor, eq. (9.1.1). Deviations from this simple formula become pronounced when J is large, or when the band's lowest value of J is $K=\frac{1}{2}$.

The departures from the ideal rotor for large J can be traced to two sources. For one, even within the picture of Peierls and Yoccoz, the simple rotor formula only emerged from the leading terms in a series expansion of the Legendre polynomial P_J of order J, which was assumed to vary slowly compared to the angular overlap functions $\mathcal{N}(\theta)$ and $\mathcal{H}(\theta)$, eq. (9.2.6). For large J, the Legendre polynomials vary more rapidly, so that higher order terms in the expansion may be necessary. We may expect that the series expansion will fail when the angular dependence of P_J becomes comparable to that of the mean-field wave functions, i.e. when $J \approx J_{max} \approx k_F R(A) \approx A^{1/3}$, the maximum angular momentum of a single nucleon.

A second source of deviations from the ideal rotor will appear if we choose the intrinsic Slater determinant ψ_K variationally to minimize the energy E_J, instead of always using the same determinant for all J. The equations determining ψ_K must be found using the rotating-frame method of Peierls and Thouless, or some other equivalent formalism. When this is done, and when other reasonable approximations are introduced, the variational equations determining the one-body wavefunctions are seen to contain, in addition to the self-consistent mean fields U^D and U^X, additional forces which can be interpreted as centrifugal and coriolis forces familiar in the classical mechanics of rotating coordinate frames. These forces modify the single-particle wave functions, and thus the Slater determinants ψ_K. These modifications are analogous to the change of the moment of inertia of a rapidly rotating deformable body as the centrifugal force changes its shape with increasing angular velocity.

The departure from the ideal rotor for bands with $K=\frac{1}{2}$ has a different explanation. Consider for example a nucleus whose symmetry axis is momentarily in the z direction, but rotating about the x-axis with angular velocity ω_x. The Hamiltonian in the rotating frame contains a coriolis-coupling term

$$H_{cor} = -\omega_x J_x = -\frac{\omega_x}{2}(J_+ + J_-). \tag{9.5.1}$$

This term will have matrix elements in first order between states of $K=+\frac{1}{2}$ and $K=-\frac{1}{2}$. Since we expect the angular frequency ω_x to be given by $J/\mathcal{I}$, we anticipate that the matrix elements of H_{cor} will be comparable to the rotational energies. Since these states were degenerate in the non-rotating mean field, it is not surprising that the ideal-rotor picture breaks down completely. In this case, it is better to think

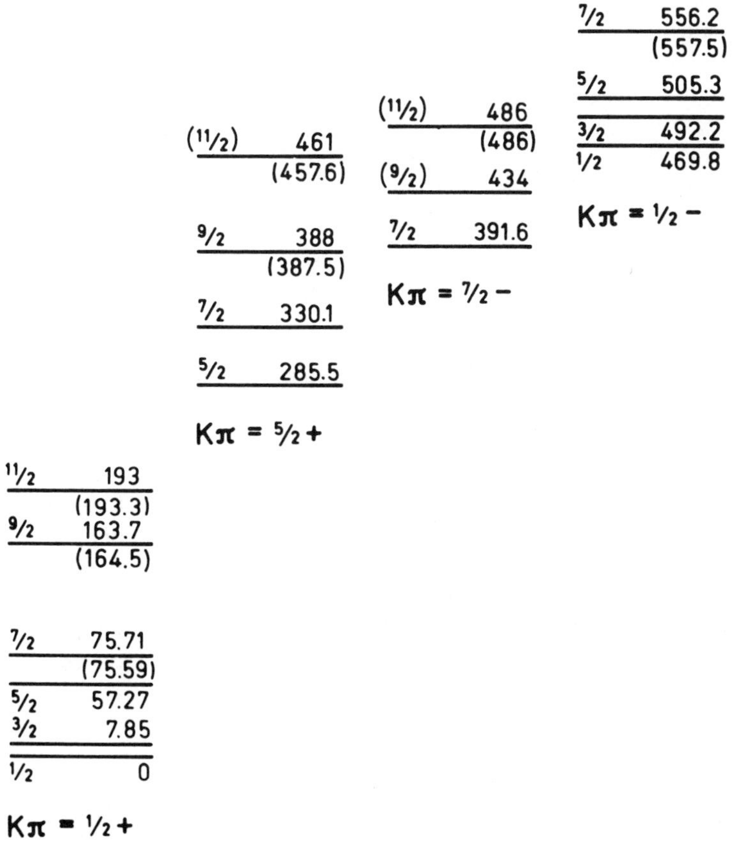

Figure 9.5   Rotational bands of ^{239}Pu based on the ground state ($K\pi=\frac{1}{2}^+$) and on several other intrinsic states. Angular momenta given at left, energies in keV at right. [from Bohr and Mottelson, vol. 2]

of the $K = \pm\frac{1}{2}$ states as a $K=0$ core with an additional nucleon in a $K=\pm\frac{1}{2}$ single-particle state (pairing enforces this, see Chapter 6). The core rotates as a $K=0$ rotor, with angular momenta $J_{core}=0, 2, 4$, etc., a corresponding rotational spectrum. The particle in the $K = \pm\frac{1}{2}$ state is then weakly coupled to the rotating core, giving a doublet of levels with $J = J_{core} \pm \frac{1}{2}$. Indeed the observed spectra (fig. 9.5) for $K=\frac{1}{2}$ nuclear states appear to have the levels grouped as suggested by this picture of particle-rotor coupling.

9.6   MATRIX ELEMENTS OF ROTATIONAL BANDS

Since the rotational states correspond to the synchronized, collective motion of many nucleons, we may expect them to be easily excited by the same types of external fields as excite the vibrational states. Indeed, the Coulomb-excitation process, in which a highly charged projectile passes far from the nucleus and is only slightly deflected by it, is extremely effective at exciting rotational states.

We saw in Sect. 8.1.C that the excitation probability is proportional to the electric transition strength $B(E\lambda)$ from the ground state to the excited level (see also Sect. 10.3). It is convenient to consider the matrix elements of the operator

$$Q_\lambda^\mu (\{\vec{r}_i\}) \equiv \sum_i eq_i r_i^\lambda Y_\lambda^\mu(\theta_i, \phi_i) \qquad (9.6.1)$$

where eq_i is the charge of the i'th nucleon. The matrix elements of $F_{\lambda\mu}^E$, eq. (8.1.29), can be constructed easily from those of Q_λ^μ. The simple generator-coordinate wave functions (9.4.5) suffice for evaluating the transition matrix elements. Since these wave functions give an inadequate estimate of the rotational energy, we may worry that they would give unreliable matrix elements; after all, we know that trial wave functions give the energy more accurately than the matrix elements of other operators. We need to remember, however, that the rotational energy, which falls with A more rapidly than A^{-1}, is very small compared to the total energy, which is proportional to A. We will see that we can estimate the matrix elements only to leading order in A; the corresponding estimate of the energy would have to be much more accurate to be useful.

We proceed to evaluate the matrix element

$$\langle \psi_{JM}^K | Q_\lambda^\mu | \psi_{J'M'}^{K'} \rangle =$$
$$\int d^3\vec{\omega} d^3\vec{\omega}' \mathcal{D}_{MK}^{J*}(\vec{\omega}) \; \mathcal{D}_{M'K'}^{J'}(\vec{\omega}') \langle \psi_K (\{\Re_{\vec{\omega}} \vec{r}_i\}) | Q_\lambda^\mu (\{\vec{r}_i\}) | \psi_{K'} (\{\Re_{\vec{\omega}'} \vec{r}_i\}) \rangle. \qquad (9.6.2)$$

Following Villars we employ the substitutions

$$\vec{r}_i' = \Re_{\vec{\omega}'} \vec{r}_i \qquad (9.6.3a)$$
$$\Re_{\vec{\omega}} \vec{r}_i = \Re_{\vec{\omega}} \Re_{\omega'}^{-1} \vec{r}_i' \equiv \Re_{\vec{\omega}''} \vec{r}_i'. \qquad (9.6.3b)$$

The multipole operator can then be calculated in the coordinate system rotated by $\vec{\omega}'$:

$$Q_\lambda^\mu(\{\vec{r}_i\}) = Q_\lambda^\mu(\{\Re_{\vec{\omega}}^{-1}\vec{r}_i'\}) = \sum_{\mu'} \mathcal{D}_{\mu\mu'}^\lambda(\vec{\omega}')Q_\lambda^{\mu'}(\{\vec{r}_i'\}). \qquad (9.6.4)$$

As in eq. (9.6.3b), the rotation $\vec{\omega}$ is decomposed into two successive rotations,

$$\Re_{\vec{\omega}} = \Re_{\vec{\omega}''}\,\Re_{\vec{\omega}'} \qquad (9.6.5a)$$

or equivalently

$$\Re_{\vec{\omega}}^{-1} = \Re_{-\vec{\omega}} = \Re_{-\vec{\omega}'}\,\Re_{-\vec{\omega}''}. \qquad (9.6.5b)$$

The corresponding representation in terms of $\mathcal{D}$ functions is [see Bohr and Mottelson vol. 1]

$$\mathcal{D}_{MK}^{J*}(\vec{\omega}) = (-1)^{M-K}\mathcal{D}_{-M-K}^J(-\vec{\omega}) = (-1)^{M-K}\sum_L \mathcal{D}_{-M-L}^J(-\vec{\omega}')\mathcal{D}_{-L-K}^J(-\vec{\omega}'')$$

$$= \sum_L \mathcal{D}_{ML}^{J*}(\vec{\omega}')\mathcal{D}_{LK}^{J*}(\vec{\omega}'').$$

$$(9.6.6)$$

Using eqs. (9.6.4) and (9.6.6) in (9.6.2) gives, after changing the integration variables $\vec{\omega}$ and $\vec{r}$ (but using their old names)

$$\langle\psi_{JM}^K|Q_\lambda^\mu|\psi_{J'M'}^{K'}\rangle = \sum_{L\mu'}\int d^3\vec{\omega}\,\mathcal{D}_{LK}^{J*}(\vec{\omega})\langle\psi_K(\{\Re_{\vec{\omega}}\,\vec{r}_i\})|Q_\lambda^{\mu'}(\{\vec{r}_i\})|\psi_{K'}(\{\vec{r}_i\})\rangle$$

$$\cdot \int d^3\vec{\omega}'\mathcal{D}_{M'K'}^{J'}(\vec{\omega}')\mathcal{D}_{ML}^{J*}(\vec{\omega}')\mathcal{D}_{\mu\mu'}^\lambda(\vec{\omega}')$$

$$(9.6.7)$$

$$= \frac{8\pi^2}{2J+1}\sum_{L\mu'}\langle J'\,M'\,\lambda\mu|JM\rangle\langle J'\,K'\,\lambda\mu'\,|JL\rangle$$

$$\cdot \int d^3\vec{\omega}\,\mathcal{D}_{LK}^{J*}(\vec{\omega})\langle\psi_K(\{\Re_{\vec{\omega}}\,\vec{r}_i\})|Q_\lambda^{\mu'}(\{\vec{r}_i\})|\psi_{K'}(\{\vec{r}_i\})\rangle.$$

Recalling that the wave functions (9.4.5) are not normalized, we realize that we have to evaluate the transition strength

$$B(E\lambda, J' \to J) = \sum_{\mu M} \frac{|\langle\psi_{JM}^K|Q_\lambda^\mu|\psi_{J'M'}^{K'}\rangle|^2}{\langle\psi_{JM}^K|\psi_{JM}^K\rangle\langle\psi_{J'M'}^{K'}|\psi_{J'M'}^{K'}\rangle}. \qquad (9.6.8)$$

It is easy to verify that this expression reduces to eq. (8.1.37) when $J'=0$ and the wave functions are normalized to unity. Here, it is straightforward to evaluate the normalizations with the technique used for the matrix elements. We find

$$\langle\psi_{JM}^K|\psi_{JM}^K\rangle = \frac{8\pi^2}{2J+1}\sum_L\int d^3\vec{\omega}\,\mathcal{D}_{LK}^{J*}(\vec{\omega})\langle\psi_K(\{\Re_{\vec{\omega}}\,\vec{r}_i\})|\psi_K(\{\vec{r}_i\})\rangle. \qquad (9.6.9)$$

The expressions so far follow exactly from the form of the wave functions (9.4.5), but unfortunately are quite complicated to evaluate. For strongly deformed intrinsic states, we can obtain a very simple result by realizing that, in that case, the overlap on the right-hand side of eq. (9.6.9) must be very small unless the rotation angle θ of $\vec{\omega} = (\phi, \theta, \psi)$ is very close to 0. For example, in the mean-field model of intrinsic states, the overlap is the product of A factors for each of the occupied single-particle orbits. Each of these factors is 1 when $\vec{\omega} = 0$, but they all fall off more or less gently as θ increases. The product will therefore be sharply peaked at $\theta = 0$,

$$\langle \psi_K(\{\Re_{\vec{\omega}} \, \vec{r}_i\}) | \psi_K(\{\vec{r}_i\}) \rangle = e^{iK(\phi+\psi) - \theta^2/\theta_0^2 + \mathcal{O}(\theta^3)} \tag{9.6.10}$$

where θ_0^2 is proportional to A^{-1}. Noting that $\mathcal{D}_{LK}^J(\phi, 0, \psi) = e^{iK(\phi+\psi)}\delta_{LK}$, we see that

$$\langle \psi_{JM}^K | \psi_{JM}^K \rangle \approx \frac{8\pi^2}{2J+1} \cdot \frac{1}{2}\theta_0^2(2\pi)^2. \tag{9.6.11}$$

We can use the same approximation to evaluate the matrix element in eq. (9.6.7), even though it also contains the multipole-moment operator. This single-particle operator is a sum of terms, each of which operates on only one nucleon; thus each term will, in the mean-field model, contain A-1 factors identical to those appearing in the normalization overlap. We conclude that the matrix element has the same angular dependence as eq. (9.6.10), with a width $\theta_0^2 \left(1 + \mathcal{O}(A^{-1})\right)$:

$$\langle \psi_K(\{\Re_{\vec{\omega}} \, \vec{r}_i\}) | Q_\lambda^\mu(\{\vec{r}_i\}) | \psi_{K'}(\{\vec{r}_i\}) \rangle \approx \langle \psi_K | Q_\lambda^\mu | \psi_{K'} \rangle e^{iK(\phi+\psi) - \theta^2/\theta_0^2} \tag{9.6.12}$$

which implies that

$$\langle \psi_{JM}^K | Q_\lambda^\mu | \psi_{J'M'}^{K'} \rangle \approx \tag{9.6.13}$$
$$\frac{8\pi^2}{2J+1} \langle \psi_K | Q_\lambda^{\mu'} | \psi_{K'} \rangle \cdot \langle J' \, M' \, \lambda\mu | JM \rangle \langle J' \, K' \, \lambda\mu' | JK \rangle \frac{1}{2}\theta_0^2(2\pi)^2.$$

Inserting eqs. (9.6.11) and (9.6.13) into eq. (9.6.8) we conclude that

$$B(E\lambda, J' \to J) = |\langle \psi_K | Q_\lambda^{K-K'} | \psi_{K'} \rangle|^2 \langle J' \, K', \lambda, K - K' | JK \rangle^2. \tag{9.6.14}$$

In the special case where $J' = 0$, as for the ground-state band of an even-even nucleus, we obtain simply

$$B(E\lambda, 0 \to \lambda) = |\langle \psi_0 | Q_\lambda^0 | \psi_0 \rangle|^2 \tag{9.6.15}$$

Actually, in the usual case where the intrinsic nuclear shape is symmetric with respect to the reflection $z \to -z$, the overlaps get contributions from the angular region $\theta \approx \pi$; the intermediate equations (9.6.10–13) then have to be modified slightly, but the results (9.6.14–15) remain valid. Note that if a similar approximation is used to calculate the energy, eq. (9.3.11), it gives just the original energy of the intrinsic wave function plus terms of order A^0 (intermediate between the intrinsic energy $\sim A$ and the rotational energy $\sim A^{-1}$); these terms would be an estimate of the kinetic energy gained by delocalizing the angular orientation of the mean-field wave packet.

Returning to the Coulomb excitation of an even-even target, we realize from eqs. (8.1.36) and (9.6.15) that since the quadrupole part of the projectile's electric field is much larger than the higher multipoles, it will excite the $\ell = 2$ rotational state much more strongly than the higher states in the band. A large, deformed nucleus has a huge quadrupole moment; thus the probability of exciting the $\ell = 2$ rotational state may become large. In this case, the lowest-order theory of Coulomb excitation will not be sufficient to describe the process, because once the $\ell = 2$ state is excited, a further excitation may occur to $\ell = 4$, and from these to $\ell = 6$, etc.

The multipole transition matrix elements between states of the same rotational band are much larger than those where the underlying mean-field states are different. They are also larger than the vibration transition strengths. Fig. 9.6 shows the observed B(E2) for the lowest $J^\pi = 2^+$ states of even-even nuclei, in units of a "single-particle" strength (compare eq. (10.3.52)

$$B_{sp}(E2, 0 \rightarrow 2) = \frac{5}{4\pi} e^2 \left(\frac{3}{5} R(A)^2 \right)^2 \qquad (9.6.16)$$

which is what you would get from exciting a single proton from $\ell = 0$ to $\ell = 2$ with constant probability densities inside the nuclear volume. The regions of large ground-state deformation are apparent.

For odd nuclei, the transition probabilities, like the moments of inertia, are much more complicated. Especially when the odd nucleon has a large angular momentum, the interference between its transition matrix elements and those of the other nucleons can give a characteristic pattern that allows detailed interpretations of its motion in terms of coriolis forces. Such studies help to show how nuclear shapes and single-particle levels are affected by rotational motion. But the details are too intricate for us to discuss here.

9.7   HIGH-SPIN STATES AND THE YRAST LINE

Since the energy and angular momentum are the two conserved quantities in a nucleus (apart from its center-of-mass motion), it is instructive to plot the nuclear states in the plane of these two thermodynamic variables. In fig. 9.7 we summarize schematically the results of our discussion in Sects. 9.3-9.5 above, for an even-even nucleus with a non-spherical ground state. The vertical axis shows the energy of the state, the horizontal axis is the angular momentum. A rotational band is seen for each "intrinsic" non-spherical Hartree-Fock-Bogolyubov state, starting with the angular momentum K of that state. Each "intrinsic" state produces a nearly-parabolic band, but since each band has a different moment of inertia, the bands may cross.

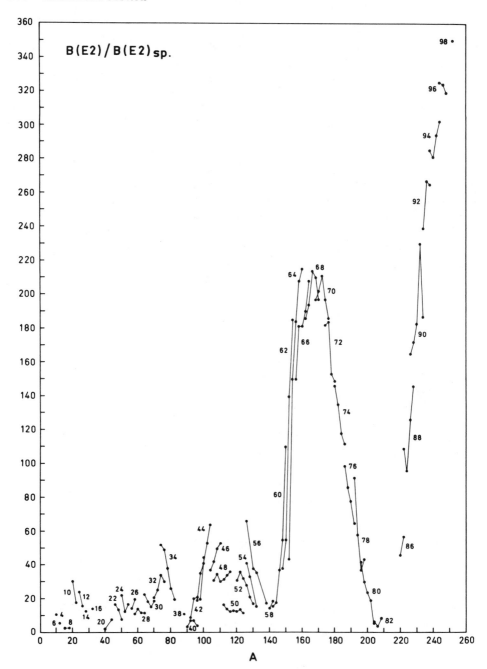

Figure 9.6   E2 transition probabilities between ground state and first excited $J^\pi = 2^+$ state in even-even nuclei, as function of A (horizontal axis) and Z (label on curves), in single-particle units (eq. (9.6.16)). [*from Bohr and Mottelson, vol. 2*]

For each J, there is a minimum energy necessary to excite the nucleus with that much angular momentum; conversely, for a given energy, there is a maximum possible angular momentum $J_{yrast}(E)$. These states form a boundary of the nuclear excitations in the E-J plane. This boundary is called the yrast line (in Swedish, the word "yr" means "dizzy"; thus the "yrast" state is the "dizziest," it has the

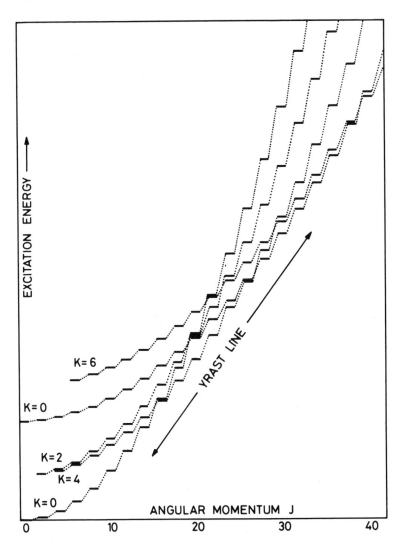

Figure 9.7   Rotational bands of five ideal rotors, with various intrinsic angular momenta, intrinsic excitation energies, and moments of inertia. For K≠0, the odd-J states are omitted for legibility.

most angular momentum for a given energy). For low J, the yrast states belong to the ground-state rotational band; for higher J, other bands with larger moments of inertia win out. Of course, where the bands cross, the Hamiltonian's eigenstates will be mixtures of the two nearby states with the same angular momentum and similar energies.

We can estimate where the first band crossing occurs by recalling that, as we mentioned in Sect. 9.2.C, pairing reduces the moment of inertia by about a factor of $\frac{1}{2}$. Since many excited states of the BCS extension of the mean-field picture occur at an excitation energy of about 2Δ, we can guess that at least one of them may have nearly the rigid moment of inertia. If so, then the band crossing would occur when

$$\frac{\hbar^2 J(J+1)}{2 \cdot \frac{1}{2}\mathcal{I}_{\text{rig}}} = 2\Delta + \frac{\hbar^2 J(J+1)}{2\mathcal{I}_{\text{rig}}}. \tag{9.7.1}$$

Using eqs. (9.1.4) and (6.6.2), we estimate

$$J(J+1) = \frac{4\Delta\mathcal{I}_{\text{rig}}}{\hbar^2}$$

$$J \approx 0.8A^{7/12}. \tag{9.7.2}$$

The first band crossing usually occurs at a somewhat larger value of J, indicating that the states with one broken pair still have less than the rigid-body moment of inertia.

Since pairing reduces the moment of inertia, the yrast states for large J are likely to have fewer pairs than those for small J. Other states with large moments of inertia may include deformed states built on different single-particle states than the ground state, with different self-consistent deformations. If the total angular momentum is very large, the collective rotational frequencies become comparable to the single-particle energies; then the rotation is no longer a collective phenomenon, but merely represents many nucleons in single-particle orbits with their angular momenta aligned along the same axis. The centrifugal forces on these nucleons push them away from the symmetry axis, so that for very high spins the nuclear shape becomes oblate.

Eventually, if too much angular momentum is put into a nucleus, the centrifugal forces will have to make it spin apart. Thus fission will become very rapid for extremely large angular momenta, and the nucleus may not be able to exist at all. These large amounts of angular momentum may be put into a nucleus by creating it in the fusion of two heavy nuclei with a large relative velocity. We will return to these exotic, short-lived nuclear systems in Chapter 11.

Fig. 9.8 shows a liquid-drop-model estimate of the angular momentum at which the fission barrier of a beta-stable nucleus with mass number A is predicted to vanish. Below the dashed curve, the fission barrier for the rotating beta-stable nuclei are higher than 8 MeV (an average neutron separation energy). The barriers are estimated by supplementing the liquid-drop energy with the rotational energy of a rigid body of the same shape and a given angular momentum. We see that neither light nor heavy nuclei can support many units of angular momentum, the former simply because their size is so small, the latter because the Coulomb energy reduces their stability. The nuclei able to support the highest angular momentum occur near $A \approx 130$ in fig. 9.8, but even in that case 100 units of $\hbar$ will make their fission barriers vanish. Experiments have not yet determined where the limit lies.

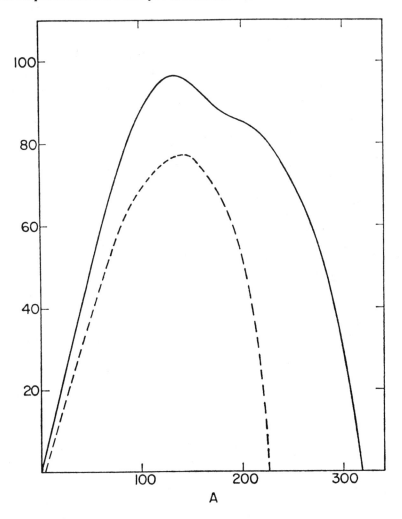

Figure 9.8   Limiting angular momentum for beta-stable nuclei of mass A. Solid line: vanishing fission barrier; dashed line: fission barrier = 8 MeV $\approx S_n$. [*from S. Cohen, F. Plasil and W. Swiatecki, Ann. Phys.* **82** *(1974) 557*]

9.8   TRANSITION FROM SPHERICAL TO DEFORMED NUCLEI IN THE INTERACTING BOSON APPROXIMATION

The characteristic rotational spectrum of eq. (9.1.1) is only a good approximation when the mean-field ground state is strongly anisotropic, as we saw in the discussion around eq. (9.3.9). Indeed, closed-shell nuclei with spherical mean-field ground states have no low-lying rotational spectrum, presenting instead the vibrational modes studied in Chapter 8. Nor do heavy even-even nuclei only a few nucleons away from a closed shell have rotational spectra, since they prefer to attain spherical symmetry through the pairing mechanism of Chapter 6 instead of through the superposition of deformed states favored by rotational nuclei. To understand in detail the emergence of a rotational spectrum away from closed shells, we need a model in which the mean field plays a less overwhelming role. A very fruitful model for this purpose is the Interacting Boson Approximation (IBA) developed by Arima, Iachello and many others [see Arima and Iachello].

To motivate the IBA we return to the discussion of Sect. 6.2 on the states of two identical nucleons in an otherwise empty spherical subshell of degeneracy $2j + 1$, interacting via a short-range attractive force. The two-nucleon states are eigenstates of total angular momentum $\vec{J}^2$, whose energy increases with the eigenvalue J. The lowest state, J=0, is non-degenerate; the next-lowest state, J=2, has fivefold degeneracy. Approximate wave functions for states with 2,4,6,8,10 or 12 nucleons may be found within a subspace of state space in which the nucleons are placed pairwise in these lowest-energy two-body states with J=0 and 2. These elementary degrees of freedom are known as **s-pairs** (J=0) and **d-pairs** (J=2).

The **interacting boson approximation** consists of treating the s-pair and d-pair degrees of freedom as if they were quantum fields obeying Bose statistics. Thus creation operators $s^+, \{d_\mu^+, \mu = -2, -1, 0, 1, 2\}$ are introduced which commute with each other, and obey the canonical Bose commutation relations with their conjugate annihilation operators

$$
\begin{aligned}
[s^+, s] &= 1 \\
[s^+, d_\mu] &= 0 = [s, d_\mu^+] \\
[d_\mu^+, d_\nu] &= \delta_{\mu\nu}.
\end{aligned}
\tag{9.8.1}
$$

Nuclei with 2N identical nucleons outside a closed shell are produced by operating on the ground state of the closed-shell nucleus with N boson creation operators. The number of s and d pairs are given respectively by the eigenvalues of the s and d number operators

$$
\begin{aligned}
n_s &= s^+ s \\
n_d &= \sum_\mu d_\mu^+ d_\mu
\end{aligned}
\tag{9.8.2}
$$

so that $N = n_s + n_d$ is the total number of pairs which is half the number of fermions outside the closed shell.

In the IBA the bosons are assumed to interact by two-boson interactions, i.e. interactions of one pair of nucleons with another pair. The most general form of the Hamiltonian, for only one type of nucleon, is

$$H_{IBA} = \varepsilon n_d + a_0 T_0^+ T_0 + \sum_{J=1}^{4} a_J \sum_{\mu} T_J^{\mu} T_J^{-\mu} \qquad (9.8.3)$$

where we have assumed conservation of angular momentum, and have ignored terms depending only on N, which would not affect the excitation spectrum. The tensors T_J^{μ}, which represent the one-body density, are given by

$$T_0 = \frac{1}{2} \sum_{\mu} \overline{d}_{\mu} \overline{d}_{-\mu} - \frac{1}{2} \overline{s} \overline{s} \qquad (9.8.4a)$$

$$T_2^{\mu} = d_{\mu}^+ s + s^+ \overline{d}_{\mu} - \sqrt{\frac{7}{4}} \sum_{\nu\nu'} \langle 2\nu 2\nu' | 2\mu \rangle d_{\nu}^+ \overline{d}_{\nu'} \qquad (9.8.4b)$$

$$T_J^{\mu} = \langle 2\nu 2\nu' | J\mu \rangle d_{\nu}^+ \overline{d}_{\nu'}, \quad J = 1, 3, 4 \qquad (9.8.4c)$$

where $\overline{d}_{\mu} = (-1)^{\mu} d_{-\mu}$ annihilates the time-reverse of the state created by d_{μ}. The labels (J, μ) on T_J^{μ} describe each tensor's behavior under rotations.

The spectrum of H_{IBA} can be computed accurately by finding the eigenvalues of a finite matrix. The parameters of H_{IBA} could then be fit to observed nuclear levels. But such a phenomenological procedure would only be realistic if it included both proton and neutron degrees of freedom as well as couplings between unlike nucleons. This approach, known as IBM-2, has been a great help in nuclear spectroscopy. For our purposes, however, it is most enlightening to consider analytically the simpler model without isospin, known as IBM-1, described above. This model can be solved exact in three limiting cases.

The first case when H_{IBA} can be solved analytically is the case when only a_1 and a_2 are non-zero. In this case H_{IBA} possesses an SU(3) symmetry beyond the O(3) symmetry of the rotational group, so that H_{IBA} can be expressed in terms of the Casimir operators of those groups. The resulting spectrum is of the form

$$E(N, \lambda, \mu, K, J, M) = BJ(J+1) - \kappa[\lambda^2 + \mu^2 + \lambda\mu + 3(\lambda + \mu)] \qquad (9.8.5)$$

where J and M are the total angular momentum and its projection, λ and μ are SU(3) quantum numbers, κ and B are constants, and K is an extra quantum number associated with a remaining degeneracy. We see that an ideal rotational spectrum is produced by the dipole-dipole and quadrupole-quadrupole interactions a_1 and a_2. We can trace the broken symmetry of the mean-field solutions to a similar interaction between the multipole moments of the nucleons' density: if some of the nucleons choose to distribute themselves in a particular multipole pattern, then the rest will gain attractive energy by redistributing their density pattern to match the others. This limiting case of H_{IBA} produces the rotational spectrum without ever introducing the mean field.

Another case when H_{IBA} can be solved analytically is the case when a_1 and a_2 vanish, leading to U(5) and O(5) symmetries beyond the rotational ones. In this case the spectrum is dominated by the term εn_d. If ε alone were non-zero, the spectrum would resemble that of a harmonic vibrator with elementary excitations of angular momentum 2: the first excited state would be a 2^+ state, then there would be degenerate 4^+ and 2^+ states from coupling the angular momenta of two 2^+

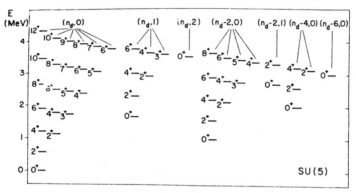

Figure 9.9   A typical spectrum with U(5) symmetry and N=6. In parentheses are the values of additional quantum numbers associated with non-rotational Casimir operators of U(5). [*from Arima and Iachello*].

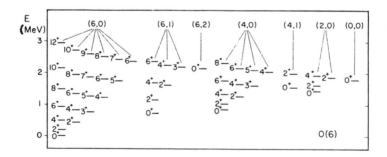

Figure 9.10   A typical spectrum with O(6) symmetry and N=6. In parentheses are the values of additional quantum numbers associated with non-rotational Casimir operators of O(6).[*from Arima and Iachello*].

vibrational quanta, etc, with all levels equally spaced. The interactions among the pairs from a_1, a_3, and a_4 break most of the degeneracies but can leave a recognizable remnant of the vibrational spectrum (see fig 9.9). This case corresponds in the mean-field picture to the case of a spherical nucleus, where the preference ε_d for a J=0 pair is large because of the rotational degeneracy; the vibrations are modified from the independent-particle model both by the attractive interactions within pairs which helps keep ε_d small, and by the interactions among pairs represented by a_1, a_3, and a_4. We note that this limiting case of H_{IBA} describes a vibrational spectrum without explicitly introducing the mean-field approximation.

The third limiting case when H_{IBA} can be solved analytically occurs when only a_1 and a_3 are non-vanishing, leading to O(6) and O(5) symmetries in addition to the rotational ones. These terms, corresponding to odd-parity changes of the nuclear shape, give spectra which have level spacings much like the ideal rotor, but with near-degeneracies more like the harmonic (see fig. 9.10). Several examples of nuclei whose spectra resemble this case have been found.

In both the IBA and the mean-field pictures, the transition from spherical to rotational nuclei is traced to the competition between pairing and shape correlations.

The IBA provides the advantage of being able to describe intermediate situations where neither correlation dominates the other. On the other hand, it has not yet been possible to establish the quantitative connection between the parameters of H_{IBA} and the nucleon-nucleon force. Vigorous efforts are underway to illuminate that connection [see Barrett], which will help us obtain a deeper understanding of collective rotational motion.

PROBLEMS

9.1   Derive eq.(9.2.6). Evaluate the cranking moment of inertia for ^{20}Ne in its ground state, using a harmonic-oscillator model without spin-orbit forces. What would be the energy and angular momentum of the lowest excited state?

9.2   Calculate the rigid body moments of inertia around the three axes of a deformed volume conserving axially symmetric homogeneous, sharp cut-off, ellipsoidially shaped density distribution. Compare the resulting rotational energies for given total angular momentum J and projection K on the symmetry axis. Consider both oblate and prolate shapes.

9.3   Derive eq. (9.2.8).

9.4   Show that ψ_{JM}, eq. (9.3.3), is an eigenfunction of the total angular momentum and its z component, with eigenvalues $J(J+1)\hbar^2$ and $M\hbar$ respectively. Hint: compare sect. 4.5.

9.5   Compute the four possible matrix elements of eq. (9.4.2) for a spin 1/2 particle.

9.6   Estimate for a strongly deformed nucleus the ratio of successive Coulomb excitation probabilities for the three first excited members of the ground state rotational band of intrinsic angular momentum $K \neq \frac{1}{2}$.

9.7   The potential energy surface for ^{240}Pu is assumed similar to that of fig. 5.4. The energy difference between second and first minimum is $E_{II} - E_I \approx 2\mathrm{MeV}$ and the moments of inertia in the two wells are $\mathcal{I}_{II}$ and $\mathcal{I}_I$, respectively. Find the spin J where the two "ground state" rotational bands cross each other. Use the result of problem 9.2 and the assumptions that $\mathcal{I}_I$ is half the rigid body value of a prolate ellipsoid of axis ratio 1.2 and $\mathcal{I}_{II}$ is equal to the rigid body value of a prolate ellipsoid of axis ratio 2.0.

9.8   Assume a liquid drop one-hump fission barrier (see fig. 5.4) and an angular momentum dependence of the deformation energy given by

$$E_D(J) = E_D(J = 0) + \frac{\hbar^2 J(J + 1)}{2\mathcal{I}_D}$$

Use rigid body values of ellipsoids for $\mathcal{I}_D$ (problem 9.2) and axis ratios of 1 and 2 for ground state and barrier deformations, respectively. What is the angular momentum where the fission barrier vanishes for A=160 and a J=0 barrier height of 30 MeV?

10
Decay of Excited States

We have seen a number of ways in which a nucleus can be produced in an excited state: stripping and pickup reactions, fission, charge exchange reactions, beta decay, Coulomb excitation, γ absorption, inelastic electron scattering. We will soon learn about other such mechanisms. In this chapter we will study what happens to such excited nuclei as they lose their excitation energy.

The various excitation mechanisms may be roughly divided into two classes, according to what kind of states they produce. In the first class, a large number of nuclear states are populated in a non-selective, random way. This produces a nucleus in or near a state of statistical equilibrium. The decay of such a state will also be described in a statistical way.

The second class of excitation mechanisms leads to specific quantum states, characterized by a selective relationship between initial and final states. These states may then also decay in a selective way, or they may instead relax into equilibrated configurations before decaying statistically.

We begin this chapter by introducing the Bohr compound nucleus as a statistical method for dealing with the decay of equilibrated nuclear excitations. Next, we consider the dominant strong-interaction decay modes of such an equilibrated system. Finally, we discuss decays mediated by the electroweak interaction.

10.1 COMPOUND NUCLEUS

The compound nucleus, introduced by Niels Bohr, is a quasi-stationary state of the nucleus where the available excitation energy is distributed statistically among all accessible nuclear degrees of freedom. The limitation of the allowed phase space is due to the ordinary conservation laws of energy, angular momentum, parity, etc. To understand this definition we must have a picture (fig. 10.1) of the formation process and a formulation of the relevent statistical model.

10.1 A. Formation

A neutron of kinetic energy ε_n collides with a nucleus. Its wavelength outside the

News and Views

Neutron Capture and Nuclear Constitution

THE new views of nuclear structure and the processes involved in neutron capture, presented by Prof. Niels Bohr in an address which appears elsewhere in this issue, were expounded by him in a lecture to the Chemical and Physical Society of University College, London, on February 11 and were illustrated by two pictures here reproduced. The first of these is intended to convey an idea of events arising out of a collision between a neutron and the nucleus. Imagine a shallow basin with a number of billiard balls in it as shown in the accompanying figure. If the basin were empty, then upon striking a ball from the outside, it would go down one slope and pass out on the opposite side with its original velocity. But with other balls in the basin, there would not be a free passage of this kind. The struck ball would divide its energy first with one of the balls in the basin, these two would similarly

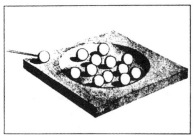

FIG. 1.

share their energies with others, and so on until the original kinetic energy was divided among all the balls. If the basin and the balls are regarded as perfectly smooth and elastic, the collisions would continue until the kinetic energy happens again to be concentrated upon a ball close to the edge. This ball would then escape from the basin and the remainder of the balls would be left with insufficient total energy for any of them to climb the slope. The picture illustrates, therefore, "that the excess energy of the incident neutron will be rapidly divided among all the nuclear particles with the result that for some time afterwards no single particle will possess sufficient kinetic energy to leave the nucleus".

Nuclear Energy Levels

THE second figure illustrates the character of the distribution of energy levels for a nucleus of not too small atomic weight. The lowest lines represent the levels with an excitation of the same order of magnitude as ordinary excited γ-ray states. According to the views developed in Prof. Bohr's address, the levels will for increasing excitation rapidly become closer to one another and will, for an excitation of about 15 million electron volts, corresponding to a collision between a nucleus and a high-speed neutron, be continuously distributed, whereas in the region of small excess energy of about 10 million volts excitation they will still be sharply separated. This is illustrated by the two lenses of high magnification placed over the level-diagram in the two above-mentioned regions. The dotted line in the middle of the field of the lower magnifying glass represents zero excess energy, and the fact that one of the levels

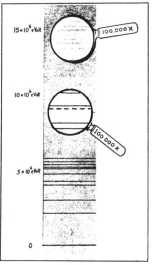

FIG. 2.

is very close to this line (about $\frac{1}{4}$ volt distant) corresponds to the possibility of selective capture for very slow neutrons. The average distance between the neighbouring levels will in this energy region be about ten volts as estimated from the statistics for the occurrence of selective capture. The diagram shows no upper limit to the levels, and these actually extend to very high energy values. If it were possible to experiment with neutrons or protons of energies above a hundred million volts, several charged or uncharged particles would eventually leave the nucleus as a result of the encounter; and, adds Prof. Bohr, "with particles of energies of about a thousand million volts, we must even be prepared for the collision to lead to an explosion of the whole nucleus".

Figure 10.1   Report of a lecture by Niels Bohr [*Nature* **137** *(1936) p. 351*]

nucleus given by

$$\lambda = 2\pi\sqrt{\frac{\hbar^2}{2m_N \varepsilon_n}} \tag{10.1.1}$$

decreases from 9 fm to 3 fm when ε_n increases from 9 MeV to 81 MeV. It seems therefore likely that a neutron with a fairly low energy must react with the nucleus as a whole and not with its constituents.

However, once the neutron is within the nucleus, it gains kinetic energy of about 40 - 50 MeV from the average potential. The wavelength then starts at about 4 fm already for $\varepsilon_n = 0$. This is still appreciable compared to the nucleon size but only about one quarter of the diameter of a heavy nucleus.

The rather abrupt change of neutron wavelength at the nuclear surface results in a rather large probability of reflection when the neutron tries to escape after having crossed the nucleus. A standing wave in the nucleus is then built up, so that sufficent time for the interaction with the nucleons is available.

The dynamical picture of the process, then, is that the incoming neutron collides with a nucleon and thereby loses some of its energy. The resulting two nucleons each continue and collide with another nucleon, again sharing the available energy. After a while the energy is distributed statistically among all the nucleons. This means that the probability of finding a given energy distribution is proportional to the phase space of that configuration. The purely statistical measure of the probability is then the number of eigenstates with the appropriate quantum numbers. Thus the nucleons obtain a distribution of energy around an average value which is lower than that of the initial neutron. It is very unlikely, but not impossible to find one nucleon with the full kinetic energy of the initial neutron.

The resulting quasi-stationary state is called the compound nucleus. In it, the energy is statistically distributed, and almost all "memory" about the initial configuration is lost. It is therefore possible to form the same compound nucleus in various other ways, for example by proton, deuteron, or α absorption. Clearly the target nucleus and the projectile energy then must be appropriately chosen to lead to the same energy, N, and Z, as well as the same angular momentum.

In the dynamical evolution sketched above, a series of intermediate states of the nucleus are populated. They are not stationary, but of course at each time they can be expanded on stationary states. This is also possible for the "last" state where all the energy is statistically distributed. Again this state is **not** stationary although it is very stable with a long lifetime τ. The width ΔE of the energies of the strictly stationary states necessary to make up the wave packet of the compound-nuclear states is related to τ by $\tau \Delta E = \hbar$.

Thus the quasi-stationary compound state can be viewed as a wave packet formed by stationary states of energy in an interval ΔE around the total energy given by energy conservation. The contributing states are also constrained by other conservation laws like that of charge, angular momentum, parity, etc. Only the states of matching quantum numbers can contribute to the formation of the wavepacket.

As we have emphasized, the compound nucleus is **not** a stationary state. Hence it will decay sooner or later in one way or another. The resulting final states can be described by a complete set of quantum numbers like emitted particles of given energy, angular momentum, etc. The rate at which the compound nucleus decays into a given final state i is denoted by $\Gamma_i/\hbar$; Γ_i is called the "partial width". The total rate of decay gives the lifetime τ of the compound nucleus,

$$\frac{1}{\tau} = \sum_i \frac{\Gamma_i}{\hbar}. \tag{10.1.2}$$

where the summation must be extended over all possible decay modes. The probability that the compound nucleus has not decayed by a time t after its formation is $e^{-t/\tau}$. We can find the total probability P_i of decay into the final state i,

$$P_i = \int_0^\infty dt \, \frac{\Gamma_i}{\hbar} e^{-t/\tau} = \frac{\Gamma_i}{\sum_k \Gamma_k}. \tag{10.1.3}$$

10.1 B.   Level Density

Bohr's picture of the formation of the compound nucleus led to the notion of the statistical distribution of energy. This means more precisely that the nucleus has a certain statistical probability of being in a given state where the nucleus' energy is well-defined. The various possible states then provide the distribution of energy. The statistical probability for being in one class of states defined by a set of quantum numbers is proportional to the number of those states.

When a specific decay rate is needed, say for neutron emission, we must calculate the probability of finding an outgoing neutron of a given energy. Clearly it is then necessary to be able to find the number of states of given characteristics. This is the number of levels in a small energy interval with given quantum numbers.

Following Bohr and Mottelson [vol. 1], we begin by defining the nuclear level density ρ for given energy E and nucleon number A as a sum of Dirac δ-functions, i.e.

$$\rho(E, A) = \sum_{i,\nu} \delta(A - A_\nu)\delta(E - E_i(A_\nu)). \tag{10.1.4}$$

The summations run over all nucleon numbers A_ν and all energy eigenvalues $E_i(A_\nu)$. The grand partition function Z given by

$$Z(\alpha, \beta) = \sum_{i,\nu} e^{\alpha A_\nu - \beta E_i(A_\nu)} \tag{10.1.5}$$

can be expressed in terms of ρ as

$$Z(\alpha, \beta) = \int dE \int dA \, \rho(E, A) e^{\alpha A - \beta E}. \tag{10.1.6}$$

The inverse relation of eq. (10.1.6)

$$\rho(E, A) = \frac{1}{(2\pi i)^2} \int \int_{-i\infty}^{i\infty} Z(\alpha, \beta) e^{\beta E - \alpha A} d\alpha \, d\beta \tag{10.1.7}$$

gives ρ as the Laplace transform of Z. The exponent $\mathcal{S}$ (in statistical mechanics called the entropy) of the integrand

$$\mathcal{S}(\alpha, \beta) = \beta E - \alpha A + \ln Z(\alpha, \beta) \tag{10.1.8}$$

is a rapidly varying function of the integration parameters. Expansion of $\mathcal{S}$ to second order around the equilibrium point and subsequent integration (the saddle-point method) then leads to

$$\rho(E, A) = \frac{e^{\mathcal{S}(\alpha, \beta)}}{2\pi \sqrt{D(\alpha, \beta)}} \tag{10.1.9}$$

where D is the determinant

$$D(\alpha, \beta) = \begin{vmatrix} \frac{\partial^2 \ln Z}{\partial \beta^2} & \frac{\partial^2 \ln Z}{\partial \alpha \partial \beta} \\ \frac{\partial^2 \ln Z}{\partial \beta \partial \alpha} & \frac{\partial^2 \ln Z}{\partial \alpha^2} \end{vmatrix} \tag{10.1.10}$$

and the saddle point (α, β) is determined by

$$\frac{\partial \mathcal{S}}{\partial \beta} = E + \frac{\partial \ln Z(\alpha, \beta)}{\partial \beta} = 0 \tag{10.1.11a}$$

$$\frac{\partial \mathcal{S}}{\partial \alpha} = -A + \frac{\partial \ln Z(\alpha, \beta)}{\partial \alpha} = 0. \tag{10.1.11b}$$

The formulation so far is general, but to continue we need a model of the nucleus. As discussed in Chapter 4, we can to a first approximation view the nucleons as moving independently in an average potential. Thus we consider the nucleus as a gas of Fermions in the nuclear volume.

The single particle spectrum is given by $\hat{\epsilon}_1, \hat{\epsilon}_2, \hat{\epsilon}_3, \ldots$. The nuclear many-body system is then characterized by the occupation numbers n_k (either 0 or 1) of these states. In particular we have

$$A = \sum_k n_k \tag{10.1.12}$$

$$E = \sum_k n_k \hat{\epsilon}_k \tag{10.1.13}$$

and the grand partition function of eq. (10.1.5), where indices i and ν now are replaced by sets of occupation numbers $\{n_k\}$, becomes

$$Z(\alpha, \beta) = \sum_{\{n_k\}} \exp\left[\sum_k n_k(\alpha - \beta\hat{\epsilon}_k)\right]$$

$$= \prod_k [1 + \exp(\alpha - \beta\hat{\epsilon}_k)] \tag{10.1.14}$$

where the last identity results from the restriction $n_k = 0$ or 1.

Defining the single-particle level density $g(\varepsilon)$ by (see fig. 10.2)

$$g(\varepsilon) = \sum_k \delta(\varepsilon - \hat{\varepsilon}_k) \tag{10.1.15}$$

we can rewrite eq. (10.1.14) as

$$\ln Z(\alpha, \beta) = \int_{U_0}^{\infty} g(\varepsilon) \ln\left[1 + \exp(\alpha - \beta\varepsilon)\right] d\varepsilon \tag{10.1.16}$$

where U_0 is an energy below which $g(\varepsilon) = 0$. By dividing the integration interval in two, i.e. $[U_0, \mu]$, $[\mu, \infty]$, where $\mu = \alpha/\beta$, and adding and subtracting the first term in the following expression, we obtain

$$\ln Z(\alpha, \beta) = \int_{U_0}^{\mu} (\alpha - \beta\varepsilon) g(\varepsilon)\, d\varepsilon$$
$$+ \int_{0}^{\infty} [g(\mu + \varepsilon') + g(\mu - \varepsilon')] \ln\left[1 + e^{-\beta\varepsilon'}\right] d\varepsilon'. \tag{10.1.17}$$

The logarithmic factor in the second integral makes the integrand small except when $\varepsilon' \leq 1/\beta$. This allows us to expand g in powers of ε' in the last term. Keeping only the lowest order, we finally have

$$\ln Z(\alpha, \beta) = \int_{U_0}^{\mu} (\alpha - \beta\varepsilon) g(\varepsilon) d\varepsilon + \frac{\pi^2}{6\beta} g(\mu). \tag{10.1.18}$$

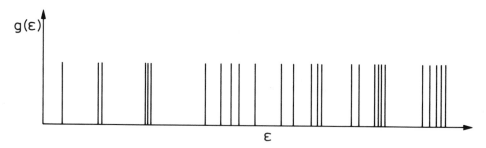

Figure 10.2   Schematic single-particle level density for a potential without degeneracy.

In the same approximation, where derivatives of g are neglected, the saddle-point equations (10.1.11) become

$$E = \frac{\pi^2}{6\beta^2} g(\mu) + \int_{U_0}^{\mu} \varepsilon\, g(\varepsilon) d\varepsilon \qquad (10.1.19)$$

$$A = \int_{U_0}^{\mu} g(\varepsilon) d\varepsilon \qquad (10.1.20)$$

where the last equation shows that the chemical potential μ may, to this order, be approximated by its value for the ground state. The excitation energy E^* is therefore obtained from eq. (10.1.19) as

$$E^* = \frac{\pi^2}{6\beta^2} g(\mu). \qquad (10.1.21)$$

We can now find approximate expressions for the entropy $\mathcal{S}$ and determinant D. Substituting (10.1.18) and (10.1.21) in (10.1.8) and (10.1.10) we see

$$\mathcal{S} = \frac{\pi^2}{3\beta} g(\mu) = 2\sqrt{\frac{\pi^2}{6} g(\mu) E^*}, \qquad (10.1.22)$$

$$\sqrt{D} = \frac{E^* \sqrt{12}}{\pi}. \qquad (10.1.23)$$

The level density then finally becomes

$$\rho(E, A) = \frac{\exp\left[2\sqrt{\frac{\pi^2}{6} g(\mu) E^*}\right]}{E^* \sqrt{48}}, \qquad (10.1.24)$$

which increases exponentially with $\sqrt{E^*}$.

It is interesting at this point to compare with the thermodynamic description of large systems. Here the grand potential Ω is related to the grand partition function by

$$\beta\Omega = -\ln Z. \qquad (10.1.25)$$

Then eq. (10.1.8) can be rewritten

$$\Omega = E - \mu A - \mathcal{S}/\beta \qquad (10.1.26)$$

which shows that $\mathcal{S}$ indeed should be interpreted as the entropy and $\beta = 1/T'$ as the inverse temperature. Calculating $d\mathcal{S}/dE$ from eq. (10.1.8) we find

$$\frac{d\mathcal{S}}{dE} = \beta + \left(E + \frac{\partial \ln Z}{\partial \beta}\right)\frac{d\beta}{dE} + \left(-A + \frac{\partial \ln Z}{\partial \alpha}\right)\frac{d\alpha}{dE} = \beta \qquad (10.1.27)$$

where the last equality is obtained by using eq. (10.1.11). This is the usual equilibrium relationship between entropy and energy

$$dE = T'd\mathcal{S}. \tag{10.1.28}$$

In statistical mechanics and thermodynamics, the probability distribution is proportional to $e^\mathcal{S}$. In our case of a finite system, this role is played by the level density which in this sense is the "nuclear entropy". The corresponding temperature T can then in analogy to eq. (10.1.28) be defined as

$$\frac{1}{T} = \frac{\partial \ln\rho}{\partial E} = \beta - \frac{1}{2}\frac{\partial \ln D}{\partial E}. \tag{10.1.29}$$

Thus the thermodynamic temperature T' equals the nuclear temperature when the energy dependence of D is neglected. This corresponds in fact to the limit of very large systems where the entropy, eq. (10.1.22), tends to infinity for fixed excitation energy whereas D, eq. (10.1.23), stays constant. Even apart from the exponential dependence on $\mathcal{S}$, its significance would completely dominate that of D for large systems.

Ordinary macroscopic thermodynamics is so similar to that of finite systems that we quite easily can exploit the close analogy in many applications. As a little example we can calculate the average occupation number $\langle n_k \rangle$ of the state "k" by, see eq. (10.1.14)

$$\langle n_k \rangle = \frac{\partial}{\partial(\alpha - \beta\hat{\varepsilon}_k)}[\ln Z] = \frac{1}{1 + \exp\left[(\hat{\varepsilon}_k - \mu)\beta\right]}. \tag{10.1.30}$$

This is the well-known Fermi distribution for a temperature $1/\beta$.

The result (10.1.24) for the level density is a continuous function of E^*. Comparing with the definition (10.1.4), we see that some averaging procedure has been applied. This clearly happened when $g(\varepsilon)$ was considered a smooth function so that it could be Taylor expanded in eq. (10.1.17). The width of the contributing energy interval is about T, which must include many single-particle contributions for g to be considered continuous. Thus $g \gg \beta$ or equivalently, from eq. (10.1.21), $g(\mu)E^* \gg 1$ which means that the excitation energy must be large compared to the energy of the first excited state. On the other hand E^* should not be too large, since then g would vary over the contributing energy interval and its derivatives could not be neglected. A closer investigation of this condition [Bohr and Mottelson volume 1] leads to the restriction $E^* \ll (\mu - U_0)A^{1/3}$. Thus the approximations should be valid for excitation energies from a few MeV up to around 100 MeV except for the lighter nuclei.

At low excitation energies, the level density has to be modified to take account of pairing. This greatly reduces the level density, since every excitation requires an energy greater than the gap Δ which is much larger than the single-particle level spacing. As the temperature becomes comparable to Δ, however, this effect on the level density becomes smaller. The value of Δ also decreases somewhat, because the gap equation (6.4.28) has to be modified to take account of the thermal occupation of quasiparticle states. Above a critical temperature T_{crit}, of the same order as Δ, the gap equation ceases to have solutions and the pairing disappears. This is why superconductors and superfluids exist only at very low temperatures. Once the temperature exceeds T_{crit}, the expressions we derive here apply, as long as the excitation energy E^* is measured relative to the ground state of the unpaired system.

The level density, eq. (10.1.24), only specifies energy and total nucleon number. A simple and straightforward extension is to distinguish between neutrons and protons. The result is quite similar and given by

$$\rho(E, N, Z) = \sqrt{\frac{g^2(\mu)}{4g_n(\mu)g_p(\mu)}} \; \frac{\exp\left[2\sqrt{\frac{\pi^2}{6}g(\mu)E^*}\right]}{\left[\frac{3}{2}g(\mu)E^*\right]^{1/4}E^*\sqrt{48}} \tag{10.1.31}$$

where g_n and g_p are the single-particle level densities for neutrons and protons and g is their sum.

Another essential quantum number is the angular momentum J. Since the single particle angular momenta are not additive (they may couple in very complicated ways), it is convenient to start with the additive projection on the intrinsic symmetry axis. The total projection component K is given in terms of the single particle projections $\hat{m}_k$, i.e.

$$K = \sum_k n_k \, \hat{m}_k \tag{10.1.32}$$

and the partition function is now

$$Z(\alpha, \beta, \gamma) = \prod_k [1 + \exp(\alpha - \beta\hat{\varepsilon}_k - \gamma\hat{m}_k)]. \tag{10.1.33}$$

The single particle level density becomes a function of both ε and m and we get

$$g(\varepsilon, m) = \sum_k \delta(\varepsilon - \hat{\varepsilon}_k)\delta_{\hat{m}_k m}, \tag{10.1.34}$$

$$\ln Z(\alpha, \beta, \gamma) = \int \int d\varepsilon \, dm \; g(\varepsilon, m) \ln[1 + \exp(\alpha - \beta\varepsilon - \gamma m)]. \tag{10.1.35}$$

Expanding to second order in γ and neglecting all derivatives of g, we obtain

$$\ln Z(\alpha, \beta, \gamma) = \ln Z(\alpha, \beta) + \frac{\gamma^2}{2\beta}g(\mu)\langle m^2\rangle \tag{10.1.36}$$

where $g(\varepsilon)$ now means

$$g(\varepsilon) = \int_{-\infty}^{\infty} g(\varepsilon, m)dm \tag{10.1.37}$$

and

$$\langle m^2 \rangle = \int_{-\infty}^{\infty} g(\mu, m) m^2 \, dm \Big/ g(\mu) \qquad (10.1.38)$$

is the average squared angular-momentum projection of the states at the Fermi energy μ.

The integral may be evaluated by the saddle-point method. The saddle-point equations, eqs. (10.1.11) are now supplemented by

$$\frac{\partial S}{\partial \gamma} = K + \frac{\partial \ln Z(\alpha, \beta, \gamma)}{\partial \gamma} = 0. \qquad (10.1.39)$$

Their solutions are

$$A = \int_{U_0}^{\mu} g(\varepsilon) d\varepsilon \qquad (10.1.40)$$

$$E = \int_{U_0}^{\mu} \varepsilon g(\varepsilon) d\varepsilon + \frac{\pi^2}{6\beta^2} g(\mu) + \frac{1}{2} \left(\frac{\gamma}{\beta} \right)^2 g(\mu) \langle m^2 \rangle \qquad (10.1.41)$$

$$K = -\frac{\gamma}{\beta} g(\mu) \langle m^2 \rangle \qquad (10.1.42)$$

which, inserted in the generalization of eq. (10.1.9), gives

$$\rho(E, A, K) = \frac{6^{1/4} g(\mu)}{24 \langle m^2 \rangle^{1/2}} \frac{\exp \left[2 \sqrt{\frac{\pi^2}{6} g(\mu) E_K^*} \right]}{[g(\mu) E_K^*]^{5/4}} \qquad (10.1.43)$$

where the reduced excitation energy is

$$E_K^* = E^* - \frac{K^2}{2 g(\mu) \langle m^2 \rangle}. \qquad (10.1.44)$$

To calculate the level density for definite angular momentum J we divide $\rho(E, A, K)$ in contributions of given J.

$$\rho(E, A, K) = \sum_{J=K}^{\infty} \rho(E, A, K, J). \qquad (10.1.45)$$

Assuming rotational invariance, so that $\rho(E, A, K, J) = \rho(E, A, K', J)$ for all $K, K' \leq J$, we finally obtain

$$\begin{aligned}
\rho(E, A, K = J, J) &= \rho(E, A, K = J) - \rho(E, A, K = J + 1) \\
&\approx -\frac{\partial \rho(E, A, K)}{\partial K} \bigg|_{K=J+1/2} \\
&= \frac{\pi(2J + 1) g(\mu)}{6^{1/4} \langle m^2 \rangle^{3/2}} \frac{\exp \left[2 \sqrt{\frac{\pi^2}{6} g(\mu) E_{K=J+1/2}^*} \right]}{48 \left(g(\mu) E_{K=J+1/2}^* \right)^{7/4}}.
\end{aligned} \qquad (10.1.46)$$

To estimate the average of m^2 for levels at the Fermi surface, we assume that the momenta of the nucleons are isotropically distributed with momentum p_F from the Fermi-gas model, and that the momenta are uncorrelated with the positions of the nucleons. Then

$$
\begin{aligned}
\hbar^2 \langle m^2 \rangle &= \left\langle \left(xp_y - yp_x \right)^2 \right\rangle \\
&\approx \langle x^2 \rangle \langle p_y^2 \rangle + \langle y^2 \rangle \langle p_x^2 \rangle \\
&\approx \frac{1}{3} p_F^2 \langle x^2 + y^2 \rangle = \frac{1}{3} p_F^2 \frac{\mathcal{I}_{\text{rig}}}{Am_N} \\
&\approx \frac{\mathcal{I}_{\text{rig}}}{g_{FG}(\mu)}
\end{aligned}
\tag{10.1.47}
$$

where

$$
g_{FG}(\mu) = 4 \frac{dN(\varepsilon)}{d\varepsilon}\Big|_{\varepsilon=\mu} = \frac{3A}{2} \left(\mu - U_0 \right)^{-1}
\tag{10.1.48}
$$

is the level density at the Fermi surface in the Fermi gas model from eqs. (4.1.6) and (4.1.7), and $\mathcal{I}_{\text{rig}}$ is the rigid moment of inertia for rotation about the z-axis.

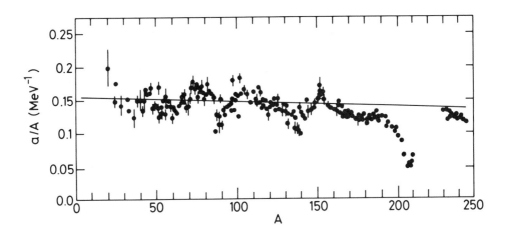

Figure 10.3   Single-particle level-density parameters $a/A = \pi^2 g\,(\mu)/6A$ for various nuclei [*from A. V. Ignatyuk in "Nuclear Theory in Neutron Data Evaluation" (IAEA–190), "Proceedings of a Consultant's Meeting, Trieste 1975," IAEA, Vienna 1976, vol. 1, p. 211.*]

The level density in eq. (10.1.46) can be generalized to two kinds of particles, neutrons and protons. For N=Z=A/2 the modification is simply a factor of $(1/24)^{1/4}$ and a change of the exponent 7/4 to 2. The main feature is still an exponential increase with available intrinsic excitation energy obtained by subtraction of the rotational energy $\hbar^2(J + 1/2)^2/2\mathcal{I}_{rig}$ from the total E^*. The other characteristic property is the $(2J + 1)$-dependence when the rotational energy is small compared to E^*, i.e. loosely speaking small J and/or large E^*.

Often the level density is only needed in the neighborhood of a particular energy E_0^*. The constant-temperature expansion is then useful, i.e.

$$\rho(E^*) \approx \rho(E_0^*)e^{(E^*-E_0^*)/T} \tag{10.1.49}$$

where T is formally given by eq. (10.1.29). It is now apparent that T is not uniquely defined, but depends on the other quantum numbers. Many different temperatures exist, for example one for each J.

The essential quantity in the level density expression is $g(\mu)$, because it enters in the exponent. It has been extracted systematically from very-low-energy neutron-scattering experiments. The compound-nuclear level density at the neutron binding energy is, on average, reproduced using a single-particle level density parameter (compare figure 10.3).

$$\frac{\pi^2}{6}g(\mu) = \frac{A}{7.5\text{MeV}}. \tag{10.1.50}$$

This should be compared to the value (A/11 MeV) of the optical potential reproducing single-particle properties. The discrepancy can be understood in terms of collective states, and perhaps surface corrections to eq. (10.1.50).

Before closing this subsection, we want to point out that the angular-momentum dependence of the level density is specified by both total and projection quantum numbers. Often the notation $\rho(E,A,J)$ is used to describe the quantity in eq. (10.1.46), which is one of the components and not the sum of all. In the latter case an extra factor of 2J + 1 would appear. Thus the $(2J+1)$ dependence does **not** arise from a summation of all directions of the spin.

10.2   STRONG INTERACTION DECAY MODES

The compound nucleus can decay in many ways. The dominant modes, when energy conservation allows, are fission and emission of neutrons, protons and α-particles. Also other light nuclei may be emitted, but the probability of such a decay decreases with the complexity of the emitted nucleus. The decay rate $\Gamma_i/\hbar$ for a particular decay "i" (specifying emitted particle, energy and angular momentum etc.) can be factorized as

$$\frac{\Gamma_i}{\hbar} = F_i \frac{1}{\tau_i} T_i \tag{10.2.1}$$

where F_i, the pre-formation factor, is the probability of finding the particle to be emitted within the nucleus; $1/\tau_i$, the knocking rate, is the frequency with which the particle hits the nuclear boundary; and T_i, the transmission coefficent, is the probability that a particle which hits the wall will penetrate it and escape.

10.2 A. The Knocking Rate

The classical picture of a particle periodically bouncing back and forth in a potential immediately gives a knocking rate. For a harmonic oscillator potential of frequency ω we have

$$\frac{1}{\tau^{\text{HO}}} = \frac{\omega}{\pi} \tag{10.2.2}$$

and for a square well potential of radius R and depth U_0

$$\frac{1}{\tau_{\text{sq}}} = \frac{\hbar K}{2\mu R} \tag{10.2.3}$$

where μ is the reduced mass and

$$\hbar K = \sqrt{2\mu(\varepsilon - U_0)} \tag{10.2.4}$$

is expressed in terms of the particle energy ε. For a neutron or proton of energy $\varepsilon = 0$ ($U_0 \approx -50\text{MeV}$), the square well estimate is $\tau_{\text{sq}} \approx 10^{-22}\text{sec}$. We have neglected the effect of centrifugal and Coulomb potentials, since they are small for the dominant cases of low charge and low angular momentum. Further the knocking rate is usually not the most important of the factors in eq. (10.2.1) and if it is, the crude estimates of eqs. (10.2.2) or (10.2.3) have to be significantly improved.

To do this we consider the quasi stationary compound nucleus characterized by a set of quantum numbers like total energy, angular momentum, parity etc. The wave packet Ψ describing the state is then a linear combination of the stationary states ψ_n around the given energy and with the other conserved quantum numbers, i.e.

$$\Psi = \sum_n a_n \psi_n(\vec{r}_1, \vec{r}_2, \dots \vec{r}_A) e^{-iE_n t/\hbar} \tag{10.2.5}$$

where the time dependence is now explicitly displayed. The energies E_n of the states ψ_n are, in the narrow energy interval around the total energy E_o, very well represented by a uniform distribution of levels,

$$E_n = E_o + \frac{(n - n_o)}{\rho_c} \tag{10.2.6}$$

where ρ_c is the level density of the compound-nuclear states of the right angular momentum and parity. Inserting eq. (10.2.6) into eq. (10.2.5) shows that Ψ is periodic and reproduces itself after the recurrence time τ_r

$$\tau_r = \hbar \cdot 2\pi\rho_c. \tag{10.2.7}$$

Thus $1/\tau_r$ could equally well be considered to be the knocking rate. Indeed, this would be reasonable when only one final state is available. Otherwise during the time τ_r, the state Ψ passes a number of configurations, each corresponding to a decay into different final states. Each final state consists of an emitted particle plus a daughter nucleus. The proper total knocking rate is therefore

$$\frac{1}{\tau_c} = \rho_d \frac{dE}{\tau_r} = \rho_d \frac{dE}{2\pi\hbar\rho_c} \tag{10.2.8}$$

where $\rho_d\,dE$ is the number of levels in the daughter nucleus in the energy interval dE around the appropriate total energy, with the right angular momentum and parity. Comparing with eq. (10.2.1), we see that the knocking rate, and therefore the decay rate, is proportional to the width of the energy band whose decays are included in the specification of "i".

For neutron emission from a state of total energy slightly above the neutron binding energy, essentially only the ground state of the daughter can be reached. The appropriate level density is then typically $\rho_c \approx 1/10$ eV, compare (10.1.46), and the resulting recurrence time is $\tau_r \approx 10^{-16}$ sec. This is in agreement with observations of the widths, see eq. (10.2.1), but a million times longer than τ_{sq}. These were the facts which in the first place led Niels Bohr to suggest the model of the compound nucleus.

10.2 B.   Transmission Coefficient

A wave packet moving towards a potential barrier has a finite transmission probability. This is a well-defined quantum-mechanical problem which of course can be solved numerically. To understand the main structure we shall rely on analytical solutions for simple potentials and the WKB approximation for the general case. First we shall relate T_i more generally to the optical model introduced in Chapter 3.

The transmission coefficient is defined as that fraction of the incoming current which is transmitted. For particles incident on a nucleus surrounded by a potential, this definition implies that the particles don't appear again, neither by being reflected nor by penetrating the nucleus. Thus the fraction in question is precisely the part which in the optical model language has been absorbed. In other words the transmission coefficient for the ℓ'th partial wave in the optical model is given by (see eqs. (3.3.5) and (3.1.2))

$$T_\ell^{OM} = 1 - |\eta_\ell|^2 \tag{10.2.9}$$

We imagine that the motion is described by a coordinate Q, which might be the position of a neutron or proton, or more generally a collective coordinate as in Chapter 8, with a mass or generalized inertia M_Q. If the barrier is described by an inverted harmonic-oscillator potential with a barrier height U_B

$$U^{HO}(Q) = U_B - \frac{1}{2}M_Q\Omega^2 Q^2, \tag{10.2.10}$$

then the transmission coefficent for an energy E is given by [see Hill and Wheeler]

$$T^{HO} = \frac{1}{1 + \exp\left[\frac{2\pi}{\hbar\Omega}\left(U_B - E\right)\right]}. \tag{10.2.11}$$

The two energies U_B and $\hbar\Omega$ characterize T^{HO}. For an energy equal to the barrier height, $E = U_B$, $T^{HO} = 1/2$. Below and above, T^{HO} approaches 0 and 1 respectively. Increasing the barrier's height or thickness or the mass, all reduce the transmission coefficent. These features also characterize the general case of an arbitrary barrier.

If the effective one-dimensional potential U_{eff} is smoothly varying with the coordinate, we can apply the WKB approximation for energies below the barrier height. The resulting transmission coefficent T^{WKB} is given by [Fröman and Fröman]

$$T^{WKB} = \frac{1}{1 + e^{2S}} \tag{10.2.12}$$

$$S = \int_{Q_1}^{Q_2} \sqrt{\frac{2M_Q}{\hbar^2}(U_{eff} - E)}dQ \tag{10.2.13}$$

where M_Q is again the appropriate mass parameter and Q_1 and Q_2 are the classical turning points defined by $E = U_{eff}(Q)$. When $U_{eff}(Q) = U^{HO}(Q)$, the integral in eq. (10.2.13) can easily be performed and the resulting transmission coefficent turns out to equal that of the exact calculation, $T^{WKB} = T^{HO}$.

For particle transmission, we start with a three-dimensional Schrödinger equation which, assuming spherical symmetry, is reduced to one dimension for the radial coordinate. The effective potential is then a sum of centrifugal U_ℓ, Coulomb U_C and nuclear U_N potentials, i.e.

$$U_{eff}(r) = U_\ell(r) + U_C(r) + U_N(r) \tag{10.2.14}$$

This potential can now be used in eqs. (10.2.12) and (10.2.13) to give the WKB transmission coefficents.

The barrier height enters in the exponent, and is therefore decisive for T. Its contributions from U_C and U_ℓ are roughly given by their values at a distance slightly outside the nuclear radius ($R_B \approx 1.4\text{fm } A^{1/3}$). For a nucleus (Z,A) and a particle with charge Z_p we have outside the nucleus

$$U_C(r) = \frac{Z_p Z e^2}{r} \tag{10.2.15}$$

$$U_\ell(r) = \frac{\hbar^2 \ell(\ell + 1)}{2\mu r^2} \tag{10.2.16}$$

where ℓ is the angular momentum and μ the reduced mass.

The factor Ze^2/R_B (see figure 10.4) varies from 4 to 15 MeV as A increases from 27 to 240. Therefore adding one charge to the emitted particle increases the

Coulomb barrier height substantially, and the resulting decay probability is reduced dramatically. Thus, due to the Coulomb barrier, only very light particles can be expected to be emitted from the compound nucleus.

The centrifugal barrier for a nucleon with $\ell = 1$ decreases (see figure 10.4) from 2.4 to 0.6 MeV when A increases from 27 to 240. The increase with ℓ, according to the factor $\ell(\ell + 1)$, is substantial, so we expect that only the lowest partial waves will contribute. However, when the centrifugal potential in the region of interest ($r \approx R_B$) is small compared to the Coulomb potential, it is less important. The ℓ dependence is rather weak for such cases (heavy nuclei). This trend is enhanced for larger charge and mass of the emitted particle, because then U_C increases and U_ℓ decreases. Thus for α emission we can expect contributions from higher partial waves than for nucleons.

Let us now make quantitative estimates of T_ℓ. We begin with $\ell = 0$ by using

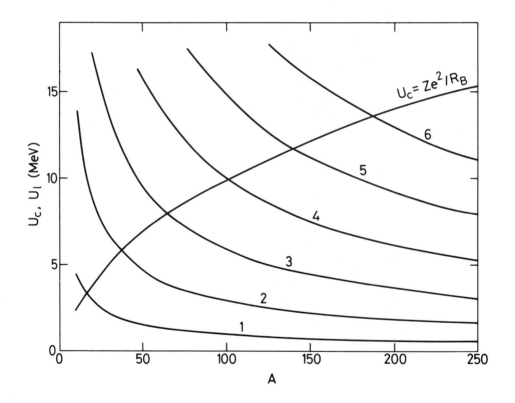

Figure 10.4   Coulomb and angular-momentum barriers, eqs. (10.2.15–16) for $r = R_B$. The decreasing curves for U_ℓ are labeled by their value of ℓ.

the Coulomb potential in eq. (10.2.15) and a square well potential for U_N. The turning points are then the nuclear radius R on the inside, and $R_t = Z_p Z e^2 / E$ on the outside. The integral in eq. (10.2.13) becomes

$$
\begin{aligned}
S &= \sqrt{\frac{2\mu}{\hbar^2}} \int_R^{R_t} \sqrt{\frac{Z_p Z e^2}{r} - E}\, dr \\
&= \sqrt{\frac{2\mu R^2}{\hbar^2 E}} U_C(R) \left[\arctan \sqrt{\frac{R_t}{R} - 1} - \frac{R}{R_t} \sqrt{\frac{R_t}{R} - 1} \right].
\end{aligned}
\tag{10.2.17}
$$

For $E \ll U_C(R)$ or equivalently $R \ll R_t$ we have approximately

$$
S \approx \sqrt{\frac{2\mu R^2 U_C(R)}{\hbar^2}} \left[\frac{\pi}{2} \sqrt{\frac{U_C(R)}{E}} - 2 \left(1 - \frac{1}{6} \frac{E}{U_C(R)} \cdots \right) \right]
\tag{10.2.18}
$$

which then necessarily is large. Therefore eq. (10.2.12) can be written

$$
\ln T_{\ell=0} \approx -2S.
\tag{10.2.19}
$$

An extension of these estimates to the case of $\ell \neq 0$ is difficult in general and we shall not attempt to do it here. Instead we turn to the opposite extreme of neutron emission where the Coulomb potential is zero. As we have already seen this process is likely to dominate, at least for the heavier nuclei. It is therefore a very important special case.

The WKB approximation is unfortunately not very useful for the neutron transmission coefficients. This is seen already from the s-waves, where the barrier has vanished. On the other hand an exact solution can be found for a square-well potential U_N. The result is [see Blatt and Weisskopf]

$$
T_\ell = \frac{4kKV_\ell(kR)}{K^2 + V_\ell(kR)\left(2kK + k^2 V_\ell'(kR)\right)}
\tag{10.2.20}
$$

where the wave numbers inside and outside the nucleus are

$$
k = \sqrt{2m_N E / \hbar^2}
\tag{10.2.21}
$$

$$
K = \sqrt{2m_N (E - U_0)/\hbar^2}.
\tag{10.2.22}
$$

The functions V_ℓ and V_ℓ' are expressed in terms of the regular and irregular Coulomb wave functions F_ℓ and G_ℓ:

$$
V_\ell(x) = \frac{1}{F_\ell^2(x) + G_\ell^2(x)}
\tag{10.2.23}
$$

$$
V_\ell'(x) = \left(\frac{dF_\ell}{dx} \right)^2 + \left(\frac{dG_\ell}{dx} \right)^2.
\tag{10.2.24}
$$

When the energy is small compared to the depth of the nuclear average potential $(E \ll |U_N|)$ or equivalently $k \ll K$, the transmission coefficient simplifies

$$T_\ell \approx 4kV_\ell(kR)/K. \qquad (10.2.25)$$

Since $V_0(x) = V_0'(x) = 1$ we have

$$T_{\ell=0} = \frac{4kK}{(k+K)^2} \qquad (10.2.26)$$

which reduces to eq. (10.2.25) for $k \ll K$. The function $V_\ell(x)$ behaves like $x^{2\ell}$ for small x, so that T_ℓ is proportional to $E^{\ell+1/2}$ for small E (i.e. $kR \ll 1$).

The general conclusion that the importance of the partial waves decreases with ℓ still holds for neutrons. If necessary, it is easy to calculate T_ℓ also for larger ℓ from the known expressions for V_ℓ and V_ℓ'. For our general discussion we don't need any more details which in any case would be inaccurate for the schematic nuclear potential employed. Some examples of transmission coefficients for a square-well U_N are shown in figure 10.5.

10.2 C.   Nucleon Emission

We shall now quantitatively investigate the emission of nucleons from the compound nucleus. The foundation is eq. (10.2.1), where the preformation factor F is unity since nucleons are the constituents of nuclei.

We consider the decay of the system with energy E and angular momentum J. The knocking rate for daughter states of angular momentum J' and energies in the interval $d\varepsilon$ about $E - S_N - \varepsilon$ is then, see eq. (10.2.8),

$$\frac{1}{\tau_c} = \frac{\rho_d(E - S_N - \varepsilon, J')d\varepsilon}{2\pi\hbar\rho_c(E, J)} \qquad (10.2.27)$$

where we have introduced the shorthand notation $\rho_d(E,J) \equiv \rho(E, A_d, K_d, J)$ for the level density of the daughter nucleus, and a similar notation for the compound nucleus. S_N is the nucleon binding energy, and ε is the kinetic energy of the emitted nucleon. The corresponding transmission coefficient then must be that of a nucleon whose angular momentum j coupled to J' gives J. In turn, j is obtained by coupling the orbital angular momentum ℓ to the spin of 1/2. Thus in principle the transmission coefficient $T_\ell(\varepsilon)$ may depend on both ℓ and j. We may obtain the energy distribution of the emitted nucleon by adding the spectra from all of the allowed angular-momentum combinations. The rate of emission per unit nucleon energy is

$$\frac{1}{\hbar}\frac{d\Gamma_N(E, J, \varepsilon)}{d\varepsilon} = \sum_{J'} \sum_{j=|J-J'|}^{J+J'} \sum_{\ell=j-1/2}^{j+1/2} \frac{\rho_d(E - S_N - \varepsilon, J')T_{\ell j}(\varepsilon)}{2\pi\hbar\rho_c(E, J)} \qquad (10.2.28)$$

where ε is in the interval $0 \leq \varepsilon \leq E-S_N$. In the previous discussion we have seen how to calculate all of the quantities on the right-hand side.

We can now derive an amusing and historically important relationship between the compound-nucleus emission spectra, and the cross section σ_{abs} for a hypothetical reaction in which the compound nucleus is created by bombarding the excited

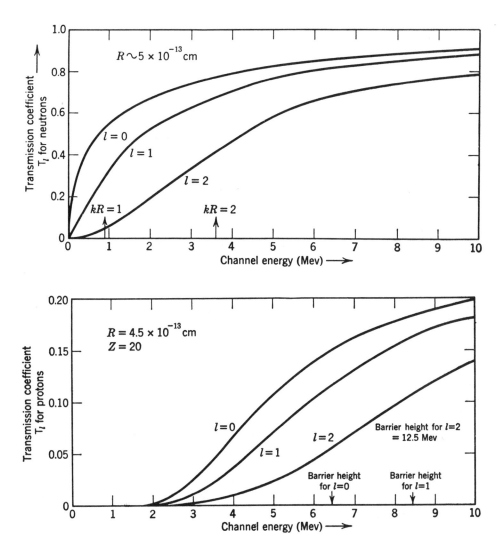

Figure 10.5   Transmission coefficients for nucleons in a square-well potential [*from J. Blatt and V. Weisskopf*]

daughter nucleus (which is clearly impractical since the daughter nucleus itself will decay in less than a femtosecond). We begin with the simplifying assumption

$$\rho(E, J) = (2J + 1)\rho(E, 0) \tag{10.2.29}$$

which is valid for small values of J (compare eq. (10.1.46)). If furthermore $T_{\ell j}(\varepsilon)$ is only a function of ℓ and ε, we get easily

$$\frac{d\Gamma_N(E, J, \varepsilon)}{d\varepsilon} = \frac{1}{2\pi\rho_c(E, 0)} \sum_{\ell=0}^{\infty} 2(2\ell + 1)T_\ell(\varepsilon)\rho_d(E - S_N - \varepsilon, 0). \tag{10.2.30}$$

We now introduce the inverse cross section $\sigma_{abs}(\varepsilon)$ for absorption of a nucleon by the excited daughter nucleus to produce the compound-nuclear state,

$$\sigma_{abs}(\varepsilon) = \frac{\pi}{k^2} \sum_{\ell=0}^{\infty} (2\ell + 1)\left(1 - |\eta_\ell|^2\right). \tag{3.3.5}$$

The absorption probability $\left(1 - |\eta_\ell|^2\right)$ for the partial wave ℓ may be estimated by the transmission coefficient T_ℓ, eq. (10.2.9). With this relation we obtain

$$\frac{d\Gamma_N(E, J, \varepsilon)}{d\varepsilon} = \frac{2\mu}{\pi^2\hbar^2\rho_c(E, 0)}\varepsilon\sigma_{abs}(\varepsilon)\rho_d(E - S_N - \varepsilon, 0) \tag{10.2.31}$$

where the kinetic energy is in the interval $0 \leq \varepsilon \leq E - S_N$.

Eq. (10.2.31) can also be derived by application of the general principle of detailed balance: the transition probability P_{ab} from a state a to a state b is related to the time reversed transition probability P_{ba} from the state b to the state a by

$$\rho_a P_{ab} = \rho_b P_{ba} \tag{10.2.32}$$

where ρ_a and ρ_b are the densities of states of a and b, respectively. Let a be the compound nucleus state of energy E and angular momentum J with level density $\rho_c(E, J)$. The probability per unit time of emitting a nucleon of energy between ε and $\varepsilon + d\varepsilon$ from this state is $\frac{1}{\hbar}d\Gamma_N(E, J, \varepsilon)$. The daughter nucleus has energy $E - S_N - \varepsilon$, an angular momentum J and a cross section $\sigma_{abs}(\varepsilon)$ for absorbing nucleons of energy ε into the compound state. The probability of capture of these nucleons by the daughter nucleus is $v_N\sigma_{abs}/\Omega_r$, where v_N is the nucleon velocity and Ω_r is the volume of the box enclosing the system. The nucleon density of states $\rho_N(\varepsilon)$ per unit energy is given by

$$\rho_N(\varepsilon)d\varepsilon = 2 \cdot \frac{4\hbar p(\varepsilon)^2 dp}{(2\pi\hbar)^3}\Omega_r$$

where $p(\varepsilon)$ is the nucleon momentum and the factor 2 is due to the spin degeneracy. The total level density of state b is then $\rho_N(\varepsilon)\rho_d(E - S_N - \varepsilon, J)$, the product of nucleon and daughter level densities with the proper energy and angular momentum values. The principle of detailed balance, eq. (10.2.32) then immediately results in eq. (10.2.31) by use of $d\varepsilon/dp = v_N$ and the assumption in eq. (10.2.29).

A simple and useful estimate of σ_{abs} is obtained by assuming that the nucleon is absorbed when it reaches a distance of $(R + 1/k)$ from the center of the nucleus. The term $1/k$ allows for the diffractive effect when the nucleon's wavelength is long. This simple picture gives the correct absorptive cross section for a black uncharged sphere of radius R in both limits of $kR \ll 1$ and $kR \gg 1$:

$$\sigma_{abs}(\varepsilon) \approx \pi R^2 \left(1 + 1/kR\right)^2 \qquad (10.2.33a)$$

for a neutron; for a proton, a crude correction for the Coulomb interaction is to multiply (10.2.33a) by $(1 - U_C/\varepsilon)$ if ε is above the Coulomb barrier U_C, and to assume no absorption below the barrier(see eq. 11.1.7):

$$\sigma_{abs}(\varepsilon) \approx \pi R^2 \cdot \max(0, 1 - U_C/\varepsilon) \qquad (10.2.33b)$$

The spectrum of the emitted nucleon is given by eq. (10.2.32) where the level density strongly (exponentially) favors small ε, while σ_{abs} even more strongly inhibits $\varepsilon < U_C$. Thus the bulk of the particles will appear with energies slightly above U_C. For this region we find, after approximating ρ_d by (10.1.49),

$$\rho_d(E - S_N - \varepsilon, 0) \approx \rho_d(E - S_N - U_C, 0)e^{(U_C - \varepsilon)/T_d}, \qquad (10.2.34)$$

that the spectrum for $\varepsilon > U_C$ is proportional to

$$\left(1 + \frac{1}{kR}\right)^2 (\varepsilon - U_C)e^{(U_C - \varepsilon)/T_d}. \qquad (10.2.35)$$

For protons, the first factor varies more slowly than the others, so the distribution is essentially a Maxwell distribution displaced in energy by U_C. For neutrons, the distribution is a slightly distorted exponential. Typical spectra are shown in figure 10.6.

It is also possible to calculate the total nucleon emission rate or the total width $\Gamma_N(E,J)$ for nucleon emission,

$$\Gamma_N(E, J) = \int_0^{E - S_N} \frac{d\Gamma_N(E, J, \varepsilon)}{d\varepsilon} d\varepsilon. \qquad (10.2.36)$$

The approximation in eq. (10.2.34) applied to eq. (10.2.32) then gives, after a variable change to $\varepsilon' = \varepsilon - U_C$ and the definition $\varepsilon_0 = \hbar^2/2\mu R^2$

$$\Gamma_N(E, J) = \frac{\rho_d(E - S_N - U_C, 0)}{\pi \varepsilon_0 \rho_c(E, 0)} \int_0^{E - S_N - U_C} \left(1 + \sqrt{\frac{\varepsilon_0}{\varepsilon' + U_C}}\right)^2 \varepsilon' e^{-\varepsilon'/T_d} d\varepsilon'. $$

$$(10.2.37)$$

Since $\varepsilon' e^{-\varepsilon'/T_d}$ has a maximum for $\varepsilon' = T_d$, we use that value in the first factor of the integral to find

$$\Gamma_N(E, J) = \frac{\rho(E - S_N - U_C, 0)}{\pi \varepsilon_0 \rho_c(E, 0)} \left(1 + \sqrt{\frac{\varepsilon_0}{T_d + U_C}}\right)^2 T_d^2 \left(1 - e^{(S_N + U_C - E)/T_d}\right).$$

$$(10.2.38)$$

The exponential term is much smaller than unity as soon as E exceeds $(S_N + U_C)$ by T_d, so it can usually by neglected. If the constant-temperature approximation $\rho_i(E, 0) \approx \rho_i(0, 0) e^{-E/T_i}$ is valid for all energies for both ρ_c and ρ_d, we get

$$\Gamma_N(E, J) = \frac{\rho_d(0, 0)}{\pi \varepsilon_0 \rho_c(0, 0)} \left(1 + \sqrt{\frac{\varepsilon_0}{T_d + U_C}}\right)^2 T_d^2 e^{E\left(\frac{1}{T_d} - \frac{1}{T_c}\right) - \frac{S_N + U_C}{T_d}} \qquad (10.2.39)$$

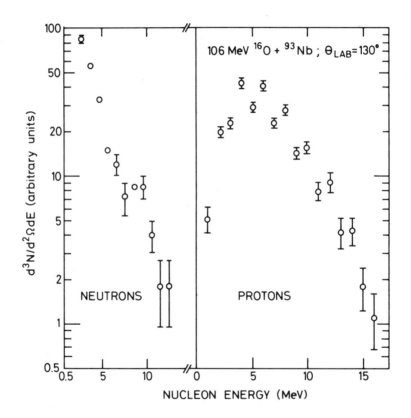

Figure 10.6 Spectra of neutrons and protons emitted at a laboratory angle of 130° from 208 MeV ^{16}O incident on ^{93}Nb [*data from R. L. Ferguson, C. F. Maguire, G. A. Petitt, A. Garron, D. Hensley, D. Horen, F. Obenshain, F. Plasil, A. Shell, G. Young, K. Geoffroy and D. Sarantites, private communication*]

where the temperatures may be different. The further, **very** crude approximations, $T_d \approx T_c$, $\rho_c(0,0) \approx \rho_d(0,0)$ and $\varepsilon_0 \ll T_d + U_C$ then finally give

$$\Gamma_N(E, J) = \frac{T^2}{\pi \varepsilon_0} e^{-(S_N + U_C)/T}. \tag{10.2.40}$$

The branching ratio between protons and neutrons is, again using the same temperature

$$\frac{\Gamma_p}{\Gamma_n} \approx e^{(S_n - S_p - U_C)/T} \tag{10.2.41}$$

showing that for the same binding energies $S_n = S_p$, the proton-emission probability is reduced compared to that of neutrons by $e^{-U_C/T}$.

Inserting typical parameter values in eq. (10.2.40) for the neutrons, we find with $T = 1 \text{MeV}$ a typical value of $\Gamma_n \approx 1$ keV. The proton width is down from this value by several orders of magnitude due to $e^{-U_C/T}$. Of course, this latter statement depends very strongly on the relative separation energies. In fact, in certain regions of N and Z, proton emission is preferred over neutron emission. A more detailed discussion of the compound-nuclear model for particle decay is given by Ericson.

10.2 D. Alpha Decay

The semiempirical mass formula is able to predict the separation energies for α particles. Using the experimental values for the α-particle binding energy, we find negative separation energies for $A \geq 145$ along the valley of β stability. These nuclei are therefore α radioactive in their ground states. However, they are prevented from decaying by the Coulomb and centrifugal barriers. The α particles can be emitted with an energy up to the absolute value of the separation energy. For $A = 250$ this is about 6 MeV with the given parameters in the liquid-drop energy, eqs. (4.3.2–3). We can calculate the α-decay lifetime as $\tau_\alpha = \hbar/\Gamma_\alpha$ where Γ_α is given by eq. (10.2.1). With τ_{sq} of eq. (10.2.3) and T_ℓ from eq. (10.2.19) we get from the first two terms of eq. (10.2.18) the rate for emission of an α paricle of energy E_α

$$\ln \tau_\alpha = -\ln F_\alpha + \ln \tau_{sq} - \ln T_{\ell=0}$$
$$= -\ln F_\alpha - \frac{1}{2} \ln \left(\frac{E_\alpha - U_0}{2\mu R^2} \right) + \pi e^2 \sqrt{\frac{2\mu}{\hbar^2}} \frac{ZZ_\alpha}{\sqrt{E_\alpha}} - 4\sqrt{\frac{2\mu R}{\hbar^2}} \sqrt{ZZ_\alpha e^2}. \tag{10.2.42}$$

which is derived in the WKB approximation. This is sufficiently accurate for our purpose although the steep, almost discontinuous, potential at the nuclear surface certainly violates the condition of validity. An improvement is obtained by including an extra term, $\ln(\frac{1}{16}(U_C - E_\alpha)/(E_\alpha - U_0))^{1/2}$, from the reflection at the dicontinuity. This modification only involves the pre-exponential factor and consequently leaves unchanged the dominant terms in eq. (10.2.42). Inserting the reduced mass of an

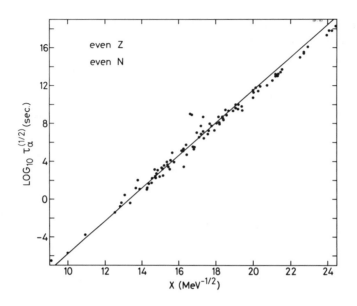

Figure 10.7a   The logarithm of measured α-decay half-lives $\tau_\alpha^{1/2} = \tau_\alpha \ln 2$ as function of X for even-even nuclei. The line is from eq. (10.2.43) with $F_\alpha = 1$, $U_0 = -200$ MeV, $R = 7.25$ fm, $U_C = 30$ MeV and $E_\alpha = 6$ MeV. [The data are from A. Rytz, Atomic Data and Nuclear Data Tables **23** (1979) 507 with a few modifications of S. Hofmann, W. Faust, G. Münzenberg, P. Armbruster, K. Güttner, and H. Ewald. Zeitschrift f. Physik **A291** (1979) 53.]

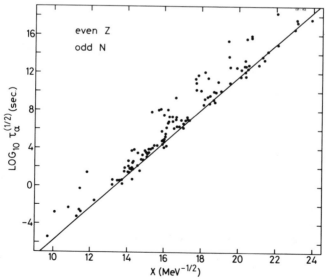

Figure 10.7b   The same as fig. 10.7.a for even Z, odd N nuclei.

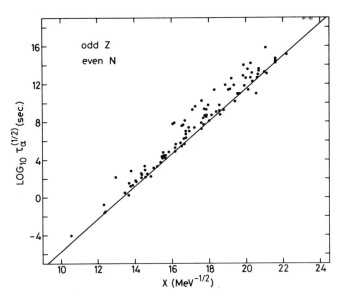

Figure 10.7c   The same as fig. 10.7.a for odd Z, even N nuclei.

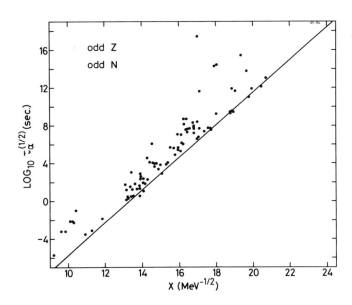

Figure 10.7d   The same as fig. 10.7.c for odd-odd nuclei.

α-particle we obtain

$$\ln \tau_\alpha = -\ln F_\alpha + \frac{1}{2}\ln\left[\frac{(U_C - E_\alpha)R^2 A m_N}{2(E_\alpha - U_0)^2(A + 4)}\right]$$
$$+ X \cdot 4\pi e^2 \sqrt{2m_N/\hbar^2} \tag{10.2.43}$$

where

$$X = \left(\frac{Z}{\sqrt{E_\alpha}} - \sqrt{\frac{8ZR}{\pi^2 e^2}}\right)\sqrt{\frac{A}{A + 4}} \tag{10.2.44}$$

and Z and A are, respectively, the number of protons and nucleons of the daughter nucleus. The $\ell=0$ α-decay lifetime therefore depends on nucleus and α-particle energy only through the combination X of eq. (10.2.44), except for the weak dependence from the modified knocking rate and the preformation factor.

The α decay need not go directly to the ground state of the daughter nucleus. Indeed more than one α particle energy is often seen in experiments. Coincidence experiments show all but the highest-energy α particles which make it possible to infer the final state's excitation energy, spin and parity. The theory applies to each of these α emissions, but the relative angular momenta may of course be different. This is a noticable effect, but not large, as argued in the derivation of transmission coefficients.

Thus plots of the logarithm of the lifetime as function of X should approximately result in a straight line, which indeed is seen in fig. 10.7a where R = 1.25 fm $A^{1/3}$ was used. The range of lifetimes is 24 orders of magnitude, but the observations fall on essentially the line given by eq. (10.2.43) with $F_\alpha=1$. It may be shifted a little depending on the actual parameter choice. The slope remains, however, and one can say that whatever the value of F_α it is about constant for all even-even nuclei. The agreement between theory and measurements is remarkable and a convincing argument for the theory.

In figs. 10.7.b, c and d are shown the measured α-decay half-lives for various parities of the neutron and proton numbers. As for even-even nuclei the main feature is a straight line. However, the scatter is now larger and almost all points lie higher than the straight-line prediction. Thus odd- odd and odd-A nuclei apparently have much longer (10-1000 times) α-decay lifetimes than the neighboring even-even systems.

It is tempting to ascribe this difference to the preformation factor, which intuitively might be expected to be less than unity when odd nucleon numbers are involved. The problem is to understand the "observed" preformation factors in terms of nuclear structure properties. This is (extremely) difficult in general, since very detailed knowledge is required both about single-particle and collective motion and about the residual interactions like the pairing force.

The excited compound nucleus can also emit an α particle. The energy distribution is given in complete analogy to eq. (10.2.28):

$$\frac{\mathrm{d}\Gamma_\alpha(E, J, E_\alpha)}{\mathrm{d}E_\alpha} = \sum_{J'} \sum_{\ell=|J-J'|}^{J+J'} \frac{\rho_\mathrm{d}(E - S_\alpha - E_\alpha, J')}{2\pi\rho_\mathrm{c}(E, J)} T_\ell(E_\alpha) \qquad (10.2.45)$$

where S_α is the binding energy of an α particle in the decaying nucleus. As in nucleon emission, we can go through the assumptions and arguments leading to eqs. (10.2.32) and (10.2.35). Also the total width can be estimated like eq. (10.2.40). The comparison to nucleon decay then is

$$\frac{\Gamma_\alpha}{\Gamma_\mathrm{N}} = 4 \exp\left[(S_\mathrm{N} - S_\alpha - U_\mathrm{C}(\mathrm{N}) + U_\mathrm{C}(\alpha))/T\right]. \qquad (10.2.46)$$

This result, like eq. (10.2.41), is very crude and only useful for not-too-large angular momenta of the decaying nucleus. Increasing the nuclear spin favors α emission, since its larger mass makes it easier for the α particle to decrease the nuclear angular momentum.

10.2 E. Fission

As discussed in Chapter 5, all nuclei of $Z \geq 34$ are unstable against fission, even in their ground states. The apparent stability is entirely due to the often substantial fission barrier. The need to penetrate the barrier results in large lifetimes which may even be comparable to the lifetime of the solar system.

We picture the fission process as the two approximately equally large parts of the total moving under the influence of their mutual interaction. The space of the trajectories is spanned by the deformation parameters which necessarily must include the limiting case of the relative distance between the two separated fragments. The relative generalized potential can be approximated by the Strutinsky energy, see Chapter 5. In this way we have conceptually reduced the fission process to a series of one-particle problems, i.e. one for each pair of fragments. The preformation factor F in eq. (10.2.1) is unity in this picture.

Let us first consider cold or spontaneous fission, meaning an excitation energy much less than the fission barrier. The fission decay rate is then

$$\frac{\Gamma_\mathrm{f}}{\hbar} = \frac{\Omega_0}{\pi} T = \frac{\Omega_0}{\pi} \frac{1}{1 + e^{2S}} \qquad (10.2.47)$$

where Ω_0 is the frequency in the ground-state motion of the two fragments-to-come, and the action integral S is given in eq. (10.2.13). The effective potential U_eff could be the Strutinsky deformation energy, and the mass parameter M_Q for the fission degree could conveniently be that of the cranking model, see Chapter 8. The deformation parameter Q increases along a one-dimensional path in the many-dimensional deformation-parameter space. The Q-trajectory is then varied until the highest fission rate occurs.

A simple and in many ways convenient phenomenological parameterization is obtained with $T = T^{HO}$, eq. (10.2.11). The fission rate is then expressed by the weak Ω_0-dependence and the crucial parameters $\Omega = \Omega_B$ and $U_B = B_f$ for the fission barrier. Roughly $\hbar\Omega_0 = \hbar\Omega_B \approx 0.5 MeV$, and B_f varies as described in Chapter 5.

The barrier-penetration factor controls low-energy fission. This becomes gradually less pronounced as the excitation energy is increased. The transmission coefficient approaches unity, and the "particle" can now pass over the barrier. The fission probability must now be described statistically like the particle emission of the compound nucleus.

The many-dimensional deformation parameter - and related conjugate momentum space constitute the relevent total phase space. The nucleus has, in this phase space, a statistical probability distribution proportional to the level density. The potential energy in deformation space (the Strutinsky energy) has minima at the ground state deformation and for the separated fragments. Total energy conservation then dictates the maxima of the available phase space at the same points. In between are one or more points of minimum phase space, i.e. at the barriers. This bottleneck has to be passed for fission to occur, and therefore controls the rate of decay.

As argued for particle emission, we obtain analogously the decay rate per unit kinetic energy of the fragments, E_f,

$$\frac{d\Gamma_f(E, J, E_f)}{dE_f} = \frac{\rho_B(E - B_f - E_f, J)}{2\pi\rho_c(E, J)} T(E_f) \qquad (10.2.48)$$

where $-B_f < E_f < E - B_f$, ρ_B is the level density on top of the barrier and $T(E_f)$ is the transmission coefficient through the barrier. Note that the relative kinetic energy E_f is negative for genuine barrier penetration required in case of $E < B_f$. Since the "particle" still has to roll down the outer Coulomb barrier before the nucleus has decayed, the function $\Gamma_f(E, J, E_f)$ does not describe the fragment kinetic energy distribution. This is much more complicated, depending on both the shape of the outer part of the barrier and the coupling between the fission degree of freedom and the intrinsic system.

The connection between high-energy ($E > B_f$) and low-energy ($E < B_f$) fission is now clear. On top of the barrier are a lot of levels, which each can be used on the way to fission. The lowest-lying level has the smallest fission barrier, and therefore the largest and perhaps only significant fission probability for $E < B_f$, due to the exponential behavior of $T(E_f)$.

The total fission rate is directly related to measurable quantities. It is simply obtained by integrating eq. (10.2.48)

$$\Gamma_f(E, J) = \int_{-B_f}^{E-B_f} \frac{d\Gamma_f(E, J, E_f)}{dE_f} dE_f \qquad (10.2.49)$$

For $E > B_f$ all "penetrations" occur above the barrier and a good approximation is then

$$\Gamma_f(E, J) = \frac{1}{2\pi\rho_c(E, J)} \int_0^{E-B_f} \rho_B(E - B_f - E_f, J) dE_f. \qquad (10.2.50)$$

The constant-temperature expansion around $E_f = E - B_f$ gives

$$\Gamma_f(E, J) = \frac{\rho_B(E - B_f, J)}{2\pi\rho_c(E, J)} T \left[1 - e^{(B_f - E)/T}\right] \qquad (10.2.51)$$

where the last exponential may be ignored for $E - B_f \geq T$. Using the constant-temperature expression for all energies with the same temperature for barrier and ground state, and assuming $\rho_B(0, J) = \rho_c(0, J)$ we obtain

$$\Gamma_f = \frac{T}{2\pi} e^{-B_f/T}. \qquad (10.2.52)$$

The most direct experimental information is the fission probability P_f, i.e. the ratio Γ_f/Γ, where Γ includes all possible decay modes. For heavier nuclei and $E > S_n$, the neutron emission and the fission decay dominate. Thus we can estimate P_f from eqs. (10.2.40) and (10.2.52)

$$P_f = \frac{\Gamma_f}{\Gamma_n + \Gamma_f} = \frac{1}{1 + 4(m_N/\hbar^2)R^2 T \, e^{(B_f - S_n)/T}}. \qquad (10.2.53)$$

For $B_f = S_n$ in a heavy nucleus, $P_f \approx 1/6$. This estimate clearly depends strongly on the relative values of B_f and S_n. It is, as stated in the derivation, at best valid under the assumption of excitation energies larger than both S_n and B_f. From energies varying from below to above the fission barrier, the fission probability increases exponentially due to barrier penetration and then bends over approaching a fairly constant value for E slightly above B_f, until there is enough energy to fission after emitting a neutron. Thus the energy dependence of P_f carries direct information about the fission barrier.

The description in terms of one barrier is of course inappropriate for the actinide nuclei with the double-hump fission barrier. However, the modification is quite simple in the limit of strong damping in the second well. This amounts to two successive compound fission decays: one through the first barrier into the second well, where all memory of its formation is lost; then through the second barrier but now in competition with a return back to the ground state. The total fission width is therefore

$$\Gamma_f(E, J) = \frac{\Gamma_f^{(1)}(E, J)\Gamma_f^{(2)}(E, J)}{\Gamma_f^{(1)}(E, J) + \Gamma_f^{(2)}(E, J)} \qquad (10.2.54)$$

where the individual fission widths are obtained by treating one barrier alone and completely neglecting the presence of the other.

The validity of the crucial strong-damping assumption is very good for not-too-low energies. The critical quantity is the width W of the intermediate state in the second well, divided by the level distance D. For low energies between the bottom of the second well and the barrier heights, the condition is in general not fulfilled. Here resonance conditions for states in the two wells may lead to enhanced fission for specific subbarrier energies. These observed structures in the fission probability are very strong indications for the interpretation in terms of a double-hump fission barrier.

The description of the fission width in terms of the barrier level density, see eq. (10.2.48), is called the transition-state method. It was originally formulated by Bohr and Wheeler for fission, and was later applied successfully for chemical decay rates. As Kramers discussed, the method is only valid provided the coupling between the fission and intrinsic degrees of freedom is neither too small nor too large. The parameter quantifying this statement is

$$q_K = 2\sqrt{BC}/\gamma \qquad (10.2.55)$$

where B is the fission mass parameter, C the curvature of the potential surface and γ is the friction coefficient describing the mentioned coupling. All three parameters depend on the place on the fission path. The transition-state method assumes quasi-stationary equilibrium, and the fission rate is simply the one-way flow across the barrier. If the friction at the barrier is too large, the net flow towards fission will be smaller than the one-way flow, because "particles" which already passed the barrier may come back. On the other hand, if friction is too small, the assumed statistical equilibrium around the barrier cannot be maintained. The "particle" disappearing over the barrier are not replaced fast enough. Thus the conditions are

$$q_K(\text{barrier}) \geq 1$$
$$q_K(\text{ground state}) \leq B_f/T \qquad (10.2.56)$$

where the first inequality refers to the barrier and the second to the ground state deformation.

The amount of damping in the region of the ground state may be estimated heuristically from the fact that nuclear vibrational motion is underdamped; from the ratios of vibrational energies to their widths in Chapter 8, we see that q_K (ground state)<0.3. Similarly, the amount of damping outside the barrier may be estimated heuristically from the observed kinetic-energy distributions of the fission fragments. A total absence of damping would mean that the relative motion of the fission fragments would contain all the potential energy attained when the system was at the fission barrier. Empirically, the kinetic energies are substantially less than the Coulomb energy of touching spheres, indicating that q_K (beyond barrier)>1.

The damping of fission motion may be estimated theoretically by way of the linear-response theory of Sect. 8.6. The expression in eq. (8.6.8a), which is zero at zero excitation energy, gives a finite result when the nuclear temperature is comparable to the single-particle level spacing. The approximate magnitude of the friction coefficient may be understood by comparison with a simplified geometry, namely nuclear matter contained in one direction by a slowly-moving plane wall represented by a potential $\hat{U}(z - Z(t))$ where $Z(t)$ is the position of the wall. $\hat{U}$ rises from a negative value U_0 for $z \to -\infty$ to a value greater than the Fermi energy μ for $z \to +\infty$, changing rapidly and monotonically in the neighborhood of $z=Z(t)$. It is convenient to contain the nuclear matter in a large box of linear dimension $L \to \infty$, in order to obtain a discrete spectrum of states characterized by their momentum $\vec{p}$ so that we can use the formulas of Chapter 8. The dissipative part of the polarizability is evaluated from eq. (8.2.10)

$$\chi_{ZZ}^{ipm''}(\omega) = \frac{\pi}{\hbar} \sum_{\vec{p}_i} \sum_{\vec{p}_k} \Theta(|\vec{p}_i|^2 \leq p_F^2)\Theta(|\vec{p}_k|^2 > p_F^2)$$

$$|\langle \vec{p}_i | F_Z | \vec{p}_k \rangle|^2 (\delta(\omega - \omega_{ki}) - \delta(\omega + \omega_{ki})) \tag{10.2.57}$$

where the wavefunctions $\langle \vec{r} | \vec{p} \rangle = u_{p_x}(x)u_{p_y}(y)u_{p_z}(z)$ are orthonormal in the box of dimension L, and we have temporarily neglected spins. Since $F_Z = -\partial\hat{U}/\partial z$ depends only on z, the perpendicular components of $\vec{p}_i$ and $\vec{p}_k$ have to be equal. Taking for example the case $\omega > 0$, we find

$$\chi_{ZZ}^{ipm''}(\omega > 0) = \frac{\pi}{\hbar} \sum_{\vec{p}_i} \sum_{p_{kz}} \Theta(\vec{p}_i^2 \leq p_F^2)\Theta(\vec{p}_{i\perp}^2 + p_{kz}^2 > p_F^2)$$

$$|\langle p_{iz} | F_Z | p_{kz} \rangle|^2 \delta(\omega - (p_{iz}^2 - p_{kz}^2)/2\hbar m_N) \tag{10.2.58}$$

where the matrix element contains only the integral over z. We can approximate the sums by integrals, replacing $\sum_p$ by $L\int_0^\infty dp(\pi\hbar)^{-1}$ for each cartesian component. Then the integrals over p_{kz} and the perpendicular components of $\vec{p}_i$ can be carried out, giving

$$\chi_{ZZ}^{ipm''}(\omega > 0) = \frac{L^4 m_N^2 \omega}{2\pi^2 \hbar^3} \int_0^\infty \frac{dp_z}{p_z} |\langle p_z | F_Z | \sqrt{p_z^2 + 2m_N\hbar\omega}\rangle|^2 \Theta(p_z^2 < p_F^2 - 2m_N\hbar\omega). \tag{10.2.59}$$

The friction coefficient is obtained from eq. (8.6.8a):

$$\gamma_{ZZ} = \frac{\partial}{\partial\omega}\chi_{ZZ}^{ipm''}(\omega)|_{\omega=0} = \frac{L^4 m_N^2}{2\pi^2 \hbar^3} \int_0^{PF} \frac{dp_z}{p_z} |\langle p_z | F_Z | p_z \rangle|^2. \tag{10.2.60}$$

The diagonal matrix elements of F_Z are readily calculated:

$$\langle p_z | F_Z | p_z \rangle = -\int_{-L}^\infty dz\, u_{p_z}^2(z)\partial\hat{U}/\partial z = 2\int_{-L}^\infty dz\frac{\partial u_{p_z}}{\partial z}\hat{U}u_{p_z}$$

$$= 2\int_{-L}^\infty dz\frac{\partial u_{p_z}}{\partial z}\left(\frac{p_z^2}{2m_N} + U_0 + \frac{\hbar^2}{2m_N}\frac{\partial^2}{\partial z^2}\right)u_{pz} \tag{10.2.61}$$

$$= \frac{\hbar^2}{2m_N}\int_{-L}^\infty dz\frac{\partial}{\partial z}(\partial u_{p_z}/\partial z)^2 = -\frac{p_z^2}{m_N L}$$

where we have used the fact that u_{p_z} solves the one-dimensional Schrodinger equation with eigenvalue $U_0 + p_z^2/2m_N$, as well as the asymptotic forms $u_{p_z}(z \to \infty) = 0$, $u_{p_z}(z \to -L) \approx (2/L)^{1/2}\sin(p_z z/\hbar)$. Substituting eq. (10.2.61) into eq. (10.2.59), we find that the friction coefficient is proportional to the area of the wall $S = L^2$ and the Fermi momentum p_F, as well as to the interior nuclear-matter density n_0 which in the absence of spin and isospin degeneracy is given by $n_0 = p_F^3/(6\pi^2\hbar^3)$:

$$\gamma_{zz} = L^2 p_F^4/(8\pi^2\hbar^3) = \frac{3}{4}n_0 p_F S \qquad (10.2.62)$$

When spins and isospins are included, each component contributes equally to the friction coefficient as well as to the density; thus the last expression in eq. (10.2.62) is valid also when the spins are taken into account.

In fact, this expression, known as the **wall formula**, was first obtained by Swiatecki [See Blocki el al] for a gas of classical particles with a velocity distribution chosen to mimic the Fermi gas. To estimate the rate of damping for fission, he averaged the energy loss per unit area of the surface of the fissioning nucleus,

$$dE/dt = \gamma_{QQ}^{\text{wall}}\dot{Q}^2 \approx \frac{3}{4}n_0 p_F \int d^2 S |v_\perp(S)|^2 \qquad (10.2.63)$$

where $v_\perp$ is the velocity of the wall when the collective shape coordinate has the velocity $\dot{Q}$. The wall formula gives a fairly good estimate of the damping when the nuclear single-particle eigenvalues are evenly distributed with a level spacing which is less than the temperature, or than the inverse lifetimes of the single-particle states. It predicts that the fission motion is somewhat overdamped, in agreement with observations for large deformations like the fission barrier. For more symmetric shapes like the ground-state deformation, the regularities of the single-particle spectrum and wavefunctions cause the damping to be much less than the wall formula estimate, leading to the observed underdamping. This change in the friction coefficient makes it possible to satisfy both conditions (10.2.56) for the validity of the transition-state method.

10.3   GAMMA DECAY

The excited states of the nucleus have so far, at least implicitly, been considered as stationary by definition. (Compound states and particle unstable states are not stationary, but can be expanded on such states). A familiar analogous case is that of the stationary states of the hydrogen atom. Both the atom and the nucleus have strictly speaking at most only their ground state as a stationary state. All other states decay; they are time dependent. We shall here describe the reason for this instability, i.e. the interaction between the charged system and the electromagnetic field around it.

10.3 A. States of the Electromagnetic Field

Until now, we have treated the electromagnetic field as a classical, static or time-dependent Coulomb field, for example in the scattering of electrons, the ground-state energy, or the Coulomb excitation process. To describe the emission and absorption of photons, however, we must have a quantal description of the electromagnetic field. Thus we have to define a state space describing its degrees of freedom.

The photons are relativistic Bose particles, and therefore must be described in a second-quantized formalism. Each photon is identified by a set of characteristics corresponding to the classical characterization of its electromagnetic field. The most obvious is the plane-wave basis, where we specify the photon's momentum $\hbar \vec{k}$ and the cartesian polarization $\vec{e}$ of its electric field, which must be perpendicular to $\vec{k}$. For each $\vec{k}$ and $\vec{e}$, we have to specify how many photons $n_{\vec{k}\vec{e}}$ are in the state. Thus a basis state would be specified by a set of quantum numbers $\{n_{\vec{k}\vec{e}}\}$; we write the state as $|\{n_{\vec{k}\vec{e}}\}\rangle$.

It is convenient, for each $\vec{k}$, to choose two polarizations $\vec{e}_a(\vec{k})$ and $\vec{e}_b(\vec{k})$, perpendicular to each other and to $\vec{k}$. We shall normalize the states by enclosing them in a large box of volume Ω_r. Then the orthonormality condition is

$$\langle\{n_{\vec{k}\vec{e}_i(\vec{k})}\}|\{n'_{\vec{k}\vec{e}_i(\vec{k})}\}\rangle = \prod_{\vec{k}}\prod_i \delta_{n_{\vec{k}\vec{e}_i(\vec{k})}, n'_{\vec{k}\vec{e}_i(\vec{k})}} \tag{10.3.1}$$

i.e. there have to be the same number of every kind of photon (specified by the discrete set of wave vectors $\vec{k}$ in the box), otherwise the states are orthogonal.

A photon of momentum $\hbar \vec{k}$ and polarization $\vec{e}$ is created by the operator $a^+_{\vec{k}\vec{e}}$, for which we adopt the relativistic Boson normalization:

$$\sqrt{\frac{k}{2\pi\hbar c}}\, a^+_{\vec{k}\vec{e}(\vec{k})}|\{n_{\vec{k}_i\vec{e}_i(\vec{k}_i)}\}\rangle = \sqrt{n_{\vec{k}\vec{e}}+1}|\{n'_{\vec{k}_i\vec{e}_i(\vec{k}_i)}\}\rangle \tag{10.3.2}$$

where

$$n'_{\vec{k}_i\vec{e}_i(\vec{k}_i)} = n_{\vec{k}_i\vec{e}_i(\vec{k})}, (\vec{k}_i, \vec{e}_i) \neq (\vec{k}, \vec{e}) \tag{10.3.2a}$$

$$n'_{\vec{k}\vec{e}(\vec{k})} = n_{\vec{k}\vec{e}(\vec{k})} + 1. \tag{10.3.2b}$$

Just as in the case of the harmonic oscillator, it is easy to see that eqs.(10.3.2) and (10.3.1) together imply commutation relations

$$[a_{\vec{k}\vec{e}}, a^+_{\vec{k}'\vec{e}'}] = \frac{2\pi\hbar c}{k}\delta_{\vec{k}\vec{k}'}, \delta_{\vec{e}\vec{e}'} \tag{10.3.3a}$$

$$[a_{\vec{k}\vec{e}}, a_{\vec{k}'\vec{e}'}] = 0. \tag{10.3.3b}$$

It is useful to express the electromagnetic field operators in terms of the photon creation operators. It is convenient to work in the transverse gauge, where the 4-vector potential $A^\mu(\vec{r}, t) = \left(A^\circ(\vec{r}, t), \vec{A}(\vec{r}, t)\right)$ satisfies the gauge condition

$$\vec{\nabla} \cdot \vec{A}(\vec{r}, t) = 0, A^\circ = 0 \tag{10.3.4}$$

and is therefore related to the electric and magnetic fields, $\vec{E}(\vec{r}, t)$ and $\vec{H}(\vec{r}, t)$, by

$$\vec{E}(\vec{r}, t) = -\frac{1}{c}\frac{\partial}{\partial t}\vec{A}(\vec{r}, t), \tag{10.3.5a}$$

$$\vec{H}(\vec{r}, t) = \vec{\nabla} \times \vec{A}(\vec{r}, t). \tag{10.3.5b}$$

The vector potential operator $\vec{A}(\vec{r}, t)$ is a linear combination of photon creation and annihilation operators:

$$\vec{A}(\vec{r}, t) = \sum_{\vec{k}\vec{e}}(\vec{A}_{\vec{k}\vec{e}}(\vec{r}, t)a_{\vec{k}\vec{e}}^{+} + \text{hermitean conjugate}). \tag{10.3.6}$$

The proportionality factors $\vec{A}_{\vec{k}\vec{e}}(\vec{r}, t)$ may be determined by considering the action of $\vec{A}(\vec{r}, t)$ on a state with only one photon of momentum $\hbar\vec{k}$ and polarization $\vec{e}$: denoting such a state by

$$|\vec{k}\vec{e}\rangle \equiv |\{n_{\vec{k}'\vec{e}'} = 0, \forall(\vec{k}', \vec{e}') \neq (\vec{k}, \vec{e}); n_{\vec{k}\vec{e}} = 1\}\rangle, \tag{10.3.7}$$

we find that, for each component $A^{i}(\vec{r}, t)$,

$$\langle\vec{k}\vec{e}|[\vec{p}, A^{i}(\vec{r}, t)]|\text{vac}\rangle = \hbar\vec{k}\langle\vec{k}\vec{e}|A^{i}(\vec{r}, t)|\text{vac}\rangle \tag{10.3.8}$$
$$= \hbar\vec{k}A_{\vec{k}\vec{e}}^{i}(\vec{r}, t)(2\pi\hbar c/k)$$

since $|\vec{k}\vec{e}\rangle$ is an eigenstate of $\vec{p}$ with eigenvalue $\hbar\vec{k}$, and the vacuum has eigenvalue 0. On the other hand the commutator is just $-i\hbar$ times the gradient of $A^{i}(\vec{r}, t)$, so that we have

$$\vec{k}A_{\vec{k}\vec{e}}^{i}(\vec{r}, t) = -i\vec{\nabla}A_{\vec{k}\vec{e}}^{i}(\vec{r}, t). \tag{10.3.9}$$

Since every term of (10.3.6) has to satisfy Maxwell's equation,

$$\left(\vec{\nabla}^2 - \frac{1}{c^2}\frac{\partial^2}{\partial t^2}\right)\vec{A}_{\vec{k}\vec{e}}(\vec{r}, t) = 0, \tag{10.3.10}$$

eq. (10.3.9) implies that the function $\vec{A}_{\vec{k}\vec{e}}(\vec{r}, t)$ has to be a plane wave with wave vector $\vec{k}$ and frequency ck. Since the photon is supposed to be moving in the direction of $\vec{k}$, we choose the plane wave

$$\vec{A}_{\vec{k}\vec{e}}(\vec{r}, t) = A(k)\vec{e}e^{i(\vec{k}\cdot\vec{r} - ckt)}, \tag{10.3.11}$$

where we have used eq. (10.3.5a) to infer that $\vec{A}$ is in the direction of $\vec{e}$.

To obtain the normalization $A(k)$, we recall that a single photon has a total energy $\hbar ck$; using the expression for the electromagnetic energy,

$$H_{\text{em}} = \int\frac{d^3\vec{r}}{8\pi}\left(\vec{E}^{+}(\vec{r})\cdot\vec{E}(\vec{r}) + \vec{H}^{+}(\vec{r})\cdot\vec{H}(\vec{r})\right), \tag{10.3.12}$$

we find using (10.3.7), (10.3.2), and (10.3.3)

$$\langle \vec{k}\,\vec{e}\,|H_{em}|\vec{k}\,\vec{e}\,\rangle = \frac{k}{4\pi\hbar c}\langle vac|a_{\vec{k}\vec{e}}H_{em}a^{+}_{\vec{k}\vec{e}}|vac\rangle$$

$$= \langle vac|\left(H_{em}a_{\vec{k}\vec{e}}a^{+}_{\vec{k}e} + [a_{\vec{k}\vec{e}}, H_{em}]a^{+}_{\vec{k}\vec{e}}\right)|vac\rangle\frac{k}{4\pi\hbar c}$$

$$= \langle vac|H_{em}|vac\rangle + \langle vac|H_{em}a^{+}_{\vec{k}\vec{e}}a_{\vec{k}e}|vac\rangle\frac{k}{4\pi\hbar c}$$

$$+ \langle vac|[a_{\vec{k}\vec{e}}, H_{em}]a^{+}_{\vec{k}\vec{e}}|vac\rangle\frac{k}{4\pi\hbar c}.$$

(10.3.13)

The second term is zero because $a_{\vec{k}\vec{e}}|vac\rangle = 0$. To evaluate the third term we use Eqs. (10.3.12), (10.3.5), (10.3.6), (10.3.11), and (10.3.3) to find

$$\langle \vec{k}\,\vec{e}\,|H_{em}|\vec{k}\,\vec{e}\,\rangle = \langle vac|H_{em}|vac\rangle + \hbar ck|A(k)|^2\Omega_r.$$

(10.3.14)

The energy that the single-photon state has is supposed to be $\hbar ck$ larger than the energy of the vacuum (which a similar computation shows to be infinite due to contributions from the zero-point fluctuations of states with large $\vec{k}$). Thus we find

$$A(k) = (\Omega_r)^{-\frac{1}{2}}.$$

(10.3.15)

In summary, we have

$$\vec{A}(\vec{r}, t) = (\Omega_r)^{-\frac{1}{2}}\sum_{\vec{k}\vec{e}} \vec{e}\left(e^{i(\vec{k}\cdot\vec{r}-ckt)}a^{+}_{\vec{k}\vec{e}} + \text{hermitean conjugate}\right)$$

(10.3.16)

for the electromagnetic potential in terms of the photon creation operators.

10.3 B. Matrix Elements of the Interaction

In our gauge, the interaction Hamiltonian can be written

$$H_{int}(t) = -\frac{1}{c}\int d^3\vec{r}\,\vec{J}(\vec{r})\cdot\vec{A}(\vec{r}, t)$$

(10.3.17)

where the current density $\vec{J}(\vec{r})$ arises from the motion and intrinsic spins of the nuclear constituents. The current contains important contributions from the mesons in the nucleus, as well as from the quark substructure of the nucleons. We will, for simplicity, ignore these effects, and assume that the nucleons are point particles with intrinsic magnetic moments as in Sect. 7.2, so we can express the current operator as

$$\vec{J}(\vec{r}) = \frac{e}{2m_N}\sum_{i=1}^{A}[q_i\vec{p}_i\delta(\vec{r}-\vec{r}_i) + q_i\delta(\vec{r}-\vec{r}_i)\vec{p}_i$$

$$-\frac{1}{2}g^{(s)}_i\vec{s}_i\times\left(\vec{\nabla}\delta(\vec{r}-\vec{r}_i) + \delta(\vec{r}-\vec{r}_i)\vec{\nabla}\right)]$$

(10.3.18)

where $q_i = 1$ or 0 for protons or neutrons respectively.

To treat the problem of emission or absorption of a photon, we will need the matrix elements of H_{int} between a state $|n\rangle$ of the nucleus in the electromagnetic vacuum $|vac\rangle$, and a state $|n'\rangle$ of the nucleus with one photon $|\vec{k}\vec{e}\rangle$ added to the vacuum. Since the operator H_{int} factorizes into nucleonic and electromagnetic operators, we can write

$$\langle n', \vec{k}\vec{e}\,|H_{int}(t)|n,\ vac\rangle = -\frac{1}{c}\int d^3\vec{r}\,\langle n'|\vec{J}(\vec{r})|n\rangle \cdot \langle \vec{k}\vec{e}\,|\vec{A}(\vec{r},t)|vac\rangle$$
$$= -\sqrt{\frac{2\pi\hbar}{ck\Omega_r}}\int d^3\vec{r}\,\langle n'|\vec{J}(\vec{r})|n\rangle \cdot \vec{e}\,e^{i(\vec{k}\cdot\vec{r}-ckt)}. \tag{10.3.19}$$

Since the nuclear states are eigenstates of angular momentum, it is useful to expand the plane wave in spherical harmonics,

$$e^{i\vec{k}\cdot\vec{r}} = 4\pi\sum_{l=0}^{\infty}\sum_{m=-l}^{l} i^l j_l(kr)Y_l^m(\theta_k,\phi_k)^* Y_l^m(\theta,\phi) \tag{10.3.20}$$

where (θ_k,ϕ_k) are the angles of $\vec{k}$, and (θ,ϕ) are the angles of $\vec{r}$. Similarly, we expand the polarization state of the photon $\vec{e}$ on a basis of spin-one states, $\vec{e}_{\nu=\pm 1} = \mp\frac{1}{\sqrt{2}}(\vec{e}_x \pm i\vec{e}_y)$ and $\vec{e}_{\nu=0} = \vec{e}_z$, with well-defined spin projections $\hbar\nu$ along the z-axis,

$$\vec{e} = \sqrt{\frac{4\pi}{3}}\left(Y_1^0(\vec{e})\vec{e}_0 - Y_1^1(\vec{e})\vec{e}_{-1} - Y_1^{-1}(\vec{e})\vec{e}_{+1}\right)$$
$$= \sqrt{\frac{4\pi}{3}}\sum_\nu (-1)^\nu Y_1^\nu(\vec{e})\vec{e}_{-\nu} \tag{10.3.21}$$
$$= \sqrt{\frac{4\pi}{3}}\sum_\nu Y_1^\nu(\vec{e})^* \vec{e}_\nu.$$

We then write the product $\vec{e}\,e^{i\vec{k}\cdot\vec{r}}$ in terms of vector eigenstates of the total angular momentum of the photon λ, its z-projection μ, and the spatial angular momentum ℓ

$$\vec{\Phi}_{\ell\lambda}^\mu = \sum_{\nu=-1}^{+1} \langle \ell(\mu-\nu)1\nu|\lambda\mu\rangle Y_\ell^m(\theta,\phi)\vec{e}_\nu. \tag{10.3.22}$$

Using eqs. (10.3.20-22) in eq. (10.3.19) we find

$$\langle n',\vec{k}\vec{e}\,|H_{int}|n,\ vac\rangle = -8\pi^2\sqrt{\frac{2\hbar}{3ck\Omega_r}}\sum_{\lambda\mu}\sum_\ell i^\ell T_{\lambda\mu}^{\ell*}(\vec{e},\theta_k,\phi_k)\langle n'|\mathcal{M}_{\lambda\mu}^\ell|n\rangle \tag{10.3.23}$$

where

$$T^{\ell}_{\lambda\mu}(\vec{e}, \theta_{k}, \phi_{k}) = \sum_{\nu} Y^{\mu-\nu}_{\ell}(\theta_{k}, \phi_{k}) Y^{\nu}_{1}(\vec{e}) \langle \ell, \mu - \nu, 1, \nu | \lambda, \mu \rangle \qquad (10.3.24)$$

$$\mathcal{M}^{\ell}_{\lambda\mu} = \int d^{3}\vec{r}\, \vec{\mathcal{J}}(\vec{r}) \cdot \vec{\Phi}^{\mu}_{\ell\lambda}(\vec{r}) j_{\ell}(kr). \qquad (10.3.25)$$

Since $T^{\ell}_{\lambda\mu}$ is the μ'th component of a tensor of rank λ, it can be evaluated by rotating the coordinate system through the Euler angles $\vec{\omega} = (\phi, \theta, \psi) = (\phi_{k}, \theta_{k}, \psi)$. The new z-axis then points in the $\vec{k}$ direction so that the new θ_{k} and ϕ_{k} are zero and we have

$$T^{\ell}_{\lambda\mu}(\vec{e}, \theta_{k}, \phi_{k}) = \sum_{\mu'} \mathcal{D}^{\lambda}_{\mu\mu'}(\vec{\omega}) T^{\ell}_{\lambda\mu'}(\vec{e}, 0, 0). \qquad (10.3.26)$$

The vector $\vec{e}$ lies in the x-y plane of the new coordinate system and therefore can be given as $(\cos v, \sin v, 0)$ corresponding to $(\theta_{1}, \phi_{1}) = (\frac{\pi}{2}, v)$. In this coordinate system we then have

$$Y^{0}_{1}(\vec{e}) = 0$$
$$Y^{\pm 1}_{1}(\vec{e}) = \mp \frac{1}{2}\sqrt{\frac{3}{2\pi}} e^{\pm iv} \qquad (10.3.27)$$

which together with $Y^{m}_{\ell}(0,0) = \delta_{mo}\sqrt{(2\ell + 1)/4\pi}$ leads to

$$T^{\ell}_{\lambda\mu}(\vec{e}, 0, 0) = -\frac{1}{4\pi}\sqrt{\frac{3}{2}(2\ell + 1)} \left(\langle \ell 011 | \lambda\mu \rangle e^{iv} - \langle \ell 01 - 1 | \lambda\mu \rangle e^{-iv} \right) \qquad (10.3.28)$$

Then, eqs. (10.3.26) and (10.3.28) immediately give

$$\begin{aligned} T^{\ell}_{\lambda\mu}(\vec{e}, \theta_{k}, \phi_{k}) = -\frac{1}{4\pi}\sqrt{\frac{3}{2}(2\ell + 1)} (&\langle \ell 011 | \lambda 1 \rangle \mathcal{D}^{\lambda}_{\mu 1}(\vec{\omega}) e^{iv} \\ &- \langle \ell 01 - 1 | \lambda - 1 \rangle \mathcal{D}^{\lambda}_{\mu-1}(\vec{\omega})) e^{-iv} \end{aligned} \qquad (10.3.29)$$

The unspecified angle ψ enters only in the combination $\psi_{v} = \psi + v$. Using

$$\sqrt{2(2\ell + 1)}\langle \ell 01 \pm 1 | \lambda \pm 1 \rangle = \left\{ \begin{array}{c} \sqrt{\lambda+1} \\ \mp\sqrt{2\lambda+1} \\ \sqrt{\lambda} \end{array} \right\} \text{ for } \ell = \left\{ \begin{array}{c} \lambda-1 \\ \lambda \\ \lambda+1 \end{array} \right\} \qquad (10.3.30)$$

we now obtain

$$T^{\lambda}_{\lambda\mu}(\vec{e}, \theta_{k}, \phi_{k}) = \frac{1}{8\pi}\sqrt{3(2\lambda + 1)}[\mathcal{D}^{\lambda}_{\mu 1}(\phi_{k}, \theta_{k}, \psi_{v}) + \mathcal{D}^{\lambda}_{\mu-1}(\phi_{k}, \theta_{k}, \psi_{v})]$$
$$T^{\lambda-1}_{\lambda\mu}(\vec{e}, \theta_{k}, \phi_{k}) = -\frac{1}{8\pi}\sqrt{3(\lambda + 1)}[\mathcal{D}^{\lambda}_{\mu 1}(\phi_{k}, \theta_{k}, \psi_{v}) - \mathcal{D}^{\lambda}_{\mu-1}(\phi_{k}, \theta_{k}, \psi_{v})] \qquad (10.3.31)$$
$$T^{\lambda+1}_{\lambda\mu}(\vec{e}, \theta_{k}, \phi_{k}) = -\frac{1}{8\pi}\sqrt{3\lambda}[\mathcal{D}^{\lambda}_{\mu 1}(\phi_{k}, \theta_{k}, \psi_{v}) - \mathcal{D}^{\lambda}_{\mu-1}(\phi_{k}, \theta_{k}, \psi_{v})]$$

We see that the matrix element, eq. (10.3.23), is determined by the matrix elements of the multipole operators conveniently defined by

$$\mathcal{M}(\mathrm{k}, \mathrm{M}\lambda\mu) = -\frac{i}{c k^\lambda}(2\lambda+1)!!\sqrt{\frac{\lambda}{\lambda+1}}\int d^3\vec{r}\,\vec{\mathcal{J}}(\vec{r})\cdot\vec{\Phi}^\mu_{\lambda\lambda}(\vec{r})j_\lambda(\mathrm{kr}) \qquad (10.3.32)$$

$$\mathcal{M}(\mathrm{k}, \mathrm{E}\lambda\mu) = -\frac{i}{c k^\lambda}(2\lambda+1)!!\sqrt{\frac{\lambda}{\lambda+1}}\int d^3\vec{r}\,\vec{\mathcal{J}}(\vec{r})\cdot \qquad (10.3.33)$$

$$[\sqrt{\frac{\lambda}{2\lambda+1}}\vec{\Phi}^\mu_{\lambda+1,\lambda}(\vec{r})j_{\lambda+1}(\mathrm{kr}) - \sqrt{\frac{\lambda+1}{2\lambda+1}}\vec{\Phi}^\mu_{\lambda-1,\lambda}(\vec{r})j_{\lambda-1}(\mathrm{kr})].$$

Using eqs. (10.3.23–25) and (10.3.31–33) we obtain

$$\langle n', \vec{k}\,\vec{e}|\mathrm{H}_{\mathrm{int}}|n, \mathrm{vac}\rangle = \pi\sqrt{\frac{2\hbar c}{k\Omega_r}}\sum_{\lambda\mu} i^{\lambda-1}\frac{k^\lambda}{(2\lambda+1)!!}\sqrt{\frac{(2\lambda+1)(\lambda+1)}{\lambda}}$$

$$[\langle n'|\mathcal{M}(\mathrm{k}, \mathrm{M}\lambda\mu)|n\rangle\,\left(\mathcal{D}^\lambda_{\mu 1}(\vec{\omega}') + \mathcal{D}^\lambda_{\mu-1}(\vec{\omega}')\right)^*$$

$$-i\langle n'|\mathcal{M}(\mathrm{k}, \mathrm{E}\lambda\mu)|n\rangle\,\left(\mathcal{D}^\lambda_{\mu 1}(\vec{\omega}') - \mathcal{D}^\lambda_{\mu-1}(\vec{\omega}')\right)^*] \qquad (10.3.34)$$

where $\vec{\omega}' = (\phi_k, \theta_k, \psi_v)$. The matrix element only involves $\mathcal{D}^\lambda_{\mu,\nu=\pm 1}$, not $\nu = 0$, reflecting the fact that the photon only has helicities ± 1, i.e., its spin projection must be along its momentum and not perpendicular to it.

The operators $\mathcal{M}(\mathrm{k}, \mathrm{M}\lambda\mu)$ and $\mathcal{M}(\mathrm{k}, \mathrm{E}\lambda\mu)$ are called the magnetic and electric multipole-moment operators, respectively. Since they carry angular momentum λ, they can connect states whose angular momentum differ at most by λ. The $\vec{\Phi}^\mu_{\ell\lambda}$ have parity $(-1)^\ell$, see eq. (10.3.22); since the current $\vec{\mathcal{J}}$ has negative parity (like the momentum $\vec{p}$), the parities of the magnetic and electric operators are $(-)^{\lambda+1}$ and $(-)^\lambda$, respectively.

The electric and magnetic multipole moment operators can be expressed explicitly in terms of the ordinary spherical harmonics instead of the vector spherical harmonics in eq. (10.3.22). To do this we employ the identities

$$i\sqrt{\lambda(\lambda+1)}\vec{\Phi}^\mu_{\lambda\lambda} = \vec{r}\times\vec{\nabla}Y^\mu_\lambda(\theta,\phi) \qquad (10.3.35)$$

$$\mathrm{k}(\lambda+1)\sqrt{\lambda}\vec{\Phi}^\mu_{\lambda-1,\lambda}j_{\lambda-1}(\mathrm{kr}) - \mathrm{k}\lambda\sqrt{\lambda+1}\vec{\Phi}^\mu_{\lambda+1,\lambda}j_{\lambda+1}(\mathrm{kr}) =$$

$$\sqrt{2\lambda+1}\left[k^2\vec{r}\,j_\lambda(\mathrm{kr})Y^\mu_\lambda(\theta,\phi) + \vec{\nabla}\left(Y^\mu_\lambda(\theta,\phi)\frac{d}{dr}(rj_\lambda(\mathrm{kr}))\right)\right] \qquad (10.3.36)$$

which, inserted into eqs. (10.3.32–33), leads after a partial integration to the formulas

$$\mathcal{M}(\mathrm{k}, \mathrm{M}\lambda\mu) = -\frac{1}{ck^\lambda}\frac{(2\lambda+1)!!}{\lambda+1}\int d^3\vec{r}\,\vec{\mathcal{J}}(\vec{r})\cdot(\vec{r}\times\vec{\nabla})(j_\lambda(\mathrm{kr})Y^\mu_\lambda(\theta,\phi)) \quad (10.3.37a)$$

$$\mathcal{M}(\mathrm{k}, \mathrm{E}\lambda\mu) = -\frac{i}{ck^{\lambda+1}}\frac{(2\lambda+1)!!}{\lambda+1}\int d^3\vec{r}\,\left(\vec{\nabla}\cdot\vec{\mathcal{J}}(\vec{r})\right)\frac{\partial}{\partial r}(rj_\lambda(\mathrm{kr}))\,Y^\mu_\lambda(\theta,\phi)$$

$$+\frac{i}{ck^{\lambda-1}}\frac{(2\lambda+1)!!}{\lambda+1}\int d^3\vec{r}\,\left(\vec{r}\cdot\vec{\mathcal{J}}(\vec{r})\right)j_\lambda(\mathrm{kr})Y^\mu_\lambda(\theta,\phi) \quad (10.3.37b)$$

10.3 C. Transition Probability for Photon Emission

Treating H_{int} in time-dependent perturbation theory leads to the transition probability $\Gamma_\gamma/\hbar$ per second from initial $|\psi_i\rangle$ to final state $|\psi_f\rangle$ (Fermi's golden rule)

$$d\Gamma_\gamma/\hbar = \frac{2\pi}{\hbar}|\langle\psi_f|H_{int}|\psi_i\rangle|^2\rho_f d\Omega_k \qquad (10.3.38)$$

where ϱ_f is the number of final states per unit energy per unit solid angle. The initial state is an excited nuclear state and the final state is a product of a nuclear state and an emitted photon state. Then ϱ_f is the number of photons of given energy $\hbar ck$ per unit photon energy per unit solid angle of the photon. Since the number of states in a momentum interval $\hbar dk$ and angular interval $d\Omega_k$ around $\hbar\vec{k}$ in a box of volume Ω_r is

$$N(\hbar k)d(\hbar k)d\Omega_k = \frac{\Omega_r}{(2\pi\hbar)^3}(\hbar k)^2 d(\hbar k)d\Omega_k \qquad (10.3.39)$$

we obtain

$$\rho_f = N(\hbar k)d(\hbar k)d\Omega_k/d(\hbar ck)d\Omega_k = \frac{\Omega_r k^2}{(2\pi)^3\hbar c} \qquad (10.3.40)$$

The total transition rate per unit time between two nuclear eigenstates, Γ_γ, is given by summing eq. (10.3.38) over two orthogonal polarizations $\vec{e}_a$ and $\vec{e}_b$ for each direction of $\vec{k}$, and integrating over the direction. Using

$$(2\lambda + 1)\int d\Omega_k \mathcal{D}^{\lambda*}_{\mu\nu}(\phi_k,\theta_k,\psi)\mathcal{D}^{\lambda'}_{\mu'\nu}(\phi_k,\theta_k,\psi) = 4\pi\delta_{\lambda\lambda'}\,\delta_{\mu\mu'} \qquad (10.3.41)$$

together with eqs. (10.3.34) and (10.3.40) we find

$$\Gamma_\gamma(n \to n') = \int d\Omega_k \sum_{\vec{e}} d\Gamma_\gamma \qquad (10.3.42)$$

$$= 8\pi \sum_{\lambda\mu} \frac{\lambda + 1}{\lambda[(2\lambda + 1)!!]^2}k^{2\lambda+1}\{|\langle n'|\mathcal{M}(k,E\lambda\mu)|n\rangle|^2 + |\langle n'|\mathcal{M}(k,M\lambda\mu)|n\rangle|^2\}$$

This result is independent of ψ_v and consequently the transition rate depends neither on the unspecified rotation angle ψ nor on the specific choice of polarization directions $\vec{e}_a$ and $\vec{e}_b$.

Since the nuclear states are eigenstates of the total angular momentum $\vec{J}^2$ and its projection J_z, with eigenvalues $J(J + 1)\hbar^2$ and $M\hbar$ respectively and degeneracy $2J + 1$ from different values of M, the total rate of photon emission with the given transition energy will be given by summing over degenerate final states

$$\Gamma_\gamma(J_n \to J_{n'}) = \qquad (10.3.43)$$

$$8\pi \sum_{\lambda} \frac{\lambda + 1}{\lambda[(2\lambda + 1)!!]^2}k^{2\lambda+1}\{B(E\lambda, J_n \to J_{n'}) + B(M\lambda, J_n \to J_{n'})\}$$

where we have introduced the multipole transition strengths

$$B(M\lambda, J_n \to J_{n'}) = \sum_{\mu M'} |\langle J_{n'} M' | \mathcal{M}(k, M\lambda\mu)|J_n M_n\rangle|^2, \qquad (10.3.44a)$$

$$B(E\lambda, J_n \to J_{n'}) = \sum_{\mu M'} |\langle J_{n'} M' | \mathcal{M}(k, E\lambda\mu)|J_n M_n\rangle|^2. \qquad (10.3.44b)$$

The Wigner-Eckart theorem ensures that $B(E\lambda)$ and $B(M\lambda)$ are independent of the projection M_n in the initial state.

The nuclear states are concentrated within the nuclear volume, so that only values $kr \leq kR$ are needed. We can then use the expansion of the Bessel functions for $kr \ll 1$:

$$j_\lambda(kr) = \frac{(kr)^\lambda}{(2\lambda+1)!!} \left(1 - \frac{1}{2}\frac{(kr)^2}{(2\lambda+3)} + \cdots\right) \qquad (10.3.45)$$

The condition is fulfilled for photon energies

$$\hbar ck \ll \frac{\hbar c}{R} \approx 165 \text{MeV} \cdot A^{-1/3}, \qquad (10.3.46)$$

i.e., for photon energies much less than about 25 MeV for the heaviest nuclei. In the limit of small kR, the multipole operators, eqs. (10.3.37), take on a simple form independent of k. The magnetic multipole operator is immediately, by use of eq. (10.3.45), given as

$$\mathcal{M}(M\lambda\mu) = -\frac{1}{c(\lambda+1)} \int d^3\vec{r}\, \vec{J}(\vec{r}) \cdot (\vec{r} \times \vec{\nabla})r^\lambda Y_\lambda^\mu(\theta, \phi) \qquad (10.3.47a)$$

Using the continuity equation for the harmonically oscillating charge density operator $\varrho(\vec{r})$ of frequency kc

$$\vec{\nabla} \cdot \vec{J}(\vec{r}) = ikc\varrho(\vec{r}) = ikce \sum_{i=1}^{A} q_i \delta(\vec{r} - \vec{r}_i) \qquad (10.3.48)$$

we rewrite the first term of eq. (10.3.37b) as

$$\mathcal{M}(E\lambda\mu) = \int d^3\vec{r}\, \varrho(\vec{r})r^\lambda Y_\lambda^\mu(\theta, \phi). \qquad (10.3.47b)$$

The last term in eq. (10.3.37b) is usually negligible since its leading contribution contains the extra small factor kr (see Sect. 10.3.D for an estimate of the magnitudes). The limiting case of eq. (10.3.47b) appeared before in eq.(8.1.37) for Coulomb excitation, see Sections 8.1 and 9.6.

10.3 D. Single-particle Decay Rates

The single-particle model is in many ways very useful either as a reasonable approximation or to gain valuable insight. We shall here benefit in both ways and simultaneously establish the traditional units for electromagnetic transition probabilities.

The many-body matrix elements of the operators in eqs.(10.3.42) and (10.3.44) are now given as a sum of single-particle matrix elements. Assuming few nucleons outside a spherically symmetric core, only contributions from the outer nucleons survive, since the multipole operators carry λ units of angular momentum. For one extra nucleon we are then left with single-particle transitions.

Let us first consider electric multipole transitions between initial, ϕ_i, and final, ϕ_f, proton states given by

$$\phi_i = r^{-1} u_i(r) Y_\ell^m(\vartheta, \phi), \quad \phi_f = r^{-1} u_f(r) Y_0^0(\vartheta, \phi). \tag{10.3.49}$$

The intrinsic proton spin is ignored for simplicity, since the first and dominant term of the operator in eq.(10.3.18) is independent of spin. The matrix element of the operator in eq.(10.3.47b) is then non-vanishing only for $(\lambda, \mu) = (\ell, -m)$ where we obtain

$$\langle \phi_f | \mathcal{M}(E\lambda\mu) | \phi_i \rangle = \frac{e}{\sqrt{4\pi}} \int_0^\infty dr \, r^\ell u_f^*(r) u_i(r). \tag{10.3.50}$$

The corresponding $B(E\lambda)$ value is then

$$B(E\ell, \ell \to 0) = \frac{e^2}{4\pi} \left| \int_0^\infty dr \, r^\ell u_f^*(r) u_i(r) \right|^2. \tag{10.3.51}$$

Assuming constant probability densities within the nuclear volume of radius $R(A)$ leads to the **Weisskopf unit**

$$B_w(E\lambda, \lambda \to 0) = \frac{e^2}{4\pi} \left(\frac{3}{\lambda + 3} \right)^2 R(A)^{2\lambda} \tag{10.3.52}$$

and its related transition rate

$$\frac{\Gamma_\gamma^w(E\lambda)}{\hbar} = \frac{e^2}{\hbar} k(kR)^{2\lambda} \frac{18(\lambda + 1)}{\lambda(\lambda + 3)^2 [(2\lambda + 1)!!]^2}. \tag{10.3.53}$$

Although these expressions are valid only for protons, they are used indiscriminately for both nucleons as units for relative comparisons.

Introducing now the spin of the proton, we replace eq.(10.3.49) by eq.(6.2.2). The matrix element can then be calculated either by using the Wigner-Eckart theorem and expressions for reduced matrix elements of coupled systems, or directly by using various identities between Clebsch-Gordon coefficients. The result is

$$\langle \Phi_f | \mathcal{M}(E\lambda\mu) | \Phi_i \rangle = \frac{e}{\sqrt{4\pi}} \langle j_i m_i \lambda\mu | j_f m_f \rangle \langle j_i \tfrac{1}{2} \lambda 0 | j_f \tfrac{1}{2} \rangle \cdot \tfrac{1}{2} \left(1 + (-1)^{\ell_i + \ell_f + \lambda} \right)$$

$$\cdot \sqrt{\frac{(2\lambda + 1)(2j_i + 1)}{(2j_f + 1)}} \int_0^\infty dr \, r^\lambda u_f^*(r) u_i(r) \qquad (10.3.54)$$

and the $B(E\lambda)$ value becomes

$$B(E\lambda, j_i \rightarrow j_f) = \frac{e^2}{4\pi} (2\lambda + 1) |\langle j_i \tfrac{1}{2} \lambda 0 | j_f \tfrac{1}{2} \rangle|^2$$

$$\cdot \left| \int_0^\infty dr \, r^\lambda u_f^*(r) u_i(r) \right|^2 \qquad (10.3.55)$$

when $\ell_i + \ell_f + \lambda$ is even, and zero otherwise. It is independent of the orientation (m_i) of the initial state. When the final-state angular momentum j_f is $1/2$, as close to zero as possible, the $B(E\lambda, j_i \rightarrow \tfrac{1}{2})$ reduces for the appropriate values of ℓ_i, ℓ_f and λ to the expression in eq.(10.3.51).

The magnetic-multipole transitions are more complicated. We shall therefore not derive the general transition probabilities, but refer to deShalit and Talmi. Their result for $B(M\lambda)$ in the important case of $\lambda = |j_i - j_f|$ is

$$B(M\lambda, j_i \rightarrow j_f) = (\mu_N)^2 \frac{(2\lambda + 1)}{4\pi} |\langle j_i \tfrac{1}{2} \lambda 0 | j_f \tfrac{1}{2} \rangle|^2$$

$$\cdot \lambda^2 \left(g^{(s)} - \frac{2g^{(\ell)}}{\lambda + 1} \right)^2 \left| \int_0^\infty dr \, r^{\lambda - 1} u_f^*(r) u_i(r) \right|^2 \qquad (10.3.56)$$

when $\ell_i + \ell_f + \lambda$ is odd and zero otherwise. This expression is, like that for $E\lambda$ transitions, independent of the orientation of the initial state. It is valid for both neutrons and protons (see Chapter 7 for the appropriate g-factors). Using the smallest possible $j_f = 1/2$ and the assumption of constant densities within R we arrive at the **Moszkowski unit**,

$$B_M(M\lambda, j_i \rightarrow \tfrac{1}{2}) = (\mu_N)^2 \left(\frac{3}{\lambda + 2} \right)^2 R^{2\lambda - 2} \frac{\lambda^2}{4\pi} (g^{(s)} - \frac{2g^{(\ell)}}{\lambda + 1})^2, \qquad (10.3.57)$$

where the selection rules still must be fulfilled. The related transition rate $\Gamma_\gamma^M(M\lambda)$ is found from eq.(10.3.43).

The simplest magnetic multipole operator has $\lambda = 1$, which for one particle may be rewritten

$$\mathcal{M}(M1\mu) = \mu_N(g^{(s)}\vec{s} + g^{(\ell)}\vec{\ell}) \cdot \vec{\nabla}(rY_\ell^\mu)$$

$$= \mu_N \sqrt{\frac{3}{4\pi}} (g^{(s)}\vec{s} + g^{(\ell)}\vec{\ell})_\mu, \qquad (10.3.58)$$

where the spherical components of the vectors $\vec{v} = \vec{s}$ and $\vec{\ell}$ are given by

$$(v_o, v_{\pm 1}) = \left(v_z, \mp\frac{1}{\sqrt{2}}(v_x \pm iv_y)\right). \tag{10.3.59}$$

Thus the magnetic dipole operators in eq.(10.3.58) are linear combinations of the magnetic dipole vector components in eq.(7.2.1). The magnetic dipole moments are then, apart from the factor $\sqrt{3/4\pi}$, obtained as the expectation values of $\mathcal{M}(M10)$.

The only non-vanishing and non-diagonal matrix elements of $\mathcal{M}(M1\mu)$ are between states differing only in orientation of $\vec{\ell}$ and $\vec{s}$. Therefore single-particle magnetic-dipole transitions occur only between spin-orbit partners, i.e., $j_f = \ell + \frac{1}{2}$ and $j_i = \ell - \frac{1}{2}$. This matrix element can then easily be calculated and the transition probability obtained. However, eq.(10.3.56) with $\lambda = 1$ is directly applicable, resulting in

$$B(M1, \ell - \tfrac{1}{2} \to \ell + \tfrac{1}{2}) = \frac{3}{4\pi}\mu_N^2 \left(g^{(s)} - g^{(\ell)}\right)^2 \left(\frac{\ell+1}{2\ell+1}\right). \tag{10.3.60}$$

where the radial wavefunctions are assumed identical. The transition probability vanishes for $g^{(s)} = g^{(\ell)}$, in which case the operator in eq.(10.3.58) would be proportional to $\vec{j}$, which of course is diagonal in $\vec{j}^2$.

10.3 E.   Preferred Electromagnetic Transitions

The electromagnetic multiple operators are spherical tensors of rank λ. The usual triangular relations between initial and final angular momentum must therefore be obeyed, i.e.

$$|J_i - J_f| \le \lambda \le J_i + J_f \tag{10.3.61}$$

The parity of $\mathcal{M}(E\lambda\mu)$ is clearly that of Y_λ^μ whereas $\mathcal{M}(M\lambda\mu)$ also contains the current; thus

$$\Pi(E\lambda) = (-1)^\lambda = -\Pi(M\lambda). \tag{10.3.62}$$

In addition to these selection rules, the photon has intrinsic angular momentum 1 and consequently $\lambda \ge 1$.

The above strict requirements lead unambiguously to one (or at the most two) dominant modes of gamma emission for each pair of states. To understand this we must estimate the relative importance of the various transitions. For this estimate, it is sufficient to use the single-particle rates. For two successive electric or magnetic transitions we get from the Weisskopf and Moszkowski estimates, within a factor of two,

$$\frac{\Gamma_\gamma^W(E\lambda+1)}{\Gamma_\gamma^W(E\lambda)} \approx \frac{\Gamma_\gamma^M(M\lambda+1)}{\Gamma_\gamma^M(M\lambda)} \approx \left(\frac{kR(A)}{2\lambda+3}\right)^2 \tag{10.3.63}$$

which for $\lambda = 1$ and a typical transition energy $\hbar kc \approx 1$ MeV is around 10^{-5}. For electric and magnetic transitions of the same multipole order we get

$$\frac{\Gamma_\gamma^M(M\lambda)}{\Gamma_\gamma^W(E\lambda)} \approx \left(\frac{\lambda \hbar g^{(s)}}{2m_Nc R}\right)^2, \qquad (10.3.64)$$

which for $\lambda=1$ varies between 0.005 and 0.05, depending on the nucleus. Thus the lowest allowed transition of one kind is always dominant. (The special case of eq.(10.3.56) is therefore especially important.)

When the selection rules allow $E\lambda$ and $M\lambda+1$ as the lowest possible transitions, the latter can be neglected. When $E\lambda+1$ and $M\lambda$ are the lowest allowed transitions, both may contribute significantly, since the electric transition may be enhanced substantially above the single-particle estimate due to collective effects (see Chapters 8 and 9). The largest of these widths is found from eq.(10.3.53) with $\lambda=1$, i.e.

$$\Gamma_\gamma^W(E1) = \frac{e^2}{4}k^3 R(A)^2. \qquad (10.3.65)$$

which for $\hbar ck=1$MeV and $R\approx 4$ fm is only about 1 eV. Thus it cannot compete with neutron emission whenever the latter is energetically possible.

A particular bound excited state may be able to decay by electromagnetic transitions to several different final states. Clearly, it will prefer to decay to states it can reach by emitting a photon of small angular momentum, i.e., E1, E2, or M1 transitions. For a given multipole order, it will also prefer to emit more energetic photons, due to the factor $k^{2\lambda+1}$ in eq.(10.3.43); this preference is stronger for high multipole order. Among states with similar energy, angular momentum and parity, transitions will be emphasized to final states whose single-particle structure resembles that of the initial state, because the mulipole operator is a very simple, single-particle operator. For example, states described by the excitation out of the ground state of two or more uncorrelated nucleons will have small transition rates among themselves or to the ground state. The biggest transition matrix elements come when the relation between initial and final states is of a collective nature, as we have seen for vibrations in Chapter 8 and for rotations in Chapter 9.

The selectivity of electromagnetic radiative decays, and their preference for collective transitions, provides a powerful tool for sorting out nuclear levels. In particular, the identification of various states with their correct rotational bands, such as in figs. 9.4 and 9.5, is made possible by studying the relative strengths of transitions among various pairs of states. A given member of a rotational band will have a strong tendency to decay to the next-lower member of the same band, which in turn will decay further down the band. Thus most of the strongest lines in a spectrum of emitted photons will correspond to transitions within a band. Once one such transition is identified, the others can be found by looking at the emitted photons in coincidence with the first transition. An example of how this technique helps to classify excited states is shown in fig. 10.8. The schemes can be checked

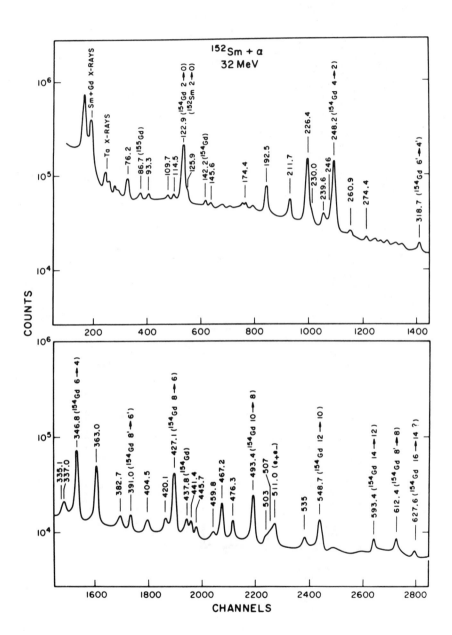

Figure 10.8a   Use of coincidence technique to understand gamma-ray spectra from reaction of 32 MeV alphas on ^{152}Sm. Singles spectrum without coincidence. [*from I. Rezenka, F. Bernthal, J. Rasmussen, R. Stokstad, I. Fraser, J. Greenberg, and D. A. Bromley*, Nucl. Phys. **A179** *(1972)51*].

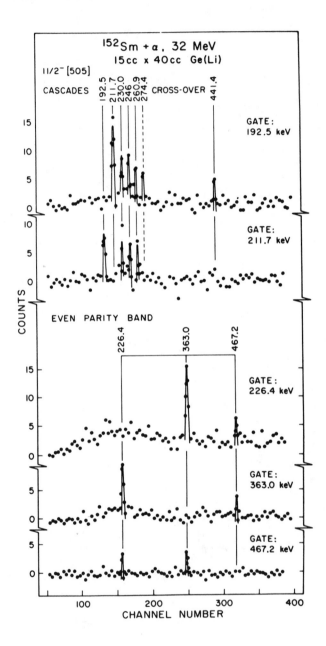

Figure 10.8b   Spectrum of photons from ^{153}Gd in coincidence with various photon energies (GATE), excited by the bombardment of ^{152}Sm with 32 MeV alphas. The coincidences with "Compton pedestals" (of photons whose energy was reduced by Compton scattering) were subtracted from the spectra shown here. Only every fourth channel was plotted in the flat portions of spectra [*from I. Rezenka et al., loc. cit.*].

by measuring the angular correlations of the emitted radiation, and by comparing the intensities for transitions within a band with transitions between bands, using eq. (9.6.14) for the transition strengths. Without these types of information about electromagnetic transitions, it would be impossible to sort out complicated spectra where several bands overlap in the same energy region.

It is worth remembering that electric dipole (E1) transitions are badly described by the independent-particle model, since the operator $\mathcal{M}(E1)$ generates an infinitesimal translation (compare Sect.8.3), which is severely misrepresented in the independent-particle model as we showed in Sect. 4.5. A dipole-type change in the motion of a single nucleon has to be accompanied by the recoil of the remaining nucleons, so that the center of mass remains fixed (the photon's momentum is negligible). This results in a substantial reduction of the transition matrix elements for single-proton transitions, and makes single-neutron transitions have electric dipole matrix elements comparable to the protons'.

10.3 F. Slow Electromagnetic Transitions

Occasionally, a nuclear excited state may have a much higher angular momentum than all the lower-energy states. In that case, the lifetime of the state may become very long, and it is called an isomeric state. For example, the first excited state of ^{243}Am has an excitation energy of 0.049 MeV and a spin-parity of 5^-, with the ground state being 1^-; it decays by E4 emission with a lifetime of 152 years. More typical examples have lifetimes of days or minutes.

We have considered the electromagnetic transitions only to leading order in the fine-structure constant $e^2/\hbar c = 1/137$. Usually, higher orders are unimportant. However, when an energetic photon is emitted from a heavy nucleus, it may produce an e^+e^- pair in the nuclear electrostatic field. Since the electrostatic field is proportional to Z, the relative importance of this process is measured by $Ze^2/\hbar c$ which need not be very small. A photon may also interact with the electrons of the atom on its way out, for example by being absorbed and using its energy to eject an electron from an atomic orbital. This process, called internal conversion, may become significant when the nuclear transition is hindered by the small energy and high angular momentum of the photon; in the internal-conversion process, the photon does not have to have a wave number proportional to its energy, but can be "off its mass shell" because it doesn't appear in the final state.

10.4 BETA DECAY

The electromagnetic interaction discussed above in Sect. 10.3 is only a part of a more general interaction, the electroweak interaction. While the electromagnetic interaction is not charge-independent, it does not change the charge of the nucleons. Other parts of the electroweak interaction are able to change the nucleon's charge.

They are responsible for the beta-decay processes discussed in Sect. 7.5, which restrict the nuclei found in nature to those near the line of beta stability. Nuclei whose ground states are not beta-stable, and many excited states of all nuclei, can in principle decay by emission of positrons or electrons together with neutrinos or antineutrinos. However, the beta-decay process is usually quite slow, and can seldom compete with electromagnetic or strong-interaction decay rates unless the latter are themselves atypically slow. We shall discuss the reason for the smallness of beta-decay rates, find expressions for them, and discuss their dependence on nuclear structure. Along the way we will show how measurements of nuclear beta decay were able to demonstrate for the first time the non-conservation of parity, one of the most fundamental discoveries of modern physics.

10.4 A.   The Effective Interaction of Beta Decay

Just as the electromagnetic interaction is mediated by the photon, the weak interaction is mediated by the vector bosons, $W^{\pm}$ and Z^0. Like the photon, the vector bosons have spin 1 and negative parity. Unlike the photon, however, they are very massive:

$$m_W c^2 = 81 \text{GeV}, m_Z c^2 = 93 \text{GeV}. \qquad (10.4.1)$$

Equally important, the W bosons carry a single unit of electric charge. Thus they are able to mediate the beta-decay interactions. While the coupling of the photon to the electromagnetic current conserves parity, the couplings of the heavy vector bosons do not; in fact, they are nearly equal mixtures of both parities. This is why beta decay does not conserve parity.

Since no nucleus has enough excitation energy to produce a W boson, these quanta appear only indirectly as virtual particles in intermediate states. Following arguments similar to those introduced in Sect. 2.5 to describe meson-exchange forces, we see that the range of the weak interaction mediated by W bosons is about $\hbar/m_W c \approx 2 \times 10^{-3}$fm, which is totally negligible on a nuclear scale. The fundamental process of beta decay is depicted in fig. 10.9.

The nucleon emits a charged W boson, which decays into a positron and neutrino or electron and antineutrino. Since the range of the effective interaction is so short, we can approximate it by a zero-range interaction. The strength of the zero-range interaction may be estimated by inspecting eq.(2.5.9): we see that the volume integral of the effective interaction is proportional to the square of the coupling constant times the square of the range of the force. Since the range of the force is inversely proportional to the boson's mass, the effective interaction becomes very weak even if the coupling constant is not small. Only the large mass of the W bosons makes the beta-decay effective interaction seem weak, because it works over a tiny volume.

To calculate the rate of beta decay, we need to find the matrix element between initial and final states corresponding to the process of fig. 10.9. The most straightforward method would be to calculate the value of the Feynman diagram; readers who are familiar with this technique are encouraged to do this. Instead, we will first derive an effective interaction of zero range which can also be applied directly to nuclear transitions between localized many-body states.

In order to derive the effective interaction, we consider the process of fig.10.10, in which an electron and proton exchange a W boson to become a neutrino and a neutron. Proceeding as in the computation of the pion-exchange potential, we start with eq.(2.4.17) for the effective interaction,

$$\Im_E = V + VG_E^{(+)}\Im_E \qquad (2.4.17)$$

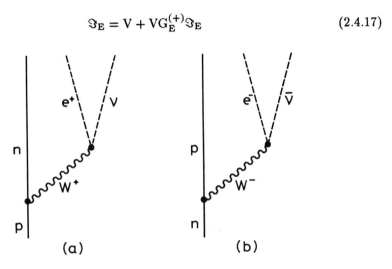

Figure 10.9 Fundamental process of beta decay for (a) positron emission, (b) electron emission.

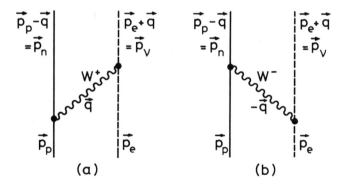

Figure 10.10 Effective lepton-nucleon interaction due to (a) W^+; (b) W^- boson exchange.

which we now use in a state space including nucleons, leptons, and charged vector bosons. Once again, we use the equation to second order

$$\Im_E = V + VG_E^{(+)}V + \dots \tag{2.5.2}$$

The fundamental interaction V is taken from the standard model of electroweak interactions, for which Glashow, Salam and Weinberg received the Nobel prize. It has matrix elements between states with fermions, represented by relativistic Dirac spinors Φ, and vector bosons, represented by their field operators: A_μ for photons, $W_\mu^\pm$ and Z_μ for the heavy bosons. The electroweak interaction is [see Halzen and Martin]

$$
\begin{aligned}
H_{ew} = e\overline{\Phi}(\vec{r},t)\gamma^\mu & \left[QA_\mu + \frac{1}{\sqrt{2}\sin\Theta_W}\frac{1-\gamma_5}{2}\left(t_+W_\mu^+(\vec{r},t) + t_-W_\mu^-(\vec{r},t)\right) \right. \\
& \left. + \frac{1}{\sin\Theta_W\cos\Theta_W}\left(\frac{1-\gamma_5}{2}t_3 - \sin^2\Theta_W Q\right)Z_\mu \right]\Phi(\vec{r},t).
\end{aligned}
\tag{10.4.2}
$$

Here $\overline{\Phi} = \Phi^+\gamma_0$, the γ's are relativistic Dirac matrices, Q is the electric charge operator, and t_+ is the fundamental fermion isospin raising operator. Θ_W is the Weinberg angle, which characterizes the relative strengths of the three parts of the interaction and has a measured value of about 30°.

The deep insight of the standard electroweak model is the universality of the interaction expressed in terms of the fundamental fermion fields Φ. These fundamental fermion fields include leptons and quarks. They all have spin and isospin $\frac{1}{2}$, and are grouped in doublets according to table 10.1. The isospin raising and lowering operators, t_+ and t_-, convert members of the doublets into each other. The up and down quarks, named for their isospin, have rest energies of a few MeV, while the other 4 quarks are heavy on the scale of nuclear energies, and play little role in nuclear physics unless one introduces them into a nucleus by a very energetic process. The nucleon consists almost entirely of u and d quarks: a neutron is udd, a proton uud. When the fundamental isospin raising operator is applied to a neutron, it gives a proton twice; thus an extra factor of two arises when $t_\pm$ are applied to nucleon spinors. In addition, the small admixture of strange quarks in the nucleon's

$$
t_3 = \begin{pmatrix} \frac{1}{2} \\ -\frac{1}{2} \end{pmatrix} : \begin{pmatrix} \nu_e \\ e \end{pmatrix}\begin{pmatrix} \nu_\mu \\ \mu \end{pmatrix}\begin{pmatrix} u \\ d \end{pmatrix}\begin{pmatrix} c \\ s \end{pmatrix}\begin{pmatrix} t \\ b \end{pmatrix}
$$

Table 10.1   Isospin doublets of fundamental fermions. Their antiparticles have the opposite sign of t_3.

constituents gives a factor $\cos\Theta_C \approx 0.97$ where Θ_C is the Cabbibo angle. Furthermore, the axial-vector current of the nucleon, corresponding to the term $\gamma_\mu\gamma_5$, has contributions from the virtual pions around the nucleon, according to the principle of the Partially Conserved Axial Current (PCAC). When the wave functions have long wavelengths compared to the nucleon's size, this effect can be included by multiplying the $\gamma_\mu\gamma_5$ term for nucleons with an additional factor, conventionally denoted g_A/g_V. The hypothesis of the Conserved Vector Current (CVC) implies that no such correction is necessary for the vector-current term γ_μ. Experimentally, in vacuum, $g_A/g_V \approx 1.25$; it may be different in nuclear matter.

Taking matrix elements of $\Im_E$ between the initial and final states of fig. 10.10, we see that the first-order term V in eq. (2.5.2) has zero matrix element because the vector-boson field operators in H_{ew} change the number of bosons, which is the same in both states. The intermediate states in the second-order term are of two types corresponding to figs. 10.10a and 10.10b. Proceeding exactly as in Sect. (2.5), we find analogous to eq. (2.5.4) that in the notation of fig.10.10,

$$\langle \vec{p}_n\vec{p}_\nu|\Im_{ew}|\vec{p}_p\vec{p}_e\rangle \approx \tag{10.4.3}$$
$$-\sum_q \left[\frac{\langle \vec{p}_\nu|H_{ew}|\vec{p}_e,\vec{q}\rangle\langle \vec{p}_n,\vec{q}|H_{ew}|\vec{p}_p\rangle}{m_W c^2} + \frac{\langle \vec{p}_n|H_{ew}|\vec{p}_p,-\vec{q}\rangle\langle \vec{p}_\nu,-\vec{q}|H_{ew}|\vec{p}_e\rangle}{m_W c^2} \right]$$

where the energies of the fermions as well as the kinetic energy of the boson have been neglected in the denominators, because they are small compared to $m_W c^2$. We have deliberately not used the fact that only one value of $\vec{q}$ contributes by momentum conservation in the matrix elements.

We can compute the matrix elements of the vector field operators as in Sect. 10.3.A; the fact that the bosons are massive means that they now have three polarization states instead of two, and that the normalization in (10.3.2) depends on the energy, approximately $m_W c^2$, instead of the photon energy $\hbar ck$, compare eq. (10.3.14). We also have to remember that the operators in the expression corresponding to eq. (10.3.16) for the field W^+ destroy W^+ bosons and create W^- bosons, and vice versa (otherwise eq. (10.4.2) wouldn't conserve charge). We find

$$\langle \vec{p}_n\vec{p}_\nu|\Im_{ew}|\vec{p}_p\vec{p}_e\rangle = \frac{-e^2\pi(\hbar c)^2}{2(m_W c^2)^2\sin^2\Theta_W}\cos\Theta_C$$
$$\sum_{q,\mu}\left[\int d^3\vec{r}\,'\, \overline{\Phi}_{\vec{p}_\nu}(\vec{r}\,',t)\gamma^\mu(1-\gamma_5)t_+\frac{e^{-i\vec{q}\cdot\vec{r}}}{\sqrt{\Omega_r}}\Phi_{\vec{p}_e}(\vec{r}\,',t)\right] \tag{10.4.4}$$
$$\cdot\left[\int d^3\vec{r}\,\overline{\Phi}_{\vec{p}_n}(\vec{r},t)\gamma_\mu(1-\frac{g_A}{g_V}\gamma_5)t_-^N\frac{e^{i\vec{q}\cdot\vec{r}}}{\sqrt{\Omega_r}}\Phi_{\vec{p}_p}(\vec{r},t)\right].$$

where the operator t_-^N in the second factor is now the ordinary nucleon isospin-lowering operator instead of the fundamental quark operator.

We now use the fact that the plane waves form a complete set on the normalization volume, Ω_r,

$$\sum_q \frac{(e^{i\vec{q}\cdot\vec{r}'})^* e^{i\vec{q}\cdot\vec{r}}}{\Omega_r} = \delta(\vec{r}' - \vec{r}) \tag{10.4.5}$$

Denoting the constant in eq.(10.4.4) by the conventional notation $G_F/\sqrt{2}$, we see that the matrix elements of the effective interaction are

$$\langle\vec{p}_n\vec{p}_\nu|\Im_{ew}|\vec{p}_p\vec{p}_e\rangle = \int d^3\vec{r}\, d^3\vec{r}' \overline{\Phi}_{\vec{p}_\nu}(\vec{r}',t) \overline{\Phi}_{\vec{p}_n}(\vec{r},t) \Im_{ew}(\vec{r},\vec{r}') \Phi_{\vec{p}_e}(\vec{r}',t) \Phi_{\vec{p}_p}(\vec{r},t) \tag{10.4.6}$$

where the effective interaction is a Dirac matrix in both lepton and nucleon variables,

$$\Im_{ew}(\vec{r},\vec{r}') = -\frac{G_F}{\sqrt{2}}\delta(\vec{r}-\vec{r}')\{\mathcal{J}_{lept}^+(\vec{r})\cdot\mathcal{J}_{nucl}^-(\vec{r}') + \mathcal{J}_{lept}^-(\vec{r}')\cdot\mathcal{J}_{nucl}^+(\vec{r})\}/c^2 \tag{10.4.7}$$

The dot product in eq.(10.4.7) is a Lorentz product,

$$(\mathcal{J}_1)\cdot(\mathcal{J}_2) = \sum_\mu \mathcal{J}_1^\mu \mathcal{J}_{2\mu},$$

and the current operators are

$$(\mathcal{J}_{lept}^\pm)^\mu = c\gamma^\mu(1-\gamma_5)t_\pm \quad \text{on leptons} \tag{10.4.8a}$$

$$(\mathcal{J}_{nucl}^\pm)^\mu = c\gamma^\mu\left(1-\frac{g_A}{g_V}\gamma_5\right)t_\pm^N \quad \text{on nucleons.} \tag{10.4.8b}$$

Actually our computation only shows the first term of eq. (10.4.7), since the matrix elements of the second term would be zero in eq. (10.4.6). We can easily obtain the second term, however, by remembering that $\Im_{ew}$ is invariant under time reversal. Thus another matrix element can be found by taking the Hermitean conjugate of eq. (10.4.6), corresponding to neutron-neutrino charge exchange. This matrix element corresponds to the second term in eq. (10.4.7).

The coupling constant $G_F/\sqrt{2}$, known as the Fermi coupling for weak interactions, is given by

$$\frac{G_F}{\sqrt{2}} = \frac{e^2\pi(\hbar c)^2\cos\Theta_C}{2(m_W c^2)^2\sin^2\Theta_W}. \tag{10.4.9}$$

Measurements of beta-decay rates determine G_F, as we will see below. Together with the measurement of the mass of the W boson, eq.(10.4.9) provides a value of the Weinberg angle $\Theta_W \approx 27°$, corresponding to $G_F = 8.7 \times 10^{-5}$ MeV fm^3.

For applications in nuclear physics we can apply an additional simplification. Since the nucleons are non-relativistic, we can simplify their part of the interaction.

Using the non-relativistic approximation for the Dirac spinors Φ in terms of the Pauli 2-component spinors Φ,

$$\Phi \approx \begin{pmatrix} \Phi \\ \frac{\vec{s}\cdot\vec{p}}{m_N c}\Phi \end{pmatrix} + \mathcal{O}\left(\frac{p}{m_N c}\right)^2 \tag{10.4.10}$$

together with the properties and explicit expressions for the Dirac matrices, we find

$$\overline{\Phi}_1 \gamma_0 t_\pm \Phi_2 \approx \Phi_1^\dagger t_\pm \Phi_2 \tag{10.4.11a}$$

$$\overline{\Phi}_1 \gamma_i t_\pm \Phi_2 \approx -\Phi_1^\dagger \frac{p_i}{m_N c} t_\pm \Phi_2 \tag{10.4.11b}$$

$$\overline{\Phi}_1 \gamma_i \gamma_5 t_\pm \Phi_2 \approx -\frac{2}{\hbar}\Phi_1^\dagger s_i t_\pm \Phi_2 \tag{10.4.11c}$$

$$\overline{\Phi}_1 \gamma_0 \gamma_5 t_\pm \Phi_2 \approx \frac{2}{\hbar}\Phi_1^\dagger \frac{\vec{s}\cdot\vec{p}}{m_N c} t_\pm \Phi_2 \tag{10.4.11d}$$

Keeping only the lowest-order terms independent of $\vec{p}/m_N c$, the effective interaction (10.4.7) simplifies greatly:

$$\Im_{ew} \approx \Im_F + \Im_{GT} \tag{10.4.12}$$

where

$$\Im_F = -\frac{G_F}{\sqrt{2}}\delta(\vec{r}-\vec{r}')[t_-^N \varrho_{lept}^+ + t_+^N \varrho_{lept}^-] \tag{10.4.13}$$

$$\Im_{GT} = -\frac{G_F}{\sqrt{2}}\frac{g_A}{g_V}\delta(\vec{r}-\vec{r}')\frac{2\vec{s}_N}{\hbar c}\cdot[t_-^N \vec{\mathcal{J}}_{lept}^+ + t_+^N \vec{\mathcal{J}}_{lept}^-] \tag{10.4.14}$$

and we have introduced the notation $\mathcal{J}^\mu = (c\varrho, \vec{\mathcal{J}})$. $\Im_F$ is known as the **Fermi interaction**, and $\Im_{GT}$ is the **Gamow-Teller interaction**.

10.4 B. Transition Rates and Electron Spectra for Beta Decay

When a nuclear state $|\psi_i\rangle = |n\rangle$ decays by beta emission to another nuclear state $|n'\rangle$, the energy released is shared by an electron positron and an antineutrino or neutrino respectively. Thus the specification of the final state $|\psi_f\rangle$ has to include the energies of both leptons as well as their directions. Since it is very inconvenient to measure neutrinos, the charged lepton's energy ε is the most useful variable. The spins of both leptons also have to be specified. Fermi's golden rule in lowest-order perturbation theory is thus

$$d^6\Gamma_{ew} = 2\pi|\langle\psi_f|\Im_{ew}|\psi_i\rangle|^2 \rho_\varepsilon \rho_{\varepsilon_\nu} \delta(\varepsilon + \varepsilon_\nu + E_n' - E_n)d\varepsilon d\varepsilon_\nu d\Omega_e d\Omega_\nu \tag{10.4.15}$$

where ε and ε_ν are the lepton energies, and $\rho_\varepsilon, \rho_{\varepsilon_\nu}$ are their densities of states per unit energy and angle in the normalization volume Ω_r

$$\rho_\varepsilon d\varepsilon = \frac{p(\varepsilon)^2 dp}{(2\pi\hbar)^3}\Omega_r = \frac{p(\varepsilon)\varepsilon d\varepsilon}{c^2(2\pi\hbar)^3}\Omega_r. \tag{10.4.16}$$

where $\vec{p}$ is the lepton's momentum. Integrating over the neutrino's energy ε_ν and angle Ω_ν, we find the rate

$$d^3\Gamma_{ew}/\hbar = \frac{\Omega_r^2}{(2\pi\hbar c)^6}(E_n - E_n' - \varepsilon)^2\varepsilon\sqrt{\varepsilon^2 - m_e^2 c^4}d\varepsilon d\Omega_e\frac{2\pi}{\hbar}\int d\Omega_\nu|\langle\psi_f|\Im_{ew}|\psi_i\rangle|^2.$$

(10.4.17)

While this expression applies for arbitrary spin projections of the leptons, the lepton current (10.4.8a) actually implies that the neutrino has negative spin projection along its momentum (positive for antineutrino), because for massless fermions $\frac{1}{2}(1 \pm \gamma_5)$ projects onto spin projection $\pm\hbar/2$ along the momentum ($\mp\hbar/2$ for antifermions). In fact the whole interaction contains the factor $(1 - \gamma_5)$ on leptons, and therefore only one polarization of neutrino can interact. As far as we know, nature contains only left-handed neutrinos. Thus there is no need to sum over neutrino spins, although it is formally convenient on occasion.

The leptons' Dirac spinors are proportional to the normalized plane waves times 4-component spin functions $\mathcal{X}_{\vec{p}}$

$$\Phi_{\vec{p}} = \mathcal{X}_{\vec{p}}'\Omega_r^{-\frac{1}{2}}e^{i\vec{p}\cdot\vec{r}/\hbar} = \mathcal{X}_{\vec{p}}\Omega_r^{-\frac{1}{2}}\left(1 + i\vec{p}\cdot\vec{r}/\hbar - \frac{1}{2}(\vec{p}\cdot\vec{r}/\hbar)^2 + \dots\right).$$ (10.4.18)

Since the nuclear wave functions restrict the spatial integrations to the nuclear volume, the plane waves are well approximated by a constant when $pR/\hbar << 1$, leading to the requirement, eq.(10.3.46), that the lepton energies be much less than 25 MeV for the heaviest nuclei, a condition that is fulfilled in all practical cases. However, while the electron's wave function is a constant inside the nuclear volume, its magnitude will be affected by the Coulomb field of the nucleus. We shall ignore this effect for now and return to consider it in Sect. 10.4.D.

In the limit of long lepton wavelengths, the matrix element in eq. (10.4.17) simplifies greatly; it becomes

$$\Omega_r\langle\psi_f|\Im_{ew}|\psi_i\rangle \approx -\frac{G_F}{\sqrt{2}}\{\overline{\mathcal{X}}_e t_+\gamma_0(1 - \gamma_5)\mathcal{X}_\nu\langle n'|t_-^N|n\rangle + \overline{\mathcal{X}}_e t_-\gamma_0(1 - \gamma_5)\mathcal{X}_\nu\langle n'|t_+^N|n\rangle$$

$$+\frac{g_A}{g_V}\sum_i[\overline{\mathcal{X}}_e t_+\gamma^i(1 - \gamma_5)\mathcal{X}_\nu\langle n'|\frac{2s_i}{\hbar}t_-^N|n\rangle + \overline{\mathcal{X}}_e t_-\gamma^i(1 - \gamma_5)\mathcal{X}_\nu\langle n'|\frac{2s_i}{\hbar}t_+^N|n\rangle]\}.$$

(10.4.19)

We see that the Fermi matrix elements are just the matrix elements of the isospin raising or lowering operator, while the Gamow-Teller matrix elements have an additional operator, the nucleon's spin. Thus, for long lepton wavelengths, the Fermi (vector) interaction leads only to transitions between nuclear states of the same spin and parity. In the same limit, the Gamow-Teller (axial-vector) interaction can cause transitions between states whose angular momenta differ by at most one, because the nucleon's spin $\vec{s}$ is a vector operator and therefore the Wigner-Eckart theorem tells us its matrix elements are proportional to Clebsch-Gordon coefficients coupling the initial and final nuclear states to angular momentum $1\hbar$. If the nuclear states have the same spin and parity, then both Fermi and Gamow-Teller matrix elements have to be added together, unless the nucleus' angular momentum is zero, in which case the Gamow-Teller part necessarily vanishes.

For the Fermi interaction, the lepton spin functions are easy to work out. Choosing therefore for simplicity the case of a pure Fermi transition, the rate (10.4.17) becomes in the long wavelength limit

$$d^3\Gamma^F_{ew}/\hbar = \frac{G^2_F}{(2\pi)^4(\hbar c)^6\hbar}(E_n - E'_n - \varepsilon)^2\varepsilon\sqrt{\varepsilon^2 - m_e^2 c^4}$$
$$\cdot(1 \pm \beta_e)|\langle n'|t^N_\pm|n\rangle|^2 d\varepsilon d\Omega_e \tag{10.4.20}$$

where $c\vec{\beta}_e$ is the electron or positron's velocity, and its spin projection along its velocity, the helicity, is $\pm\frac{1}{2}\hbar$. The fact that the rate depends on the electron's helicity shows that parity is not conserved, since the expectation value of $\vec{p}_e \cdot \vec{s}_e$, an odd-parity quantity, is different from zero.

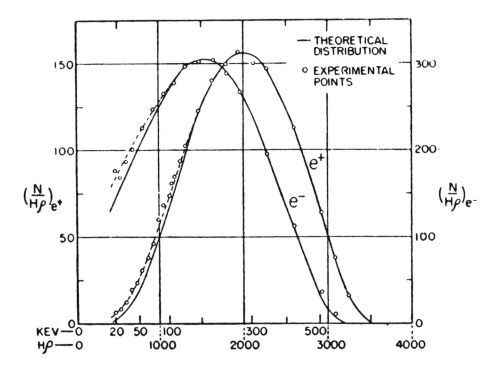

Figure 10.11   Spectrum of electrons and positrons emitted in the beta decay of ^{64}Cu [from C. S. Wu and R. D. Albert, Phys. Rev. 75 (1949) 315].

For the Gamow-Teller interaction, the lepton spin functions are quite complicated, and also lead to polarization of the charged lepton. An easier quantity to observe, however, is the correlation of the lepton's direction of motion with the nuclear angular momentum. When the initial state of the nucleus has its angular momentum polarized along a given axis, then the rate of decay, summed over electron and neutrino polarizations, neutrino direction, and final nuclear spin projection, is in the limit of long electron wavelengths [see Preston]

$$
\begin{aligned}
d^3\Gamma_{ew}/\hbar = {} & \frac{G_F^2(E_n - E_{n'} - \varepsilon)^2 \varepsilon \sqrt{\varepsilon^2 - m_e^2 c^4}}{(2\pi)^4 (\hbar c)^6 \hbar} \left\{ |\langle n'|t_\pm^N|n\rangle|^2 + \frac{g_A^2}{g_V^2} |\langle \vec{n}'||\frac{2\vec{s}}{\hbar} t_\pm^N||n\rangle|^2 \right. \\
& \mp \mathcal{P}\cos\theta\beta_e \frac{g_A^2}{g_V^2} \left[|\langle n'||\frac{2\vec{s}}{\hbar} t_\pm^N||n\rangle|^2 \lambda_{J_i J_i} \right. \\
& \left. \left. \pm 2\sqrt{\frac{J_i}{J_i+1}} \delta_{J_i J_i} \frac{g_V}{g_A} |\langle n'|t_\pm^N|n\rangle| \cdot |\langle n'||\frac{2\vec{s}}{\hbar} t_\pm^N||n\rangle| \right] \right\} d\Omega_e d\varepsilon
\end{aligned}
$$

$$(10.4.21)$$

where $\mathcal{P}$ is the fraction of polarization, θ the angle of the electron with respect to the axis of polarization, the reduced matrix elements are defined for an operator $\mathcal{O}_\nu$ of helicity component ν by

$$\langle n'|\mathcal{O}_\nu|n\rangle = \langle n'||\mathcal{O}||n\rangle \langle J_{n'} M_{n'} 1\nu|J_n M_n\rangle, \qquad (10.4.22)$$

and

$$\lambda_{IJ} = \tfrac{1}{2}[2J^2 + J + 1 - I(2J+1) + \delta_{IJ}]/(J+1). \qquad (10.4.23)$$

In this case, the violation of parity is shown by the term proportional to $\mathcal{P}\cos\theta$: different numbers of electrons are emitted along the nucleus' spin than opposite to it. For an unpolarized nucleus, the spin-averaged decay rate is independent of the electron's angle.

The spectrum of the emitted electrons or positrons, integrated over angle, is, still in the long-wavelength limit and ignoring the distortion of the electron's wave function by the Coulomb field,

$$\frac{d\Gamma_{ew}}{d\varepsilon} = \frac{G_F^2 \mathcal{M}_{ew}^2}{4\pi^3 (\hbar c)^6} \varepsilon (E_n - E_{n'} - \varepsilon)^2 \sqrt{\varepsilon^2 - m_e^2 c^4} \qquad (10.4.24)$$

where

$$\mathcal{M}_{ew}^2 = |\langle n'|t_\pm^N|n\rangle|^2 + \frac{2}{\hbar} \frac{g_A^2}{g_V^2} |\langle n'||\vec{s} t_\pm^N||n\rangle|^2 \qquad (10.4.25)$$

characterizes the nuclear states. The shape of the spectrum depends only on the maximum beta energy $E_{max} = E_n - E_{n'}$. Figure 10.11 demonstrates the excellent agreement of the theoretical and measured spectra.

The assumption in eq. (10.4.18) of plane waves is unfortunately not very accurate for the β-particles, which interact with the residual nucleons via the long-range Coulomb interaction. The necessary modification leads to a correction factor $F(Z, p_e)$

in the β-spectrum in eq. (10.4.24). In a non-relativistic treatment, the function can be estimated by the probability density of the electron or positron at the nucleus [Messiah vol. 1 app. B],

$$F(Z, p_e) = \frac{\delta}{e^\delta - 1} \tag{10.4.26}$$

$$\delta = \pm 2\pi Z \alpha \frac{\varepsilon}{c p_e} \tag{10.4.27}$$

where α is the fine structure constant, Z is the charge of the daughter nucleus, p_e is the momentum of the electron or positron, and the plus sign applies to e^+ decay and the negative sign to e^- decay.

The β spectrum is now obtained from eq. (10.4.24) by multiplying by $F(Z,p_e)$. If the quantity $((d\Gamma_{ew}/d\varepsilon)/(Fp_e\varepsilon))^{1/2}$ then is plotted as a function of ε, the result should be a straight line intersecting the abscissa at E_{max} (Figure 10.12). This is called a Fermi-Kurie plot. Its shape depends on two assumptions: the constancy of the matrix element in eq. (10.4.17), and the vanishing of the neutrino's mass. The neutrino is expected to have a small mass, which would affect the shape of the Kurie plot near the endpoint of maximum electron energy where the neutrino's kinetic energy is least. Careful measurements of the shape of this part of the spectrum have led to the limit quoted in table 1.1. Small deviations from the straight line are seen, but they are traced to the energy dependence of the matrix element, which depends on both nuclear and (for high precision) atomic wave functions.

The total decay rate cannot be found analytically with the Coulomb correction included. Equation (10.4.24) has to be multiplied by F and integrated over ε leading to the inverse lifetime. Traditionally the dimensionless function f, proportional to the decay rate, is defined by

$$f(Z, E_{max}) = \int \left(\frac{E_{max} - \varepsilon}{m_e c^2}\right)^2 F(Z, p_e) \left(\frac{p_e}{m_e c}\right)^2 \frac{dp_e}{m_e c} \tag{10.4.28}$$

and its product with the half-life, called the "ft"-value ($t_{1/2}$ is the half-life) then is found to be

$$ft_{1/2} = \frac{4\pi^3 (\hbar c)^6 \hbar \ln 2}{G_F^2 (m_e c^2)^5 \mathcal{M}_{ew}^2} \tag{10.4.29}$$

Thus the "ft"-value is a measure of the nuclear overlap matrix element when the coupling constant is known.

The experimental "ft"-values vary over many orders of magnitude and therefore $\log_{10}(ft)$ is often used with $t_{1/2}$ given in seconds. Then the $\log_{10}(ft)$ values range from about 3 to 23. The smallest values can be expected to result from the largest nuclear overlap which is unity. The estimate of G_F is then about 10^{-4} MeVfm^3. The corresponding result for a typical strong interaction matrix element is at least seven orders of magnitude larger.

It is clear that the matrix elements may vanish leading to infinite lifetime. In this case it was not justified to omit all the higher terms of Eq. (10.4.18). The next term changes the overlap into a matrix element of r which then may be different from zero. If we continue until a non-vanishing term is reached, the transitions classified in this way are denoted allowed, first forbidden, second forbidden, etc. For each step the matrix element contains an extra factor of kR, which depending somewhat on nucleus and β-particle energy, typically is about 10^{-2}. Thus the $\log_{10}(ft)$ values increase by around 4 for each step in forbiddenness.

10.4 C.   Electron Capture

Instead of emitting a positron, a nucleus may capture an electron to turn a proton into a neutron as in fig. 10.4. The decay rate is still given by Fermi's golden rule in the form

$$\frac{d\Gamma_{ec}}{\hbar} = \frac{2\pi}{\hbar} |\langle \psi(Z-1)\Phi_\nu |\Im_{ew}| \psi(Z)\Phi_e \rangle|^2 \rho_\nu d\Omega_\nu \tag{10.4.30}$$

where the neutrino's density of states ρ_ν is given by Eq. (10.3.40). Φ_ν is approximated by the first term in eq. (10.4.18) and the electron wavefunction at the origin [Messiah vol. 1 app B] is approximately

$$|\Phi_e(r = 0)|^2 \approx 4 \left(\frac{m_e c}{\hbar}\right)^3 \left(\frac{\alpha Z}{n}\right)^3 \tag{10.4.31}$$

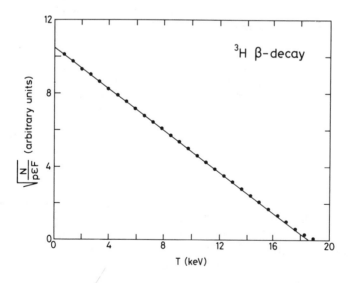

Figure 10.12   Fermi-Kurie plot of ^{3}H beta decay. Every tenth data point is plotted. The straight line is drawn to guide the eye. The data are from J. J. Simpson, private communication.

where n is the principal quantum number of the electron orbit. Integrating over the neutrino variables, we find

$$\frac{\Gamma_{ec}}{\hbar} = G_F^2 \mathcal{M}_{ew}^2 \frac{2m_e^3}{\pi\hbar^7} \left(\frac{\alpha Z}{n}\right)^3 \varepsilon_\nu^2 \qquad (10.4.32)$$

where the neutrino's energy is determined by energy conservation. Since e^+-decay always occurs in competition with electron capture, we can compare the decay rates in eq. (10.4.32) and eq. (10.4.29), i.e.

$$\frac{\Gamma_{ec}}{\Gamma_{e^+}} = 8\pi^2 \left(\frac{\alpha Z}{n}\right)^3 \left(\frac{\varepsilon_\nu}{m_e c^2}\right)^2 \frac{1}{f} \qquad (10.4.33)$$

This prediction is in fair agreement with the measured branching ratio.

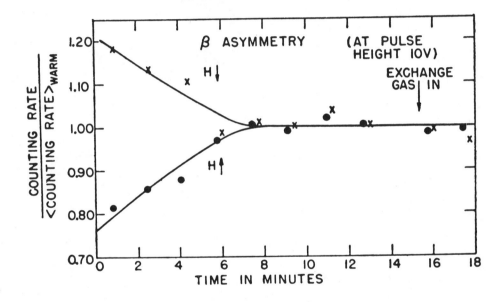

Figure 10.13   Results of β asymmetry from the polarized ^{60}Co experiment. The opposite changes of counting rate with magnetic field demonstrate the violation of parity. The polarization decreases with time as the nuclear spins come into thermal equilibrium with their surroundings. [*From C. S. Wu, E. Ambler, R. Hayward, D. Hoppes, and R. Hudson, Phys. Rev.* **105** *(1957) 1413*]

10.4 D.   Experimental Discovery of Parity Violation

We saw in eq. (10.4.21) that an allowed Gamow-Teller decay of a spin-polarized nucleus would lead to a correlation of the direction of the emitted β- particle with the direction of the nucleus' spin. This shows that parity is not a symmetry in nature: viewed in a mirror, the correlation would appear to have the opposite sign.

In 1956, C. S. Wu and her collaborators demonstrated this experimentally in the e^- decay of ^{60}Co, whose ground state of $J^\pi = 5^+$ decays into a 4^+ state of ^{60}Ni through a pure Gamow-Teller transition [see Wu and Moszkowski]. A very cold sample of ^{60}Co, at about 0.01 Kelvin, was placed in a strong magnetic field. The magnetic field aligned the spins of the cobalt atoms' electrons, whose magnetic field then aligned the spins of the nuclei. As they decayed, they emitted electrons preferentially in a direction opposite to the angular momentum, as predicted by eq. (10.4.21). Reversing the magnetic field changed the number of electrons detected by a detector placed along the axis of the magnetic field near the sample of ^{60}Co, as shown in fig. 10.13.

The discovery that parity is not a symmetry of nature was one of the most interesting contributions of nuclear physics. The nucleus continues to provide a useful system for precise studies of the electroweak interaction. Many other nuclear experiments have helped to prove the form of the electroweak interaction (10.4.2), which was determined more than a decade before the electroweak bosons were produced in high-energy particle accelerators.

PROBLEMS

10.1   Assume a compound nucleus is being formed by successive collisions initiated by an incoming neutron of given kinetic energy less than 100 MeV. Estimate the number of collisions of the incoming neutron before its extra kinetic energy is shared roughly equally between all the nucleons. Estimate then the total time of this equilibration process.

10.2   Compute approximately $\ln Z(\alpha, \beta)$ for the level spectrum $\hat\varepsilon_k = \varepsilon_0 + kd$, $k = 0, 1, 2, \ldots$. Each level can hold f particles (the degeneracy). Use the Euler-Maclaurin expansion for a discrete function g_k

$$\sum_{k=0}^{\infty} g_k = \int_0^\infty g(k)dk + \frac{1}{2}[g(0) + g(\infty)] + \frac{1}{12}[g'(\infty) - g'(0)] + \ldots$$

and neglect terms of order $e^{-\alpha + \beta\varepsilon_0}$.

Solve the saddlepoint equations and derive the level density as function of f, d and β.

Calculate the ground state energy E_0 for A particles and a fraction b $(0 \le b \le 1)$ of the full degeneracy f in the last occupied level.

Express the level density in terms of the excitation energy $E^* = E - E_0$, d, f, and b, and discuss its dependence on these quantities, especially on b.

10.3   Formulate in detail a numerical procedure to calculate the level density of eq. (10.1.9) for any given single-particle spectrum $\hat{\varepsilon}_k$ without any continuous-limit assumptions.

10.4   Derive the one-dimensional transmission coefficient for a particle hitting a potential wall of height $|U_0|$. Compare to eq. (10.2.26).

10.5   Investigate the global neutron-proton evaporation competition given by eq. (10.2.41) along the β-stability line and along lines of constant nucleon number.

10.6 ^{226}Ra decays by emission of three successive α particles

$$^{226}\text{Ra} \rightarrow\ ^{222}\text{Rn} \rightarrow\ ^{218}\text{Po} \rightarrow\ ^{214}\text{Pb}$$

Emission of a ^{12}C particle is favored since the mass of ^{12}C is less than the mass of three α-particles. Make rough estimates of the relative probability for ^{12}C and α decay of ^{226}Ra and discuss the reason for the observed decay mode. (^{12}C decay has not been observed, but ^{14}C was recently observed from ^{223}Ra with a probability of about 10^{-9} [*H. Rose and G. Jones, Nature* **307** *(1984) 245*], and other heavier products have also been found since.)

10.7   Increasing the excitation energy of a heavy nucleus allows evaporation of an increasing number of neutrons. Assume schematically that all fission barrier B_f and neutron separation energies S_n are equal to 6 MeV. Estimate roughly (eqs. (10.2.52), (10.2.53), (10.1.21), and (10.1.50)) and sketch the total fission "width" as function of excitation energy. Use an average kinetic energy for each emitted neutron and ignore all other channels.

10.8   Show eq. (10.1.31).

10.9   Use the wall formula, eq. (10.2.63), to compute the damping time for the β and γ deformation degrees of freedom as functions of the axially-symmetric equilibrium deformation. Use this result to estimate the width of the giant quadrupole resonance in heavy nuclei.

10.10   Show eq. (10.3.35) by using

$$L_z Y_\lambda^\mu = \mu Y_\lambda^\mu$$
$$L_\pm Y_\lambda^\mu = \sqrt{(\lambda \mp \mu)(\lambda + 1 + \mu)} Y_\lambda^{\mu \pm 1}$$

where

$$L_\pm = L_x \pm i L_y$$

10.11   Show eq. (10.3.36) by using the recurrence relations for the Bessel functions

$$\frac{d}{dx}j_\lambda(x) = j_{\lambda-1}(x) - \frac{\lambda+1}{x}j_\lambda(x) = -j_{\lambda+1}(x) + \frac{\lambda}{x}j_\lambda(x)$$

and the identities (see Edmonds eqs. (5.9.16) and (5.9.17))

$$\sqrt{2\lambda+1}\ \vec{r}Y_\lambda^\mu(\theta,\phi) = r\sqrt{\lambda}\vec{\Phi}_{\lambda-1,\lambda}^\mu - r\sqrt{\lambda+1}\vec{\Phi}_{\lambda+1,\lambda}^\mu$$

$$\sqrt{2\lambda+1}\ \vec{\nabla}[f(r)Y_\lambda^\mu(\theta,\phi)] = \sqrt{\lambda}\left(\frac{d}{dr} + \frac{\lambda+1}{r}\right)f(r)\vec{\Phi}_{\lambda-1,\lambda}^\mu$$

$$- \sqrt{\lambda+1}\left(\frac{d}{dr} - \frac{\lambda}{r}\right)f(r)\vec{\Phi}_{\lambda+1,\lambda}$$

10.12   An even-even nucleus has an excited state that decays to a lower excited state and then to the ground state in two successive electric quadrupole transitions. What are the possible spins and parities of these two excited states?

10.13   A hypothetical even-even nucleus with A=150 has its lowest excited-state rotational band with intrinsic angular momentum K=2 and excitation energy $E^*(K=2)=2$ MeV. Both the ground-state band and the K=2 band have a moment of inertia equal to 0.6 times the value of a rigid sphere.
(a) Find the ratio of probabilities for decaying by quadrupole radiation from the J=4 state of the K=2 band to the J=6, J=4, and J=2 states of the ground-state band.
(b) The branching ratio from the J=4 state of the K=2 band to its own J=2 state is $B_{42} \ll 1$, since the energy of this transition is so much less than the energy of the transitions to the ground-state band. Estimate the branching ratio B_{64} for decay of the J=6 state of the K=2 band to its own J=4 state.

10.14   The lithium nucleus ^{7}Li emits a 0.48 MeV gamma ray in a transition from an excited state of angular momentum and parity $\frac{1}{2}^-$ to the ground state with angular momentum and parity $\frac{3}{2}^-$. What kind of transitions are possible? Which of these is likely to dominate the emitted radiation? Estimate the lifetime of the excited state by using harmonic-oscillator wave functions. Compare to the Moszkowski estimate.

10.15   Modify eq. (10.4.17) to allow for a finite mass m_ν of the neutrino. Show that the spectrum is sensitive to m_ν only at the upper end of the spectrum.

10.16   Calculate the β-decay lifetime for the free neutron.

10.17   Calculate Γ_γ^W for an E1 transition of energy 0.78 MeV within a radius of 1 fm. Compare to the observed β-decay lifetime of the free neutron.

11

Collisions of Heavy Nuclei

Many modern accelerators can produce high-velocity beams of complex nuclei. They are true "atom-smashers," capable of causing dramatic nuclear reactions like the one in the photograph on the cover of this book. These fast, heavy nuclei, which often retain some of the atomic electrons, are commonly referred to as **heavy ions**.

A collision of a heavy-ion beam with a target nucleus is likely to have more violent consequences than the mild perturbations caused by the electroweak and few-nucleon probes discussed above. The collision shown on the cover has dozens of charged particles in the final state, while the processes we have studied in earlier chapters could never lead to more than three or four. Just as the gentler probes revealed the fundamental nuclear degrees of freedom at low excitation energy, heavy-ion collisions can show what happens to nuclear matter under extreme conditions. For example, the nuclei with high angular momentum discussed in Chapter 9 were produced by collisions of heavy nuclei.

In this chapter, we will see how collisions of nuclei with increasing velocities lead to increasingly excited states of nuclear matter. We will use concepts of macroscopic physics to describe phenomena like phase transitions and hydrodynamic flow. We have already become accustomed to interpreting our microscopic quantal descriptions of nuclei in the language of macroscopic physics: mean free path, uniform nuclear matter, superfluidity, collective normal modes, temperature. The microscopic theory of heavy-ion reactions has not yet progressed far enough to provide a detailed description of all the very complicated observations. Nevertheless, a reasonably coherent picture has emerged which enables us to begin to study nuclear matter under extreme conditions somewhat like those in a collapsing supernova or in the early universe.

We begin with the most straightforward collisions, where the relative velocity of the colliding nuclei is slower than the Fermi velocity of their nucleons. These collisions are closely related to phenomena we have already studied, including stripping and pickup reactions, Coulomb excitation, and fission. As the relative velocity approaches the Fermi velocity, we explore the breakdown of the mean-field picture when density fluctuations develop into a mixed liquid-gas phase of nuclear matter. Still higher collision velocities lead to compression of the nuclear matter, with increasing numbers of thermally- excited pions. We tell why collisions of ultrarelativistic nuclei may lead to even more exotic forms of matter in which the nucleons and mesons dissolve into a soup of their constituent quarks and gluons. By the end of this chapter we will hazard a guess at a phase diagram for nuclear matter.

11.1   COLLISIONS NEAR THE COULOMB BARRIER

Before anything more interesting than elastic scattering can happen, enough energy must be supplied for the relative motion of the projectile and target nuclei to overcome the Coulomb barrier U_C of electrostatic energy which holds them apart. For a projectile and a target with masses and charges of (A_a, Z_a) and (A_A, Z_A) respectively, the Coulomb barrier may be estimated by

$$U_C \approx \frac{Z_A Z_a e^2}{R(A_A) + R(A_a)} \approx \frac{Z_A Z_a}{A_A^{1/3} + A_a^{1/3}} \frac{e^2}{r_0} \tag{11.1.1}$$

with r_0 taken from the strong-interaction radius of the black-sphere model as approximately 1.3 fm. Typical values of U_C range from 14 MeV for ^{16}O on ^{16}O, through 86 MeV for ^{16}O on ^{208}Pb, up to 630 MeV for ^{208}Pb on ^{208}Pb, corresponding to projectile energies in the laboratory of about 28 MeV, 93 MeV and 1260 MeV respectively, necessary to produce a non-trivial reaction cross section.

These energies cannot be obtained by the electromagnetic acceleration of singly-charged ions. An electrostatic accelerator, such as a van de Graaf machine, is limited to terminal voltages of less than 10 million volts before its insulation breaks down. A 56 MeV singly-charged oxygen ion in the magnetic field of a saturated ferromagnet, which is about 2 tesla, moves in a circle of over 4 m diameter; only a few cyclotrons this large have ever been built. The need for multiply-charged ions is clear. Several electrons can be removed by an electric arc in a sputtering process (a "PIG" source), a dozen or more by an intensely excited plasma of electrons (an "ECR") gyrating in a magnetic field at their cyclotron resonance frequency $\omega_c = eB/m_e$. To obtain more highly charged ions, the partially-ionized atoms may be accelerated in a van de Graaf, cyclotron, or linear travelling-wave accelerator (LINAC), then sent at high velocity through a "stripper" foil of material to remove additional electrons. These highly-charged ions may then be further accelerated until the desired energy is obtained. For very heavy ions, the stripping-acceleration process may proceed through several stages. Since the radius of curvature in a magnetic field depends on the charge-to-mass ratio and the velocity, the capabilities of a heavy-ion cyclotron or (for relativisitic velocities) synchrotron are usually specified in terms of the **energy per nucleon** for maximally-stripped ions. When we speak of collisions near the Coulomb barrier, we mean that the ions' relative motion contains only a few MeV per nucleon.

11.1 A.   Degrees of Freedom for Low-velocity Collisions

While huge total excitation energies are available in heavy-ion collisions at a few

MeV per nucleon, the excitation energy per nucleon remains much smaller than the kinetic energy of the nucleons' internal motion. Thus the system is still highly degenerate. The degree of degeneracy of a Fermi system is characterized by its entropy per fermion. From eqs. (10.1.22) and (10.1.48) we have, in the Fermi gas model,

$$\mathcal{S}/A = \pi\sqrt{\frac{E^*/A}{\mu - U_0}} \approx \sqrt{\frac{E^*/A}{3 \text{ MeV}}}. \tag{11.1.2}$$

We see that the entropy per nucleon will be small, and the system degenerate, when the excitation energy per nucleon is small compared to 3 MeV.

As long as the nucleons are strongly degenerate, the Pauli principle will keep their mean free path long, and nuclear dynamics should be dominated by the mean field as we saw in Chapter 4. The interplay between collective and single-particle degrees of freedom will be similar to that described in Chapter 8. There we saw that there would be slow collective motion governed by equations whose coefficients—inertia, static force, friction—are determined by the single-particle motion in the mean field. The entire discussion of Chapter 8 can be carried through at finite temperature with only minor modifications. Instead of tracing over the ground state, expectation values have to be evaluated in an ensemble determined by the amount of excitation energy which has been dissipated from the collective motion into the independent-particle degrees of freedom.

Our aims are therefore very similar to those we had in Chapter 8: to identify the collective degrees of freedom, to understand their dynamic interplay, and to see how they relate to observations on the one hand, to the nucleons' underlying motion on the other.

We have already identified one important collective variable, namely the relative coordinate $\vec{R}$ of the centers of mass of the target and projectile. For collisions near the Coulomb barrier, $\vec{R}$ is the most important collective variable. It is used to classify collisions into several types, depending on its time evolution. If the nuclear surfaces never touch, we have either **elastic scattering** (ES) or **Coulomb excitation** (CE). If the surfaces barely touch, we have a **grazing collision** (GC). If the nuclei make firm contact with each other, but still separate afterwards, we have a **deeply inelastic collision** (DIC). Finally, if the nuclei become sufficiently attached to each other, they may form a compound nucleus in a **fusion reaction** (FR). If the angular momentum and total charge are both very large, this combined nucleus may not live long enough to reach its equilibrium shape before it fissions into two or more large fragments in **incomplete fusion** (IF). Of course, in all these collisions except ES, other collective degrees of freedom also come into play; in the FR and IF processes, $\vec{R}$ eventually loses not only its dominant role but even its meaning as the nuclear state evolves in time until the projectile and target nuclei can no longer be distinguished.

Before we proceed to brief descriptions of these reaction types, we pause to check whether the implied classical picture of localized wave packets is reasonable. After all, in collisions of low-velocity nucleons with nuclei we could not get far with such a picture. The large masses of the ions makes their wave packets easier to localize. The spread of energies in a wave packet of width ΔR may be estimated by

$$\Delta E = \frac{1}{2m_{tot}}\Delta p^2 = \frac{(\hbar/2\Delta R)^2}{2m_N A} \approx \frac{5 MeV\ fm^2}{A(\Delta R)^2}. \qquad (11.1.3)$$

Thus even an ion as light as oxygen may be localized to within 0.2 fm by a wave packet with an energy spread of 0.5 MeV per nucleon. For heavier ions the energy spread per nucleon decreases like A^2. We conclude that a classical description of the relative motion is appropriate.

11.1 B.   Elastic Scattering and Coulomb Excitation

At energies well below the Coulomb barrier, elastic scattering and Coulomb excitation predominate. Since the probability of Coulomb excitation remains quite small until the energy approaches the barrier, the elastic scattering is well described by Rutherford's formula

$$\frac{d\sigma_R}{d\Omega} = \left(\frac{Z_a Z_A e^2}{2\mu_\alpha v^2}\right)^2 \frac{1}{\sin^4\theta/2}, \qquad (11.1.4)$$

which is valid both in classical mechanics and in quantum mechanics for spinless, non-identical nuclei with initial relative velocity v and reduced mass $\mu_\alpha = m_N A_a A_A/(A_a + A_A)$ in the notation of Sect. 4.1.C. Other collective degrees of freedom come into play when the Coulomb field of one nucleus excites rotational or vibrational normal modes of the other nucleus. We discussed the computation of Coulomb excitation probabilities in Sect. 8.1.C.

As the projectile's energy is increased, its orbit may take it closer to the target nucleus. When its energy is great enough, it may be deflected also by the strong-interaction force between the nuclear surfaces. Then the Rutherford cross section (11.1.4) will no longer describe the data. From this argument applied to measurements of the scattering of radioactively produced alpha particles impinging upon a gold foil, Rutherford discovered that the mass of atoms was concentrated in nuclei whose radius was less than 3×10^{-12} cm.

If the projectile's energy is large enough to surmount the Coulomb barrier, inelastic processes play a large role. If the angular momentum of the relative motion is small enough for the nuclear surfaces to touch, the strong-interaction forces come into play. The distance of closest approach R_L for a Coulomb trajectory of angular momentum $L\hbar$ and energy E is given implicitly by

$$E = \frac{L(L+1)\hbar^2}{2\mu_\alpha R_L^2} + \frac{Z_a Z_A e^2}{R_L}. \qquad (11.1.5)$$

The value of L for which R_L is equal to the sum of the nuclear radii $r_0(A_a^{1/3}+A_A^{1/3}) \equiv R_{gr}$ is called the **grazing angular momentum** L_{gr}. It is given by

$$L_{gr}(L_{gr} + 1) = (E - U_C)2\mu_\alpha r_0^2 (A_a^{1/3} + A_A^{1/3})^2/\hbar^2. \qquad (11.1.6)$$

The total cross section for reactions dominated by the strong interaction may be estimated by adding the contributions to the absorptive cross section, eq. (3.3.5), from angular momenta less than L_{gr},

$$\sigma_{geom} = \frac{\pi}{k^2} \sum_{0}^{L_{gr}} (2L + 1) = \frac{\pi}{k^2}(L_{gr} + 1)^2$$

$$\approx \pi r_0^2 (A_a^{1/3} + A_A^{1/3})^2 (1 - U_C/E).$$

(11.1.7)

This is called the geometric cross section.

Trajectories with angular momenta larger than L_{gr} will still lead to Coulomb scattering. Trajectories with angular momenta near L_{gr} are likely to lead to inelastic reactions, but they may also lead to elastic scattering. The elastic scattering angle corresponding to these trajectories will be reduced by the attractive nuclear force between the target and projectile. Thus there is, classically, a maximum angle of deflection θ_r corresponding to a Coulomb trajectory with angular momentum just larger than L_{gr}. For larger L, the deflection angle is smaller because the Coulomb trajectories keep the nuclei farther from each other; for smaller L, the deflection angle is reduced because the attractive nuclear force pulls the nuclei together, reversing the deflection due to the Coulomb force. θ_r is called the **rainbow angle**. In classical mechanics, the differential cross section becomes infinite at θ_r, because many classical trajectories lead to the same scattering angle,

$$\frac{d\sigma}{d\theta}(classical) = \frac{d\sigma}{dL}\frac{dL}{d\theta_{cl}} = \frac{d\sigma}{dL}\left(\frac{d\theta_{cl}}{dL}\right)^{-1}.$$

(11.1.8)

Since the flux of incident particles is smoothly distributed in the angular momentum L, $d\sigma/d\theta$(classical) diverges where $d\theta_{cl}/dL$ vanishes, i.e., at the rainbow angle. In quantum mechanics, only a few partial waves contribute to the scattering near θ_r, so the differential cross section remains finite, but it may have a peak at θ_r, see fig. 11.1.

The inelastic excitation of collective modes in the target or projectile is also strongly affected by the nuclear forces for scattering angles near θ_r. Here, the external field provided by the passing of the other nucleus changes sign as the repulsive Coulomb force is supplanted by the attractive nuclear field. Thus the differential cross section for inelastic excitation has a minimum near the maximum of the elastic cross section, as one sees in fig. 11.1.

For angular momenta less than L_{gr}, the amplitude for elastic scattering, or for simple inelastic excitation of individual collective modes, becomes small due to competition with other, more complicated reactions.

11.1 C.   Grazing Collisions

The main reaction process for collisions near the grazing angular momentum is the transfer of nucleons between target and projectile. This process begins at even larger angular momenta, but near L_{gr} it dominates the reaction amplitude. Nucleon transfer is accompanied by the transfer of energy from the relative motion into the internal motion of the target and projectile. The amount of energy transferred is determined by kinematic considerations, which may be understood in a simple classical picture. We will, for simplicity, discuss transfer from projectile to target; of course, symmetric expressions are obtained for the reverse reaction.

Consider a nucleon in the projectile moving in a classical orbit with angular momentum projection m_i perpendicular to the plane of the relative motion of the target and projectile. As the surface of the projectile grazes the target's surface, the particular nucleon we consider happens to be at the point where the surfaces overlap. In this region, the forces from the target and projectile affect the nucleon equally, so it continues its trajectory on a straight line and finds itself moving with angular momentum projection m_f in the target nucleus (see fig. 11.2). Its velocity $-m_f/m_N R(A_A)$ relative to the target nucleus differs from its velocity $m_i/m_N R(A_a)$ relative to the projectile by the relative velocity

$$v_{rel} = (2(E_\alpha - U_C)/\mu_\alpha)^{1/2} \tag{11.1.9}$$

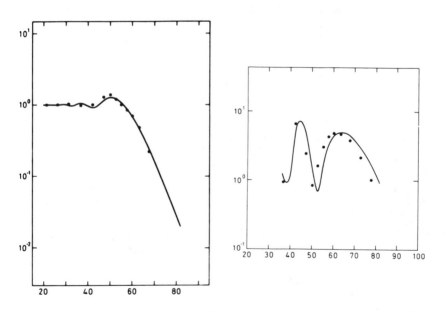

Figure 11.1   Differential cross sections for elastic scattering (left) and for inelastic excitation of the lowest 3^- state of ^{208}Pb (right) under bombardment by ^{11}B ions at 6.5 MeV per nucleon. The curves are a semiclassical approximation to the optical model and DWBA, respectively. [*From R. Broglia and A. Winther, "Heavy Ion Reactions," Benjamin, New York, 1981*]

of the target and projectile. Thus the nucleon's angular-momentum projection in the target is

$$m_{\mathrm{f}} = -\frac{R(A_A)}{R(A_a)} m_i - m_N R(A_A) v_{\mathrm{rel}}. \tag{11.1.10}$$

We see that the transferred nucleon's motion in the target is determined by its motion in the projectile together with the relative motion.

The transfer of the nucleon changes the energy of the relative motion. Since the projectile's velocity is unchanged while its mass decreases, its momentum at the moment of contact, $\vec{p}_\alpha$, is decreased by a factor $(1 - A_a^{-1})$ in addition to the loss of the transferred nucleon's internal momentum $m_i/R(A_a)$. The center-of-mass kinetic energy of the relative motion then is just the square of the projectile's momentum, divided by twice the effective mass μ_β:

$$\begin{aligned}
(\vec{p}_\beta)^2/2\mu_\beta &= \left[p_\alpha(1 - A_a^{-1}) - \frac{m_i}{R(A_a)} \right]^2 \frac{A_a + A_A}{2m_N(A_a - 1)(A_A + 1)} \tag{11.1.11} \\
&\approx (2\mu_\alpha)^{-1}(\vec{p}_\alpha)^2(1 - m_N/\mu_\alpha) - v_{\mathrm{rel}} m_i/R(A_a) + \mathcal{O}(\mu_\alpha^{-1}).
\end{aligned}$$

The change in the energy of relative motion has to match the change of the combined internal energy of the target and projectile. In the mean-field model, this change of energies is the difference of the single-particle eigenvalues ε_{f} and ε_i. Conservation of energy thus implies

$$\varepsilon_{\mathrm{f}} - \varepsilon_i \approx \tfrac{1}{2} m_N v_{\mathrm{rel}}^2 + v_{\mathrm{rel}} m_i/R(A_a) + \Delta U_C, \tag{11.1.12}$$

where ΔU_C is the change in the Coulomb energy of the relative motion. Together, the relations in eq. (11.1.10) and (11.1.12) impose a strong selectivity in the states

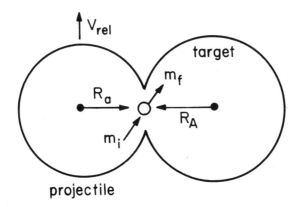

Figure 11.2   Kinematics and geometry of heavy-ion stripping reaction. See text, eqs. (11.1.9–12).

populated by grazing heavy-ion transfer reactions. These relations are known as the **Brink selection rules**. They may be derived in quantum mechanics by using the eikonal approximation for the wave functions appearing in the DWBA transition amplitude of Sect. 4.1.C. By choosing the relative velocity, various states may be preferred or suppressed, as seen in fig. 11.3. This selectivity makes grazing heavy-ion transfer reactions a useful tool for studying the structure of nuclear states.

Even when the states populated by grazing transfer reactions are too highly excited to be seen as individual peaks in the spectrum, the Brink selection rules lead to a favored energy loss, known as the **preferred Q value**. The preferred Q value follows from eq. (11.1.12) by assuming that the average initial angular-momentum projection m_i is zero, i.e., that nucleons moving forward and backward have equal chances of transferring into the other nucleus. Then the transfer of an average nucleon will cause its part of the kinetic energy of relative motion to be turned into internal energy of the target and/or projectile. The energy converted

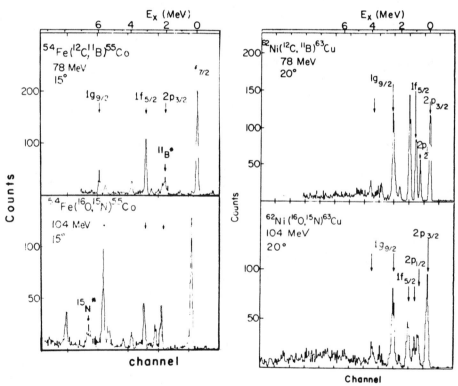

Figure 11.3   Spectra of states in ^{55}Co and ^{63}Cu populated by heavy-ion stripping reactions at different beam energies. Ex is the excitation energy. The changing ratios of different states are due to the selection rules, eqs. (11.1.9) and (11.1.12). [*from F. Becchetti, B. Harvey, D. Kovar, J. Mahoney, and M. Zisman, Phys. Rev. C10 (1974) 1846*].

in this way is the same for stripping or pickup reactions, i.e., it is independent of which direction the nucleon is transferred. If several nucleons are transferred, each one contributes equally to the average energy loss. In addition, the kinetic energy of the relative motion is diminished or enhanced by the difference in the Coulomb barriers of the incoming and outgoing channels, which is proportional to the net proton transfer $\Delta Z \equiv Z_b - Z_a$. The preferred Q value, or average change of energy of relative motion during the reaction, assuming few particles transferred compared to the masses and charges of the projectile and target, is

$$Q_{av} = E_\beta - E_\alpha = -(E_\alpha - U_C)|\Delta A|m_N/\mu_\alpha + \Delta Z(Z_A - Z_b)e^2/R_{gr} \qquad (11.1.13)$$

where $|\Delta A|$ is the total number of nucleons transferred in either direction.

Fig. 11.4 gives an example of the measured energy-loss spectrum from a grazing heavy-ion reaction, and shows how the projectile's energy loss depends on the number of transferred nucleons. We see that eq. (11.1.13) explains the dependence of the average energy loss on the number of transferred nucleons, but that there is an additional energy loss which doesn't depend on ΔA or ΔZ. This loss may be partly due to the fact that more nucleons are actually transferred than the observed net transfer: if a nucleon from the target changes places with one from the projectile there will be a net energy loss of $2 \cdot \frac{1}{2}m_N v_{rel}^2$ without increasing the apparent value of ΔA. In addition, other energy-loss mechanisms, such as the excitation of collective vibrational states, may also be present. However, the particle-transfer mechanism seems to account for most of the energy loss in grazing collisions.

11.1 D. Deeply Inelastic Collisions and Fusion

For angular momenta somewhat less than L_{gr}, more of the energy of relative motion is turned into internal excitation. The most striking experimental evidence for this is seen in the **Wilczynski plot** which displays the doubly-differential cross section $d^2\sigma_b/d\theta dE$ for the production of a particular projectile-like fragment b as a function of its energy E and deflection angle θ. Contours of equal cross section are plotted in the (E, θ) plane. We see in fig. 11.5 that, in the case of a lighter-mass projectile on a heavy target, there is a pronounced peak at $(E_{beam} - Q_{av}, \theta_{gr})$, from which a ridge extends to decreasing energy at more forward angles. Apparently the projectile slows as it swings around the target. If the angular deflection from the target's attraction is greater than the Coulomb deflection, the deflection angle becomes negative: a projectile which struck the target's right side emerges on its left. However large the deflection, a minimal amount of energy E_{min} is left in the relative motion, due to the acceleration of the Coulomb force after the nuclei separate. E_{min} is less than the Coulomb barrier for the outgoing channel Bb, because the attraction between the surfaces of the nuclei causes them to deform as they separate. Thus the centers of the nuclei are farther apart when they separate than when they first touched, leaving a smaller Coulomb energy to be gained on the way out. Because so much of the original kinetic energy is dissipated, these processes are known as **deeply inelastic collisions**.

The simplest model of deeply inelastic collisions consists of classical Newtonian equations of motion of the form derived in Sect. 8.6 for slow collective motion,

$$\mathcal{K}_\mu(\{Q_\alpha(t)\}) - \sum \gamma_{\mu\nu}\dot{Q}_\nu(t) = \sum_\nu M_{\mu\nu}\ddot{Q}_\nu(t). \qquad (8.6.6)$$

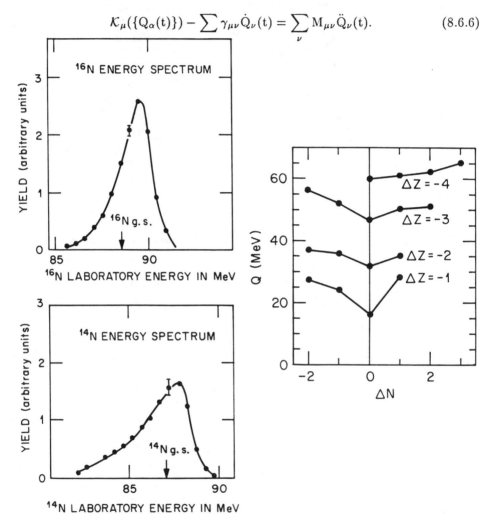

Figure 11.4   Energy loss in grazing collisions of ^{15}N on ^{232}Tl. Left: typical spectra of reaction products at 60° with beam energy 98.5 MeV. Right: average energy loss of projectile vs. change in its neutron and proton numbers, ΔN and ΔZ, at 40° with beam energy 130 MeV [*data from V. Volkov, G. Gridnev, G. Zorin, and L. Chelnokov, Nucl. Phys. A126 (1969)1, and from J. Wilczynski, private communication*].

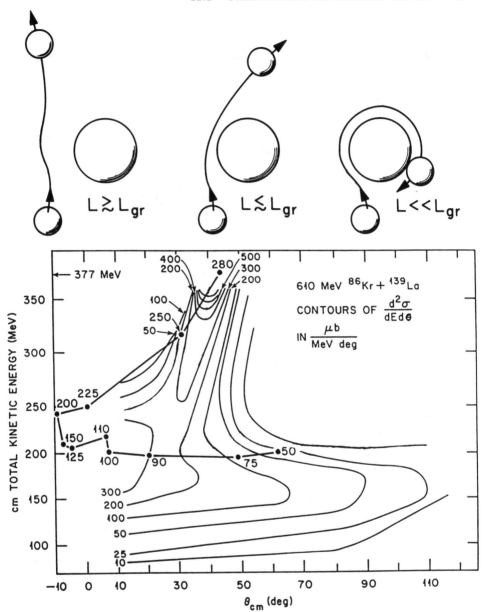

Figure 11.5 Correlation of energy loss and angular deflection for heavy ion collisions near the Coulomb barrier. Above: interpretation in terms of classical trajectories for various angular momenta. Below: Wilczynski plot gives contours of equal $d^2\sigma/d\theta dE$ for the reaction of ^{86}Kr on ^{139}La at 610 MeV laboratory energy. Broken line connects theoretical energy losses and deflection angles for various angular momentum values (solid points), for Time-Dependent Mean Field computation of fig. 11.8. [*data R. Vandenbosch, M. Webb, P. Dyer, R. Pugh, R. Weisfield, T. Thomas, and M. Zisman, Phys. Rev.* **C17** *(1978) 1672*].

Both the inertial masses $M_{\mu\nu}$ and the friction coefficients $\gamma_{\mu\nu}$ may depend on the values of the coordinates as well as the amount of internal excitation or temperature. The collective degrees of freedom need to include not only the separation of the nuclear centers of mass, but also the nuclear deformations in order to explain the small exit-channel Coulomb energy. To conserve angular momentum, the nuclear rotational degrees of freedom also have to be included. This rather large set of degrees of freedom implies many transport coefficients $\gamma_{\mu\nu}$ and $M_{\mu\nu}$ as well as a complicated static force $\mathcal{K}_\mu$ or equivalently an energy surface of which $\mathcal{K}_\mu$ is the gradient. These functions of the collective coordinates have to be adjusted to fit the data on energy-angle correlations from a variety of target-projectile combinations and initial beam energies. The observations are insufficient to constrain the interpretation to a unique set of parameters and functional shapes. Nevertheless a number of interesting conclusions may be drawn from the extensive phenomenology of deeply inelastic collisions within the Newtonian approximation.

The most remarkable results concern the friction tensor $\gamma_{\mu\nu}$. It varies rapidly with the collective coordinates, depending mainly on the distance between the nuclear surfaces where they are closest, and falling off rapidly as the surfaces are drawn apart. As the nuclei approach, their relative radial motion is damped out very rapidly, during a time when the nuclei move only one or two fm closer. The angular motion is more slowly damped, allowing the nuclei to swing around each other for up to a radian or sometimes even more before the angular momentum is shared proportionately between the relative motion and the internal rotations of the two nuclei. Apparently the radial motion is more strongly damped than the angular motion.

Eventually, if the attraction between the surfaces succeeds for a long enough time in holding the nuclei together in spite of the Coulomb and centrifugal forces, the whole system rotates almost rigidly with one angular velocity. At this point much of the original kinetic energy has disappeared from the relative motion of the nuclei: all the radial motion and part of the angular motion has been dissipated, and the remaining rotational kinetic energy is shared with the internal rotational motion of the two nuclei. The fraction of the rotational kinetic energy left in the relative motion is estimated by comparing the original moment of inertia for the centers of mass, $\mathcal{I}_{\text{initial}} = (R_a + R_A)^2 m_A m_a / (m_A + m_a)$, to the moment of inertia for rigid co-rotation of two touching spheres, $\mathcal{I}_{\text{final}} = (R_b + R_B)^2 m_B m_b / (m_B + m_b) + m_B R_B^2 / 5 + m_b R_b^2 / 5$. For equal target and projectile, $\mathcal{I}_{\text{initial}} / \mathcal{I}_{\text{final}} = 5/6$ if the masses are unchanged, while for $m_A = 10 m_a$, $\mathcal{I}_{\text{initial}} / \mathcal{I}_{\text{final}} \approx 0.5$. The fraction of the original rotational kinetic energy left at the end is $\mathcal{I}_{\text{initial}} / \mathcal{I}_{\text{final}}$. The fraction left in the relative rotational motion is $(\mathcal{I}_{\text{initial}} / \mathcal{I}_{\text{final}})^2$, which is also the factor by which the centrifugal force is reduced if the nuclei remain spherical throughout. The deformation of the nuclei causes further reductions in the rotational kinetic energy and centrifugal force of the relative motion. Evidence for the internal rotation of the excited nuclear fragments comes by observing that they emit more nucleons, alphas and gammas in the reaction plane than perpendicular to it, as expected for the decay of a spinning compound nucleus. The direction of rotation is determined by measuring the circular polarization of the emitted photons. This polarization changes sign as the energy loss increases corresponding to negative-angle scattering (fig. 11.6).

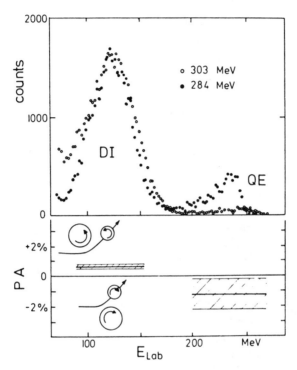

Figure 11.6   Polarization of photons from target remnants in the reaction of 284 MeV (dots) and 303 MeV (circles) ^{40}Ar on Ag. The count-rate asymmetry (PA) measures the circular polarization of the photons, shown as a function of the laboratory energy of the ejectiles. The ejectiles, whose charges range from Z = 11 to 21, are detected at 35° in the laboratory, about 10° outside the grazing angle. Their spectrum is shown in the upper part of the figure. [*from W. Trautman, J. de Boer, W. Dünnweber, G. Graw, R. Kopp, C. Lauterbach, H. Puchta, and U. Lynen, Phys. Rev. Lett. 39 (1977) 1062*].

Since the relative motion of the nuclei is so strongly damped, they may become trapped by the attractive forces between them. Thus the Newtonian orbits of low angular momentum show that the nuclei never separate. It is natural to interpret these orbits as a description of **fusion reactions**, in which the target and projectile form a compound nucleus which then decays by the mechanisms discussed in chapter 10. Indeed, the observed fusion cross sections may be quantitatively explained in this way. Often the results are parametrized in terms of a **critical angular momentum** for fusion, L_{cr}. Collisions with less angular momentum than L_{cr} lead to fusion; the others lead to deep-inelastic reactions or to grazing reactions. These three reaction types together account for the geometrical cross section. The fusion cross section is

$$\sigma_{\text{fusion}} = \frac{\pi}{k^2} L_{\text{cr}}(L_{\text{cr}} + 1). \tag{11.1.14}$$

Measurements of σ_{fusion} show that L_{cr} usually depends on which target and projectile are used, even when the mass, charge, and energy of the compound nucleus are held fixed. Thus L_{cr} is usually not a property of the compound nucleus, but reflects the reaction dynamics leading to fusion. In some cases, L_{cr} appears to be independent of the reaction channel; in those cases, it may be interpreted as the maximum angular momentum possible for a nucleus of a given excitation energy.

Another important degree of freedom in deeply inelastic collisions is the transfer of nucleons between the target and projectile. As soon as the nuclear surfaces touch, the average neutron-to-proton ratios of the target- and projectile-like fragments become equal, mainly because protons move from the lighter to the heavier nucleus. While at first it may seem paradoxical that the protons flow against the gradient of the Coulomb potential, this observation is easy to understand in terms of the particle-transfer mechanism discussed for grazing collisions. Before the collision, the Fermi energies (chemical potentials) of all stable nuclei are roughly equal, about -8 MeV. In the heavy nucleus, the protons' larger Coulomb potential is compensated by the deeper nuclear potential due to the isovector force from the neutron surplus, together with the smaller Fermi kinetic energy due to the lower density of the protons. As the nuclei approach each other, the heavy nucleus' Coulomb field raises the Fermi energy of the light nucleus' protons by more than the light nucleus affects the heavy nucleus' protons; thus the light nucleus has a higher chemical potential for protons than the heavy nucleus, to which the protons migrate because they find unoccupied states of the same energy. The neutrons, unaffected by Coulomb forces, retain their original chemical potentials of about -8 MeV and therefore have no incentive to move until the protons' migration has proceeded far enough to change the isovector potential.

Once the initial transfer of protons has established charge equilibrium, the flow of nucleons becomes quite slow, the slowest of the collective degrees of freedom to play an important role in deeply inelastic collisions. This is why the target and projectile may retain vestiges of their identity even when they have rotated through a large angle. Nevertheless the transfer of nucleons is very important to the collision dynamics, since it is the main mechanism for the damping of the angular orbiting motion. We saw in grazing collisions that each nucleon transfer causes a loss of energy from the relative motion. The random diffusion of nucleons between target and projectile thus forms a dissipative mechanism. Extension of the classical arguments invoked to explain grazing transfer reactions leads to a contribution to the friction coefficient γ_ν which is sufficient to explain the observed damping of the angular motion. This picture is corroborated by the correlation of the energy loss with the width of the distribution of the projectile fragment's mass. In a random walk, the width of a distribution grows with the square root of the number of steps; since each nucleon-transfer step leads to the loss of a fixed amount of the relative kinetic energy, the width of the mass distribution should be proportional to the square root of the energy loss. Such a correlation is seen in fig. 11.7. Similar correlations are observed for the width of the energy distribution as a function of energy loss.

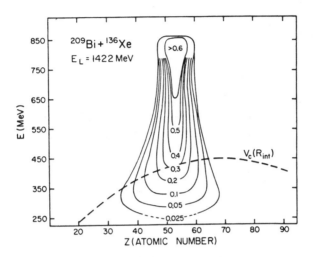

Figure 11.7   Correlation between width of projectile-fragment charge distribution and energy loss for the reaction of 1422 MeV ^{136}Xe on ^{209}Bi. Contours show regions of equal differential cross section $d\sigma/dE$ in mb per MeV as a function of fragment center-of-mass energy E and atomic number Z. Dashed line shows Coulomb energy of touching spheres. [*from H. Wollersheim, W. Wilcke, J. Birkelund, J. Huizenga, W. Schröder, H. Freiesleben, and D. Hilscher, Phys. Rev. **C24** (1981) 2114*].

The theoretical description of deep-inelastic heavy-ion collisions is still incomplete. In principle the forces and inertias appearing in the Newtonian equation of motion could be calculated using the linear-response method of Chapter 8, but realistic computations have not yet been carried out for the very asymmetric nuclear shapes and the many collective degrees of freedom necessary to describe these reactions. However, enough has been learned to cast doubt upon the Newtonian approximation for the collective motion, which rests on the assumption that the mean-field potential changes little during the relaxation time of the independent-particle degrees of freedom. This condition is violated at the moment when the nuclear surfaces touch, forming a window through which nucleons can freely pass between the nuclei. The time it takes for this window to open is roughly one fermi divided by the relative radial velocity; at one MeV per nucleon radial kinetic energy, this time is about 7×10^{-23} sec, about the time it takes for a nucleon to go five or six fm at the Fermi velocity. The time for the window to open is thus less than the relaxation time, so that the collective motion is not well described by a differential equation: the memory of earlier collective motion is not lost in the random motion of the nucleons.

The best theoretical method for treating the memories stored in the motion of the individual nucleons during large changes of the mean field is the **time-dependent mean field** theory or Time-Dependent Hartree-Fock (TDHF). A density-dependent effective interaction allows the wave function to be modeled by a time-dependent Slater determinant $|\psi(t)\rangle$ of the form (4.2.5) at each time. Instead of minimizing the energy, the time-dependent determinant is chosen to minimize the average action

$$\mathcal{A} = \int dt \langle \psi(t) | (H - i\hbar \partial / \partial t) | \psi(t) \rangle. \tag{11.1.15}$$

Note that $\mathcal{A}$ would be zero if ψ exactly solved the Schrödinger equation. The variational equations lead to time-dependent Schrödinger equations for the single-particle factors in the Slater determinant, similar to eq. (4.2.11):

$$i\hbar \frac{\partial}{\partial t} \phi_i(\vec{r}, t) = \left(-\frac{\hbar^2}{2m_N} \vec{\nabla}^2 + U^D(\vec{r}, t) \right) \phi_i(\vec{r}, t) - \int d^3 \vec{r}' U^X(\vec{r}, \vec{r}', t) \phi_i(\vec{r}', t) \tag{11.1.16}$$

where the self-consistent mean fields U^D and U^X are given at each instant by equations (4.2.26a–b). The initial conditions are obtained by starting with solutions $\phi_i^{gs}(\vec{r})$ of the static mean-field equations for the target and projectile, then boosting each by its initial velocity $\vec{v}_A$ or $\vec{v}_a$, similar to eq. (4.5.7):

$$\phi_i^A(\vec{r}, t \to -\infty) = e^{i m_N \vec{v}_A \cdot \vec{r} / \hbar} \phi_i^{gs(A)}(\vec{r} - \vec{v}_A t) e^{-i(\epsilon_i + m_N v_A^2 / 2)t / \hbar}. \tag{11.1.17}$$

Such computations (see fig. 11.8) exhibit features of energy loss, angular deflection, angular momentum, internal excitation energy, etc. which are very similar to those of the Newtonian phenomenological equations.

While the time-dependent mean field theory gives a satisfactory description of average properties, it is not able to describe the large fluctuations which develop in heavy-ion collisions near the Coulomb barrier. The reason for this failure is easy to understand: since there is only a single mean field at each time, the growth of a fluctuation is inhibited by the contributions of opposing fluctuations to the mean field which governs their evolution. For example, TDHF possesses a stationary solution in which a fissioning nucleus remains forever perched on top of its fission barrier in unstable equilibruim. In fact, Brack showed that the restriction to a single Slater determinant imposes an upper bound on the possible fluctuations of any one-body observable; when this result is applied to the division of nucleons between target and projectile, the resulting bound is much less than the width of the observed distributions.

The observed fluctuations apparently include not only the quantum fluctuations associated with the uncertainty principle, but also statistical fluctuations. The presence of statistical fluctuations is not surprising, since the heavy-ion reactions often lead all the way to the statistical compound nucleus. Indeed, such fluctuations necessarily accompany any irreversible process, of which the damping of the ions' relative motion is a clear example.

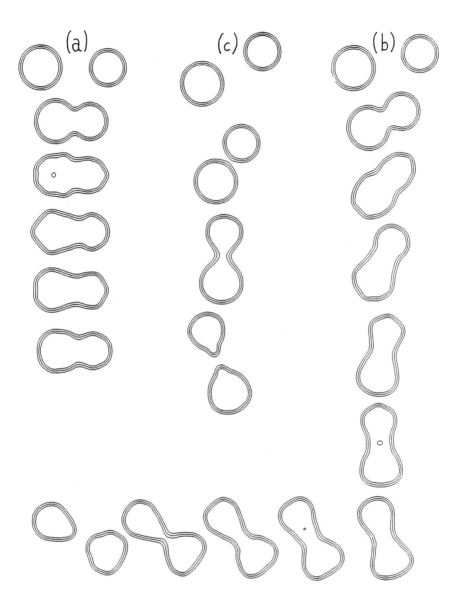

Figure 11.8   Time evolution of a collision of ^{86}Kr on ^{139}La at $E_{lab}/A = 7$MeV, viewed in the center of mass. Contours show equal densities $(0.04, 0.08, 0.12$fm$^{-3})$ in the reaction plane Z=0, calculated by the TDHF method. Successive pictures are separated by time intervals of 4×10^{-22}sec. (a) head-on collision L=0; (b) deep-inelastic collision L=110; (c) grazing collision L=250. [*courtesy M. Strayer, method after K. T. R. Davies and S. E. Koonin, Phys. Rev.* **C23** *(1981) 2042*].

The collective picture of nuclear dynamics may be extended to include fluctuations. The time evolution of the system can no longer be described by the average values of collective variables. Instead, a density matrix is needed. If $\rho(t)$ is the density matrix in the state space containing both microscopic and collective variables, then the collective motion is described by the reduced density matrix $d(t)$ defined by

$$d(t) = \mathrm{tr}_{\text{intrinsic}}\,\rho(t). \tag{11.1.18}$$

$d(t)$ is still an operator in the collective variables. An expecially useful representation of the operator is the **Wigner representation**

$$d(\{Q_k\}, \{P_k\}, t) = \int d\{Q_k'\} \langle \{Q_k - \tfrac{1}{2}Q_k'\}|d(t)|\{Q_k + \tfrac{1}{2}Q_k'\}\rangle e^{i\Sigma_k Q_k' P_k}. \tag{11.1.19}$$

The Wigner representation is convenient because when integrated over the momenta, it gives the probability distribution in the coordinate representation; conversely, integration over the coordinates yields the probability distribution in momentum space. It is tempting to think of the Wigner distribution function as the quantum analog of the joint probability distribution in phase space of classical statistical mechanics. This picture is intuitively helpful, but the analogy must be taken cautiously: for example, $d(\{Q_k\}, \{P_k\}, t)$ may be negative for some values of its arguments, which could never happen for a true probability distribution.

The dynamics of the collective fluctuations are determined by the von Neumann equation for the density matrix,

$$i\hbar \frac{d}{dt}\rho(t) = [H, \rho]. \tag{11.1.20}$$

Using a Hamiltonian like those described in Chapter 8, eq. (11.1.20) is solved approximately in perturbation theory, then averaged over the non-collective variables [see Hofmann and Jensen]. With assumptions about the relaxation of the microscopic degrees of freedom, analogous to those introduced in Sect. 8.6, differential equations for $d(\{Q_k\}, \{P_k\}, t)$ are obtained [see Hofmann and Siemens] of the form

$$\frac{\partial d}{\partial t} = -\sum_{\mu}\left[\mathcal{K}_\mu \frac{\partial}{\partial P_\mu} + \sum_\nu \left(\frac{\partial}{\partial Q_\mu} M_{\mu\nu}^{-1} - \frac{\partial}{\partial P_\mu}\Gamma_{\mu\nu} \right) P_\nu \right] d$$
$$+ \sum_{\mu\nu} \frac{\partial}{\partial P_\mu}\left[D_{\mu\nu}^0 \frac{\partial}{\partial P_\nu} + D_{\mu\nu}^1 \frac{\partial}{\partial Q_\nu} \right] d, \tag{11.1.21}$$

where $\Gamma_{\mu\nu} \equiv \sum_\sigma M_{\mu\sigma}^{-1}\gamma_{\sigma\nu}$ and $\mathcal{K}_{\mu\nu}, \gamma_{\mu\nu}$, and $M_{\mu\nu}$ are moments of response functions defined in Sect. 8.6, while $D_{\mu\nu}^0$ and $D_{\mu\nu}^1$ are new coefficients defined in terms of the **correlation functions**

$$\tilde{\mathcal{C}}_{\mu\nu}(t) \equiv \frac{1}{2\hbar}\langle \{e^{iH_0 t/\hbar} F_\mu e^{-iH_0 t/\hbar}, F_\nu\} \rangle \tag{11.1.22}$$

where the curly brackets denote an anticommutator, and the average is now over the ensemble describing the non-collective degrees of freedom. The coefficients $D_{\mu\nu}^n$ are moments of $\tilde{C}_{\mu\nu}(t)$:

$$D_{\mu\nu}^n \equiv \int_0^\infty dt\, t^n \tilde{C}_{\mu\nu}(t) \qquad (11.1.23)$$

analogous to eq. (8.6.7) for $M_{\mu\nu}$ and $\gamma_{\mu\nu}$.

To understand the diffusion equation for d, eq. (11.1.21), it is instructive to see how to derive Newton's equation (8.6.6) from it. Multiplying by Q_μ and integrating over all $\{P_\nu\}$ and $\{Q_\nu\}$, all terms vanish except two, which give

$$\frac{d}{dt} \sum_\nu \langle M_{\mu\nu} Q_\nu \rangle = \langle P_\mu \rangle. \qquad (11.1.24a)$$

Similarly, multiplying by P_μ and integrating gives

$$\frac{d}{dt} \langle P_\mu \rangle = \langle \mathcal{K}_\mu \rangle + \sum_\nu \langle \Gamma_{\mu\nu} P_\nu \rangle. \qquad (11.1.24b)$$

Substituting eq. (11.1.24a) into (11.1.24b) yields Newton's equation if we assume the distributions are narrow enough that we can ignore the variations of $\mathcal{K}_\mu, \Gamma_{\mu\nu}$, and $M_{\mu\nu}$ when evaluating the mean values.

The coefficients $D_{\mu\nu}^n$ do not affect directly the motion of the mean values, but do help the probability distributions to spread out. They are related to the Newtonian coefficients because of the **fluctuation-dissipation theorem**, which relates the Fourier transform of $\tilde{C}_{\mu\nu}(t)$ to the response function's Fourier transform

$$\mathcal{X}_{\mu\nu}''(\omega) = \mathcal{C}_{\mu\nu}(\omega)\tanh(\hbar\omega/2T) \qquad (11.1.25)$$

for the case when the microscopic variables are described by a density matrix of temperature T. When the frequency of the collective motion is slow compared to the temperature, then the diffusion coefficients are related to the Newtonian ones by the Einstein relation

$$D_{\mu\nu}^0 = T\gamma_{\mu\nu} \quad \text{for } \hbar\omega \ll T \qquad (11.1.26a)$$

$$D_{\mu\nu}^1 = TM_{\mu\nu} \quad \text{for } \hbar\omega \ll T. \qquad (11.1.26b)$$

Extensions of the phenomenology of the Newtonian classical trajectories to include approximate solutions of the diffusion equation (11.1.21) yield a reasonable description of the observed probability distributions, such as the Wilczynski plots, for many different heavy-ion reactions. However the fundamental basis of the diffusion equations, the rapid relaxation of the microscopic degrees of freedom, suffers from the same doubts as apply to the Newtonian equations for the mean values. On the other hand, the fact that the diffusion equations are linear in the probability density (unlike the TDHF equation) means that they permit various fluctuations to grow independently of each other.

The collective variables are not the only possibility for describing the probability distribution. For example, Kadanoff and Baym derive analogous equations for the one-body density matrix in terms of the single-particle coordinates and momenta. Randrup has adapted this method to the phenomenological description of heavy-ion collisions with good success. The most general method, applied by Agassi, Ko, and Weidenmüller, describes the system's evolution in the space of the eigenstates of the many-body Hamiltonian; for practical applications, however, this method has to be approximated in ways that resemble the collective-coordinate description, leading to the conclusion that corrections due to the memory of the system's earlier state may modify the rates of probability diffusion by up to a factor of two at the crucial time when the window opens between the nuclei.

11.2   COLLISIONS NEAR THE FERMI VELOCITY

When the relative velocity of the colliding nuclei becomes comparable to the velocity of the nucleons' motion inside the nucleus, a host of new phenomena become possible. We have seen that gentle collisions in the neighborhood of the coulomb barrier could be understood by fairly straightforward extensions of the self-consistent mean-field picture, which was first developed for slow, small-amplitude collective motion. When the relative motion has enough energy to dissociate the nuclei into their constituent nucleous, the mean-field picture must be questioned. Experimentally, final states with many separated nucleons or clusters demand much more complicated detector systems, as well as new ways of analyzing the multivariate measurements.

The key problem for understanding high-velocity heavy-ion collisions is **event characterization**. We have seen that many, very different types of reactions – elastic scattering, Coulomb excitation, grazing transfer reactions, deeply inelastic collisions, fusion – may all result from the same projectile incident on a given target at a fixed energy near the Coulomb barrier. We must be prepared for a comparable variety of reactions to take place when even more energy is supplied. In low-energy collisions, the various mechanisms could be distinguished by a relatively simple **inclusive measurement** of a single reaction product. In that case a single heavy fragment told the story: if it was like the projectile, than its angle and energy, located in a Wilczynski plot, could characterize the reaction; if the heavy fragment was like the combined projectile-target system, then a fusion reaction had occurred (unless the compound system, or the excited projectile, fissioned, in which case both fission fragments would have to be observed). Such a tell-tale signature of the reaction's type must be found for each new regime of incident velocities. A priori it is by no means clear that any inclusive measurement will suffice.

The ideal solution to the problem of event characterization would be to measure the masses, charges, and momenta of all the reaction products in the final state of each individual collision. Such complete information would surely give the best chance of correctly recognizing patterns typifying various sorts of reactions. Unfortunately a complete determination of a final state containing dozens or hundreds of products is beyond the capabilities of current experimental techniques. Two main difficulties must be met. First, the large number of final-state products demands the collection and analysis of huge amounts of information. If the information is collected from electronic detectors then it must be sorted and recorded very rapidly; if it is collected by optical techniques, the sorting and processing problem is merely postponed for later analysis. Second, no single type of detector can accurately identify and measure all the enormously different reaction products, which span the gamut from heavy target residues recoiling with ranges measured in microns to high-energy photons producing meter-long electromagnetic showers. The best detector systems are built up by layers, with low-mass detectors (gas-phase ionization detectors or very thin semiconductors) nearest the target to detect the shortest-range products, backed by thicker layers to stop the more penetrating ejecta.

The experimental information presently available (Spring 1987) on Fermi-velocity nuclear collisions is extensive, but not yet sufficient to convincingly construct a unified scenario like the one we discussed above for lower-velocity collisions (or one we will describe below for higher velocities). Therefore we must turn to theoretical speculation for clues about what to expect from this rapidly developing field.

First, we need to understand why we need to look beyond the mean field for our scenarios of high-velocity collisions. Two main difficulties challenge the validity of the mean field:

(1) The Pauli supresion of collisions must be reduced as the relative velocity increases, since the larger velocity permits the colliding nucleons to find unoccupied states more easily. Indeed, eventually the nucleons' mean free path ought to become comparable to the interparticle spacing, as we estimated by ignoring Pauli effects in Sect. 3.3.

(2) The large excitation energy causes the nuclear system to become unstable with respect to density fluctuations, a situation which the mean-field picture describes poorly as we discussed in Sect. 11.1.D above.

It is not hard to extend the mean-field picture to allow for collisions of nucleons, but the more serious problem of the dynamics of density fluctuations has not yet been solved. To understand the nature of this problem, we start in Sect 11.2.A by considering the thermodynamics of nuclear matter at excitation energies comparable to the binding energy. We will see that nuclear matter has a first-order phase transition in this region of excitation. Next, we explore in Sect. 11.2.B the consequences of the phase structure of nuclear matter in a simple dynamic model, the hydrodynamic picture.

11.2 A. Liquid-Gas Phase Transition in Nuclear Matter

The possibility of a liquid-gas mixture exists for any self-bound Fermi system interacting through short-range forces. Consider first homogeneous matter of uniform density and zero temperature. At very low densities, the pressure will be positive, due to the Fermi degeneracy pressure. At higher densities, the attractive forces counteract the thermal pressure, and eventually overcome it: the pressure becomes negative because the forces try to decrease the volume. At still higher densities, short-range repulsion and other saturation mechanisms drive the pressure positive again. For each small, positive pressure, there are three densities of uniform matter which have the same pressure (see fig. 11.9). For example, zero pressure is a common characteristic of the ground state of the liquid – i.e. nuclear matter at saturation density n_0 – and of the zero-density gas, with a third state of intermediate density sharing this distinction.

At a small but finite temperature, the trend will be similar, but the pressure will be larger at each density. If the temperature is larger than the flash temperature T_F, the pressure will stay positive at all non-zero densities, but there will still be a range of positive presures shared by three densities. Increasing the temperature further will narrow this range of pressures, until above the critical temperature T_C the minimum disappears and the pressure-density relation becomes unique.

We know that many of the homogeneous states we have described are thermodynamically unstable: Their entropy can be increased, at fixed volume and energy content, by allowing them to separate into part liquid, part gas. For these two phases to be in equilibrium, they must have the same pressure, temperature, and chemical

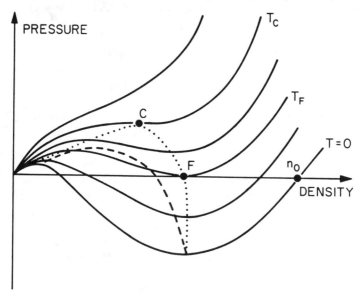

Figure 11.9   Schematic isotherms (full curves) of nuclear matter at various temperatures. C is the critical point, F the flash point. Isothermal (dotted line) and isentropic (dashed line) spinodals are also shown, see Sect. 11.2.B.

potential, or Gibbs free energy G per particle. The true isotherms have, instead of mixima and minima, a region of constant pressure in which the proportion of liquid and gas varies with the volume of the container from pure liquid at the high-density end to pure gas at the low-density boundary of the constant-pressure region (see fig. 11.10).

The mixed-phase state is characterized not only by its temperature and density, but also by the spatial distribution of the interface between the liquid and gas. At high densities just below the liquid density, the gas will be found in isolated bubbles. Near half the liquid density, the gas-liquid interface will have a complicated geometry, with interconnecting regions of dense liquid and less-dense gas interpenetrating much like the voids and lumps of a sponge. At low densities, the liquid

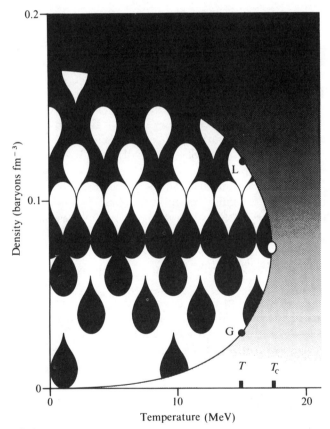

Figure 11.10   Schematic phase diagram for nuclear matter. Scales of temperature and density are approximate. An open circle marks the critical point. Points G and L illustrate the gas and liquid densities, respectively, for a temperature T not far below the critical temperature T_C. [*from P. J. Siemens, Nature* **305***(1983)410*].

will be dispersed as droplets in the gas. Since the final state of a nuclear collision is observed at low density, we will concentrate on the description of the low-density mixture. Concentrating on small droplets, we neglect Coulomb forces.

The probability P_A of finding a cluster of A particles in a fluid is given by the difference of free energies of the cluster and gas:

$$P_A = \exp - (G_A - AG_G)/T \tag{11.2.1}$$

The Gibbs free energy $G = E - T\mathcal{S} + PV$ appears because the appropriate thermodynamic variables the pressure P and temperature T rather than the volume V and entropy $\mathcal{S}$. Following Fisher, we suppose that, for sufficiently large A, the cluster free energy G_A possesses a liquid-drop expansion

$$G_A \approx G_V A + \sigma S(A) + \kappa T \ln A \tag{11.2.2}$$

where G_V is the free energy per particle in the liquid phase, σ is the surface tension, S(A) is the mean surface area of a droplet of A particles, and κ is a critical exponent. The first two terms are a generalization of the liquid-drop model to infinite temperatures; the last would become important near T_C. Comparing eqs. (11.2.1) and (11.2.2), we inspect first the leading behavior in A. Three cases arise:

(1) $G_V < G_G$. Then the biggest clusters are favored: the whole system would like to be in one big cluster. We identify this as the liquid phase.

(2) $G_V > G_G$. Then the cluster probability falls rapidly, like x^A with $x < 1$. We identify this as the gas phase.

(3) $G_V = G_G$. Then the cluster distribution is determined by the free energy for forming surfaces. We recognize the mixture phase.

Thus we see that the statistical description reproduces the phase-equilibrium condition known from thermodynamics.

In the mixed phase, the droplet distribution is determined principally by the surface energy σS. Since the surface area

$$S \approx 4\pi r_0^2 A^{2/3} \tag{11.2.3}$$

is easily estimated from the liquid density $n_L = (\frac{4}{3}\pi r_0^3)^{-1}$, the surface tension σ is the main ingredient. We know that for $T \to 0$, $\sigma = 1.14$ MeV/fm^2 from the semiempirical mass formula; we also know that σ vanishes as $T \to T_C$, because the densities of the liquid and gas phases approach each other so the surface becomes a mere ripple. Thus, at the critical temperature, the droplet distribution is determined by the critical exponent κ, and assumes a power-law form

$$P_A \sim A^{-\kappa} \tag{11.2.4}$$

which implies that $\kappa > 2$ since the distribution must be normalizable ($\Sigma A \cdot P_A$ is finite). At and below the critical temperature, the droplet distribution has the form

$$P_A = \text{const} \cdot A^{-\kappa} \exp\left(-b(T)A^{2/3}\right) \tag{11.2.5}$$

where $b(T) = 4\pi r_0^2 \sigma(T)/T$ decreases monotonically towards zero as $T \to T_C$.

We conclude that, as the temperature increases below the critical temperature, the probability of finding large droplets increases with increasing temperature. This counter-intuitive behavior is a signature of the liquid-gas equilibrium: one might have expected the droplets to be harder to hold together as thermal agitation increases, but the vapor around them reverses the trend. As the temperature increases above T_C, the probability of finding heavy droplets falls off again, since the leading term proportional to A in eq. (11.2.1) becomes non-zero. By seeing where the large-droplet formation is most probable, we could identify the critical temperature; the distribution of droplet sizes at T_C reveals the critical exponent κ. Below T_C, the droplet distributions are determined by the surface tension σ of the liquid-gas interface.

11.2 B. Limits of Nuclear Stability in the Hydrodynamic Model

From a thermodynamic viewpoint, an isolated nucleus can only be stable in its ground state. A drop of warm nuclear matter in a large container would have to evolve to a state of greater entropy by filling the space around it with vapor and smaller droplets. We realize now that the compound-nucleus decay mechanisms of Chapter 10 were expressions of the thermodynamic instability. Indeed, when Coulomb forces are taken into account, even the ground state of a very large nucleus is unstable!

The consequences of this instability for the time evolution of the nuclear system depend both on the excitation energy and on how it is introduced. A small amount of excitation energy, gently deposited, may leave a statistically excited compound nucleus corresponding to the lowest non-zero temperature among the isotherms of fig. 11.9. In the absence of external pressure, the nucleus settles into a **hydrostatic equilibrium** with its surroundings: their pressures must be the same, otherwise matter would flow and the boundary of the nucleus would move. It could only come to rest at an equilibrium density given by the point where the corresponding isotherm crosses the axis P=0, a little below n_0. Once the nucleus is in hydrodynamic equilibrium, it can slowly evaporate nucleons and other particles until it cools to the temperature of its surroundings. The time scale for reaching hydrostatic equilibrium is represented by the times for nuclear vibrational motion, much shorter than the evaporation time required for reaching full thermodynamic equilibrium.

If the nucleus has too much excitation energy, however, it can never reach hydrostatic equilibrium. Inspecting fig. 11.9, we realize that there is a maximum **flash temperature** T_F above which a statistical nucleus cannot exist at zero pressure: greater excitation causes the pressure to be always positive, because the increased thermal motion of the nucleons exceeds the attractive capability of the nuclear force. Stocker estimated the flash temperature of nuclear matter as about 12 MeV, which is reduced by Coulomb forces in nuclei. Indeed, thermal mean-field computations, in which the density matrices are obtained from thermal ensembles of Slater determinants, also fail to find solutions above excitation energies roughly equal to the binding energy [see Bonche, Levit, and Vautherin]. Nor have compound nuclei been reported experimentally above 5 MeV temperature, although that lower bound is rapidly rising as new experimental results become available.

The equality of pressure is a necessary condition for hydrostatic equilibrium. The existence of such an equilibrium is not sufficient for it to be attained practically: the equilibrium must also be stable with respect to small perturbations of its configuration. In the hydrodynamic model, this requirement is formulated by the condition for **hydrodynamic stability**

$$\partial P / \partial n \geq 0. \qquad (11.2.6)$$

This condition assures that a small inhomogeneity will be smoothed out by the flow of matter from the higher-pressure region, where the density is greater, to the lower-pressure region, where the density is less. Small density fluctuations are thereby discouraged, so that the nucleus may wait long until there is a fluctuation large enough to lower the free energy by separating the liquid and gas phases. On the other hand, if the condition (11.2.6) is not fulfilled, then a local accumulation of density would lead to a lower pressure, causing more matter to flow into the high-density region.

The limit of hydrodynamic stability, $\partial P / \partial V = 0$, is known as the **spinodal line** in the phase diagram: it separates the unstable region, where small fluctuations in the uniform medium can grow into a mixed phase, from the **metastable** region where only large fluctuations can grow. If the density fluctuations are assumed to occur without associated fluctuations of the temperature, the boundary of instability is the **isothermal spinodal** $\partial P / \partial n|_T = 0$; if the fluctuations are isentropic, like sound waves, the metastable region is expanded to the **isentropic spinodal** $\partial P / \partial n|_S = 0$. The isothermal spinodal passes through both the flashpoint and the critical point (see fig. 11.9); the isentropic spinodal encompasses a much smaller region of the state variables, restricted to temperatures well below T_C. The hydrostatically stable statistical nuclei, with pressure zero and densities between n_0 and the flash point, are all hydrodynamically stable both at constant temperature and at constant entropy.

Whether statistical nuclei at the flash temperature can be formed in nuclear collisions depends on the damping of the collective motion describing the collision. Since the colliding nuclei start out at normal density n_0, they can only form a lower-density hydrostatic nucleus by expanding. The degree of damping of the collective expansion determines the result of that expansion.

If the collective motion were strongly overdamped, the expansion would be **isoergic** because most of the energy would remain in the thermal and compressional internal energy: the work PdV done to expand the matter would be quickly turned from collective motion into heat, with little collective kinetic energy. Overdamped isoergic expansion would permit formation of compound nuclei right up to their maximum temperature T_F. If more energy were available than needed to reach the flash point, the expansion would slowly continue, driven by the positive pressure, until the spinodal line of instability was reached. Since the expansion would be slow, thermal conduction would probably be able to provide isothermal density fluctuations allowing the liquid and gas phases to separate at the isothermal spinodal.

If the collective expansion were undamped, it would be more difficult to make equilibrated nuclei near the flash temperature. Then the work PdV done by the internal pressure would be preserved as kinetic energy of the outward-flowing matter, so that the expansion would continue past the equilibrium P = 0, after which the outward flow would be slowed by the negative pressure inside the matter. If the stored energy were small enough, the outward flow would soon be stopped, after which the negative pressure would cause the matter to contract again (fig. 11.11a).

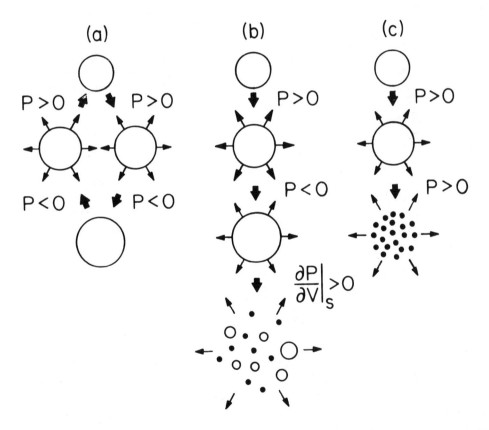

Figure 11.11   Scenarios for nuclear hydrodynamic expansion: a) Monopole vibration at low excitation; b) Expansion to liquid-gas phase mixture; c) Expansion to gas at very high excitation. [*from P. J. Siemens, Nucl. Phys.* **A428***(1984)189c*].

We recognize the oscillatory cycle of the giant monopole vibration described in Sect. 8.5.C. However, a sufficiently large amount of expansion energy would lead to a different conclusion: if the vibration's amplitude were large enough to reach the spinodal instability, then a phase separation would terminate the oscillation after a half cycle as the matter disintegrated into a liquid-gas mixture (fig. 11.11b). Since the mixed phase would have positive pressure, it would continue to expand. For undamped expansion, the relevant instability is isentropic, since undamped motion is irreversible and therefore can't produce entropy. If the internal excitation were large enough, the matter would be so hot that it would never reach the spinodal instability, but would pass smoothly from the liquid to the gas phase (fig. 11.11c) without going through a phase separation.

The degree of damping of collective flow in exploding nuclear matter will be one of the most interesting conclusions we can hope to draw from studying Fermi-velocity nuclear collisions. We know that the small-amplitude collective vibrations are underdamped at zero temperature. The degree of damping observed at higher temperatures and larger amplitudes may reflect the mean free paths of the nucleons in the warm matter: long mean free paths permit ready diffusion of fast nucleons from rapidly-moving or hot regions into nearby slower-moving or cooler regions. The resulting quick equilibration of flow patterns and thermal gradients reflects the large viscosity and heat conductivity of a degenerate Fermi liquid. These irreversible processes generate entropy as they damp the collective motion. In an especially intricate scenario noted by Pethick and Ravenhall, the thermal conduction is estimated to be most effective for short-wavelength fluctuations, which are also argued to grow most rapidly; for some region of the excitation energies, the long-distance-scale overall expansion might be almost undamped until the density reaches the isothermal spinodal, at which time the short-wavelength density fluctuations grow to produce the phase separation.

11.3   COLLISIONS NEAR THE SPEED OF LIGHT

Surprisingly, much more is known about nuclear collisions with beam velocities from 0.5c to 0.95c than has been observed about collisions around the Fermi velocity. Beams of nuclei from 100 to 2000 MeV per nucleon have been available since the mid 1970's, long before accelerators could produce heavy beams in the 20–100 MeV range. The experimental identification of multifragment final states is easier at higher energies because most of the reaction products are moving quite rapidly in more or less the direction of the beam. This eases the dynamic-range requirements for detector systems, which also benefit from the convenience of having most of the fragments appear in the forward hemisphere.

11.3 A.   Classification of Reactions and Products

Our way of thinking about nuclear collisions in the energy regime of a few hundred MeV per nucleon has been strongly shaped by models and ideas borrowed from classical physics. The de Broglie wavelengths of nucleons at these energies are much shorter than the radius of a nucleus, so that classical physics can provide a reasonable orientation. In a classical picture, the initial conditions for a collision are specified by the baryon numbers of the target and projectile A_t and A_p, their charges Z_t and Z_p, the projectile velocity $\vec{\beta}_0 c$, and the impact parameter $\vec{b}$. If b is greater than the sum of the nuclear radii, the collision will be rather gentle, and the projectile will continue after the collision with approximately its initial speed. For small impact parameters, on the other hand, nucleons from the target and projectile are expected to scatter strongly, leading to the production of many nucleons, pions and perhaps other products with speeds very different from either target or projectile. Thus the signature of a peripheral collision is a heavy fragment near 0° with a velocity close to $\vec{\beta}_0 c$, and a mass close to the projectile (often somewhat less, because even a small excitation may cause the projectile to emit nucleons). The signature of a central collision, on the other hand, is the detection of many nucleons with velocities very different from the projectile's.

The classification of collisions into peripheral and central is neither exhaustive nor exclusive. Indeed, for intermediate impact parameters, portions of the projectile and target may be strongly scattered, while other portions are relatively undisturbed. A simple description of these collisions is provided by the participant-spectator model. **Participant nucleons** are those which have suffered a large momentum transfer, **spectators** have received little or no change of momentum. This distinction may be considered as an operational definition. Remarkably, the numbers of participant and spectator nucleons for a given target-projectile combination is, within errors, independent of the beam energy from 400 MeV to 2 GeV per nucleon. This suggests

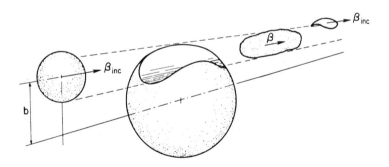

Figure 11.12   Visualization of a neon nucleus incident on a uranium nucleus, with velocity $\beta_{inc}c$ and impact parameter b. The swept-out nucleons from the projectile and target, called the fireball, move with velocity βc [*from G. Westfall, J. Gosset, P. Johansen, A. Poszkanzer, W. Meyer, H. Gutbrod, A. Sandoval and R. Stock, Phys. Rev. Lett.* **37** *(1976)1202*].

a geometrical interpretation: the spectators are those nucleons in the target or projectile whose straight-line trajectories would miss the other nucleus, the rest are participants (see fig. 11.12). Simple geometry then leads to the prediction

$$\sigma_{\mathrm{part}} = \pi r_0^2 (Z_p A_t^{2/3} + Z_t A_p^{2/3}) \tag{11.3.1}$$

for the total cross section for participant protons, in good agreement with experiment (fig. 11.13) for collisions where the target and projectile have similar sizes. In comparing experiment to equation (11.3.1), one must of course remember to count those protons which appear in deuterons and heavier fragments.

How can some of the nucleons be so little influenced by these high-energy collisions? After all, they are much more violent than the low-energy collisions of Sect. 11.1 where no spectators are observed. When the relative motion is much faster than the motion of the nucleons inside the nuclei, the projectile remnant can go right past the target before any nucleons from the target can reach it to carry the impact of the collision. Thus we expect the participant-spectator separation to appear when the beam velocity exceeds the Fermi velocity. This means that the emergence of the participant-spectator division is tied up with the liquid-gas phase transition of

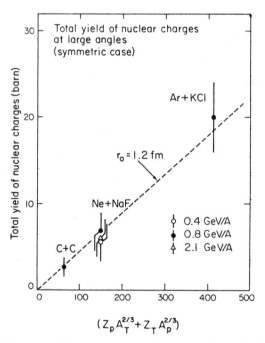

Figure 11.13   Total integrated cross sections of nuclear charge for high-energy particles emitted at large angles (participants). [from S. Nagamiya and M. Gyulassy, Advances in Nuclear Physics, J. Negele and E. Vogt eds. vol 13 (Plenum, New York, 1984) p. 201].

Sect. 11.2. In the relativistic regime, the spectator liquid remains cool, as evidenced by its narrow distribution in transverse momentum, while the participants become a hot gas well above the critical temperature for forming a mixed phase.

It is perhaps surprising that equation (11.3.1) gives such a good prediction of the number of participants, when we consider what is left out of this very simple picture. On the one hand, participant nucleons can scatter into the spectator matter, where they can knock out more nucleons to increase the number of participants. On the other hand, some of the projectile nucleons would be expected to go right through the target without much momentum transfer, particularly at higher energies where the nucleonic cross section is strongly forward peaked. Evidently these two effects approximately cancel when the target and projectile are about the same size and not too heavy. If the target is much larger than the projectile, we should expect the number of participants to be greater than given by equation (11.3.1). For such unequal systems, the distinction between participants and spectators is blurred, as each of the original participants may have to share its momentum with a number of the surrounding spectator nucleons.

11.3 B. Spectator Matter

One of the earliest systematic observations of high-energy heavy-ion reactions was the distribution of the energies of projectile-like fragments, which have a gaussian velocity distribution centered about a velocity slightly less than the beam's. The widths of these distributions were explained by Goldhaber using a simple picture in which the target slices off a piece of the projectile, containing nucleons whose momenta are randomly selected from the Fermi sea. The same picture explains the width in transverse momentum, when proper account is taken of the Coulomb deflection. The shift in the mean longitudinal momentum is less than the width of the distributions, and is probably due to the occasional participant nucleon — which has less longitudinal velocity on the average — lodging in or scattering from the projectile fragment. It seems that the effect of the participants on the spectator matter is indeed rather small.

11.3 C. Participant Nucleons

The patterns of emission of reaction products are most easily comprehended when they are presented in a two-dimensional plot of cross section as a function of the product's velocities along (longitudinal) and transverse to the beam. It is convenient to use the relativistically **invariant cross-section**

$$\sigma_I = \frac{1}{|\vec{p}c|}\frac{d^2\sigma}{d\Omega dE} = \frac{E}{c^2}\frac{d^3\sigma}{dp^3} \tag{11.3.2}$$

where $\vec{p}$ and $E = \sqrt{m^2 + \vec{p}^2 c^2}$ are the relativistic momentum and energy of the product of mass m. Instead of the longitudinal velocity one uses the **rapidity**

$$y = \frac{1}{2}\ln\frac{E + p_{\parallel}c}{E - p_{\parallel}c} \qquad (11.3.3)$$

where $p_{\parallel}$ is the component of the momentum along the beam direction. The rapidity y has the convenient property that it may be transformed to another Lorentz frame with a different velocity $v_0 = \beta_0 c$ along the beam direction by merely adding the rapidity $y_0 = \frac{1}{2}\ln((1+\beta_0)/(1-\beta_0))$ of the moving frame's origin, so that $y' = y - y_0$

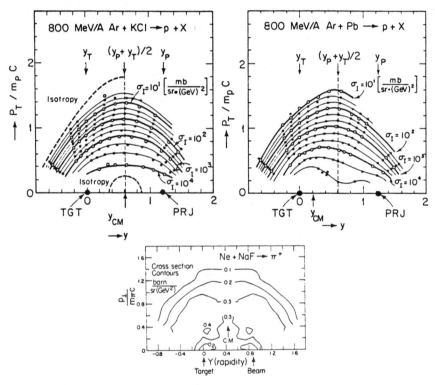

Figure 11.14   Contour plots of inclusive proton (top) and pion (bottom) spectra. Contours indicate equal invariant cross sections σ_I versus rapidity y and transverse momentum p_T or $p_\perp$. The rapidities of the target, projectile and center of mass are shown in each plot. Top: protons from 800 MeV/A Ar on a target of KCl(left) and Pb(right). Bottom: pions from 380 MeV/A Ne on NaF target. [*proton figures from S. Nagamiya, M. Lemaire, E. Moeller, S. Schnetzer, G. Shapiro, H. Steiner and I. Tanihata, Phys. Rev.* **C24** *(1981) 971; pions from J. Sullivan, J. Bisterlich, H. Bowman, R. Bossingham, T. Buttke, K. Crowe, K. Frankel, C. Martoff, J. Miller, D. Murphy, J. Rasmussen, W. Zajc, O. Hashimoto, M. Koike, J. Peter, W. Benenson, G. Crawley, E. Kashy, and J. Nolen, Phys. Rev.* **C25** *(1982) 1499*].

is the rapidity of the particle in the moving frame. Some examples of contour plots of σ_I as a function of y and $p_\perp$, the transverse momentum (which is the same in all frames moving in the beam direction) are shown in fig. 11.14.

These plots show that products with large transverse momentum are mainly produced at rapidities between the target and projectile rapidities, with additional cross section for low-transverse-momentum products near the target and projectile rapidities. There appear to be three distinct sources of nucleons and pions. We identify these sources as the spectator remnants of the target and projectile, and the broadly-spread participants.

The participant nucleons' momentum distribution goes smoothly out to very large perpendicular momenta, far beyond the kinematic limit for two-nucleon scattering even when the Fermi motion of the nucleons is taken into account. This suggests that the participant nucleons undergo many collisions with each other. This is not suprising since the Pauli principle would not be expected to lengthen the mean free paths much at these high energies where the Fermi spheres occupy only a small portion of the available momentum space for the scattered nucleons.

At high energies, the nucleons' elastic and inelastic cross sections are forward peaked. Individual nucleon-nucleon collisions involve, on average, momentum transfers of only a few times $m_\pi c$. In this case the mean free path is not an adequate measure of the rate at which a nucleon shares momentum and energy with the others, because most collisions transfer only a modest part of the nucleon's momentum. A better measure is the **stopping distance**

$$\lambda_s = |\langle\vec{p}\rangle|/|\vec{\nabla}\cdot\langle\vec{p}\rangle| \tag{11.3.4}$$

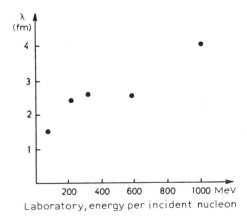

Figure 11.15   Stopping distance λ_s for a nucleon in cold nuclear matter at its saturation density, as a function of the nucleon's kinetic energy. [*from M. Sobel, P. Siemens, J. Bondorf, H. Bethe, Nucl. Phys.* **A251** *(1976) 502*].

where $\langle \vec{p} \rangle$ is the average momentum of the nucleon moving through the matter. The stopping distance λ_s is larger than the mean free path by the ratio of the average longitudinal momentum transfer to the total momentum. This ratio is unity for isotropic cross sections but is smaller when the cross sections are forward peaked. Its values in nuclear matter, estimated from free scattering by ignoring the Pauli principle, are shown in fig. 11.15. They have a plateau value of about 3 fm just above 400 MeV per nucleon, due to the onset of pion production, then rise slowly for higher energies because the average momentum transfer increases more slowly than the momentum.

The simplest picture of the participants is the **fireball model** in which the initial kinetic energy of the relative motion is distributed statistically among all the participant nucleons. In fact the observed inclusive spectra look rather like a thermal Boltzmann distribution,

$$\frac{\mathrm{d}^3\sigma_T}{\mathrm{d}p^3} = \mathrm{const} \cdot e^{-E/T}. \tag{11.3.5}$$

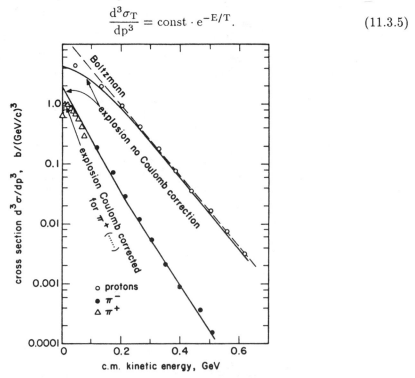

Figure 11.16 Inclusive cross sections $\mathrm{d}^3\sigma/\mathrm{d}p^3$ at $90°$ in the C.M. system for ^{20}Ne on NaF at 800 MeV/A laboratory kinetic energy. Open circles: protons; closed circles: π^-; triangles: π^+. Dashed line: Boltzmann distribution, eq. (11.3.4); Solid lines: exploding fireball with T= 44 MeV, $\beta = 0.373$ has nearly equal energy in thermal motion and radial flow. Dotted curve: exploding fireball with Coulomb correction for π^+. [from P. Siemens and J. Rasmussen, Phys. Rev. Lett. **42** (1979) 880].

Fig. 11.16 shows that while the inclusive distributions of protons and pions resemble eq. (11.3.5) they have a shoulder at low energies. Also the apparent temperature T obtained by fitting to the form (11.3.5) is surprisingly small. For the example in fig. 11.16, the apparent temperatures for protons and pions are 70 MeV and 52 MeV respectively while the center-of-mass energy per nucleon is 182 MeV. These features may be explained by supposing that the fireball is exploding. The flow of matter outward uses a portion of the energy which is not available for thermal excitation and therefore cools the system. The apparent temperature T_{app} is then hotter than the actual temperature because the hot matter is moving towards the observer, just as a far-away galaxy appears cooler because it is moving away:

$$T_{app} \equiv - \left(\frac{d}{dE} \ln \frac{d^3\sigma}{dp^3} \right)^{-1} \approx T\gamma^{-1}(1 - \beta E/pc)^{-1} \qquad (11.3.6)$$

where βc is the expansion velocity and $\gamma = (1 - \beta^2)^{-1/2}$. The pions appear cooler than the protons, even though they have the same temperature and expansion

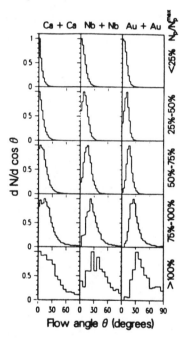

Figure 11.17 Distribution of flow angles for various symmetric target-projectile combinations at E/A = 400 MeV as a function of fractional multiplicity, the ratio of observed participant charges to the total charge of projectile plus target. This may exceed 100% because pions are created. [from J. Harris, in Intersections Between Particle and Nuclear Physics, D. Geesaman ed., AIP Conference Proceedings 150 (AIP, New York, 1986) p. 835].

velocity, because they have a larger velocity pc^2/E. The radial expansion can also cause the shoulder in the spectrum (see fig. 11.16), because the distribution is pushed toward the average radial velocity βc: in the limit when the expansion velocity is greater than the thermal velocity, the distribution is peaked at the energy corresponding to the expansion velocity.

While the exploding-fireball picture can explain the inclusive spectra, more detailed observations show that the flow is not spherically symmetric. For example, by observing most of the charged participants from a single collision event, a direction of preferred flow θ can be established in which nucleons are most likely to be moving. This angle increases with the number of participants, suggesting that the collisions with smallest impact parameter (and therefore most participants) cause the most flow away from the beam direction (fig. 11.17). These features of the flow can be qualitatively understood in a hydrodynamic model, in which the nuclear matter is treated as a compressible fluid. Its time evolution is studied by numerically following the time evolution of two liquid droplets as they smash into each other. Fig. 11.18 shows an example of the results of a hydrodynamic computation.

The final-state momentum distribution at the end of a nuclear collision does not directly tell about the spatial structure of the fireball. One way of obtaining spatial information is by **hadron interferometry**, which measures the correlations of pairs of nucleons or pions with nearly the same momenta. If the difference Δp between their momenta is less than $\hbar/R$ where R is the size of the source, then the hadrons' wave functions will interfere with each other constructively or destructively for Bose or Fermi statistics respectively. After taking account of the interactions of the observed pair with each other, the size of the source may be inferred. In principle both the shape of the source, its time duration, and its flow could be deduced from hadron interferometry. So far limitations due to the statistical fluctuations of the samples observed (pairs with small momentum differences are rare) prevent the full exploitation of this method; preliminary observations show that, by the time the nucleons have finished scattering from each other, the system expands to a density substantially less than that of equilibrium nuclear matter.

Another way to find out about the spatial extent of the system is by measuring the probability of finding the nucleons in a bound cluster. This probability depends on the temperature, because of the binding energy of the cluster. But it also depends on the density. An easy way to see this is provided by the deuteron, whose binding energy is negligible in relativistic collisions. A deuteron is a single quantum state consisting essentially of a neutron and a proton with the same momentum and position, in the same h^3 element of 6-dimensional phase space $d^3\vec{r}\,d^3\vec{p}$. Thus the ratio of deuterons to protons tells the probability that a given phase-space element is occupied. Since we can measure the momentum distribution we can infer information about the spatial distribution. In a dilute gas this information is equivalent to the entropy, given by [Siemens and Kapusta]

$$\mathcal{S} = 3.95 - \ln{(N_d/N_p)} \qquad (11.3.7)$$

where N_d and N_p are the numbers of deuterons and protons respectively. In denser systems, other clusters have to be taken into account too. The observed numbers of the lightest clusters seem to imply that the system continues to interact until the density is many times less than the equilibrium density of cold nuclear matter.

The entropy is an especially interesting quantity to know because it has the possibility of carrying information about the early stages of the collision as well as the final state. We can understand this remarkable property best within a transport

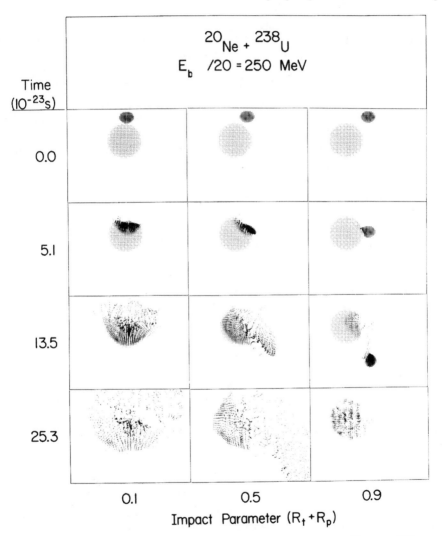

Figure 11.18 Hydrodynamic reaction mechanism for collision of ^{20}Ne on ^{238}U at 250 MeV per nucleon. [*computation by A. Amsden, G. Bertsch, F. Harlow and J. Nix, private communication, see also Phys. Rev. Lett.* **35** *(1975) 905*].

theory for the single-particle degrees of freedom. Boltzmann, who invented transport theory, showed that the rate of entropy production due to two-body collisions is proportional to the square of the distribution function's deviation from local equilibrium. On the other hand, Liouville's theorem shows that in the absence of two-body collisions the mean phase space density is conserved. Thus the entropy will mostly be created by the initial process of thermalizing the energy of relative motion; later evolution, such as an explosion, will not change the entropy very much. Both hydrodynamic and transport-theory model simulations of relativistic nuclear collisions confirm that little entropy is created after the initial stopping of the relative motion. Thus the entropy provides a trace of this early stage of the reaction. Since the model simulations show that the density of the matter may exceed twice that of normal nuclei, the entropy contains information about nuclear matter at high density and high temperature.

The main obstacle in the way of a quantitative theoretical interpretation of relativistic nuclear collisions is the difficulty of treating the pions, which are produced copiously. We know from Sect. 3.6 that pions have a large potential energy in nuclear matter, and that they are absorbed mainly by way of the Δ resonance. The creation of pions can absorb a lot of energy from the nucleons, which is returned to the nucleons if the pions are reabsorbed. Thus the pion and nucleons must influence each other in a profound and complicated way. When the theory of pion-nucleus scattering is incorporated into the transport theories needed for heavy-ion reactions, we can hope to obtain quantitative information about nuclear matter at very high excitation. Among the nuclear-matter properties we may hope to learn are how its energy varies with temperature and density, which are related to the equation of state. But other properties are also important, especially the mean free paths of pions and nucleons in hot nuclear matter.

11.4   ULTRARELATIVISTIC NUCLEAR COLLISIONS

It has recently become possible to accelerate nuclei to energies of hundreds of GeV per nucleon by injecting them into high-energy accelerators meant for protons. While hardly any experiments have been performed yet, theorists expect that collisions in this energy regime will lead to the formation of a radically new phase of nuclear matter in which the hadrons melt together into a soup of quarks and gluons.

To see why a phase transition ought to occur, imagine first what would happen if nuclear matter were compressed adiabatically without heating it. When the distance between the baryons becomes much less than the distance between the quarks inside a baryon, it will become impossible to identify the quarks as belonging to particular baryons. Instead, each quark will be able to move freely from one region to another without taking any companions along with it. The quarks are no longer confined in clusters (identified as hadrons), but are said to be **deconfined**. This deconfinement phase transition is analogous to the ionization transition in which a gas of neutral atoms and molecules dissociates into electrically charged constituents. In the case of nuclear matter, the fundamental charge of the strong interaction, called **color**, become manifest explicitly instead of being hidden inside the color-neutral hadrons. The deconfined state is often called a **color plasma** or **quark-gluon plasma**.

A similar effect must occur at high temperatures. Imagine now what would happen if nuclear matter were heated to a temperature much larger than $m_\pi c^2$. Then huge numbers of pions would be produced, so that the density of hadrons would rise. Once again we see that the association of quarks into hadrons would become unsustainable.

The theoretical description of the deconfinement phase transition is hampered by the lack of a quantitative theoretical model for how the color forces bind quarks into hadrons. The best theoretical computations deal with the baryon-free vacuum at elevated temperature, where the phase transition should also occur for the same reasons as in nuclear matter. These computations, performed within the context of lattice gauge theory, indicate a rapid rise in the vacuum's specific heat at a particular temperature, as would be expected at a phase transition. The long-range correlations of color charges also appear to change qualitatively at the same transition temperature, as would be expected when the hadrons dissolve.

The deconfinement transition appears in its simplest form in the **bag model** of hadrons. In the bag model, the physical vacuum is thought to be impermeable to any color charges — a color insulator for the strong charge. This viewpoint is motivated by many studies of the vacuum state in quantum chromodynamics. The region inside the hadron, on the other hand, is supposed to be a color conductor in which the quarks move freely. This conducting state of vacuum has a higher energy than the usual vacuum, which of course is the ground state in the absence of quarks; the difference in energy density B between the conducting and non-conducting vacua is the most important phenomenological constant of the bag model, and has a value of roughly 200 MeV fm^{-3}. When account is taken of the effective interactions among the quarks inside the bag due to exchange of gluons, the masses of hadrons (except the pion) can be reproduced by an appropriate choice of 2 constants: the energy density B, and the quark-gluon coupling constant; the bag radius is chosen to minimize the hadron's energy, which also includes the kinetic energy of the quarks in their quantum states inside the infinitely high square well of the bag. The plasma state is then simply this color-conducting excited state of the vacuum, filled with massless quarks and gluons.

In a region where the net baryon-number density is negligible, we can easily see how the phase transition occurs in the bag model by comparing the pressure of the plasma with that of a gas of hadrons. At equal temperatures, the state with the largest pressure has the lowest free energy and thus will be thermodynamically preferred. The pressure of the plasma state is given by [see Satz]

$$P_{\text{plasma}} = g_{\text{plasma}}\frac{\pi^2}{90}T^4/(\hbar c)^3 - B \qquad (11.4.1)$$

where the first term is just the pressure of a high-entropy gas of massless particles (like photons) in plane-wave states of degeneracy g_{plasma}. The pressure of a non-interacting gas of relativistic pions at the same temperature $T > m_\pi c^2$ would be

$$P_\pi = g_\pi \frac{\pi^2}{90} T^4 / (\hbar c)^3. \qquad (11.4.2)$$

where $g_\pi = 3$ for the three charge states of the pion. At low temperatures $P_\pi > P_{plasma}$ because of the bag constant B, which represents the energy penalty for making the conducting vacuum. However, the extra degrees of freedom of the plasma phase make its pressure grow more rapidly with temperature: there are eight gluons (each gluon carries an SU(3) octet color charge) each with two spin states ($1\hbar$, like the photon) giving $g_{glue} = 16$; then there are three colors each of quark and antiquark, each with spins $\pm \hbar / 2$, and isospin up or down. Thus the plasma has about thirteen times as many degrees of freedom as the pion gas, and will be thermodynamically preferred when

$$T > [90B(\hbar c)^3 / \pi^2 (g_{plasma} - g_\pi)]^{1/4} \approx 140 \text{MeV}. \qquad (11.4.3)$$

Of course this estimate of the transition temperature is very crude; fancier computations give transition temperatures above 200 MeV.

At these very high energies, the time evolution of heavy-ion collisions is also expected to change qualitatively. Each nucleon has enough kinetic energy to make

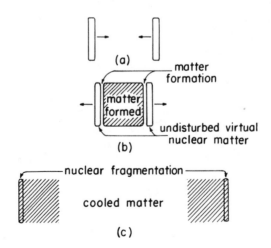

Figure 11.19   Imagined scenario of ultrarelativistic heavy-ion collision. The incident nuclei are shortened by Lorentz contraction. [*from L. McLerran, Revs. Mod. Phys.* **58**(1986) 1021].

many pions, while the average number of pions created rises much more slowly than the available energy. Thus it seems likely that two heavy nuclei could pass right through each other. Of course they would be slowed down a lot, and excited enormously, but a large part of the original relative motion might be left. Huge amounts of energy would be deposited in the region between the nucleons. That energy would eventually materialize as pions, but initially it could form a high-temperature color plasma. The baryon-rich remnants of the original nuclei might also be excited into the plasma phase. These regions would then expand, decreasing their energy densities until they fell below the temperature of the deconfinement phase transition. Then they would turn back into hadrons — mostly nucleons and pions — as the quarks and gluons grouped themselves into color-neutral clusters. Fig. 11.19 sketches this imagined scenario.

The first generation of experiments faces staggering difficulties, because the hundreds of reaction products all go forward into a very tiny angular regime due to the large center-of-mass velocity. The difficulties of this situation in the much simpler case of nucleon-nucleon collisions have led particle physicists to prefer experiments with colliding beams of equal-velocity projectiles, where the center of mass is at rest in the laboratory. Perhaps colliding-beam experiments will be needed before we will be able to study in detail the deconfined color plasma, which will resemble conditions in the first microsecond of the Big Bang.

11.5   PHASE DIAGRAM OF NUCLEAR MATTER

We are now in a position to piece together a picture of the pase diagram of nuclear matter. As state variables we will choose the baryon number density n_B and the temperature T. Actually there is a third state variable, the ratio of neutrons to protons: we will concentrate on isospin-symmetric matter with equal numbers of neutrons and protons. The results of the discussion are summarized in fig. 11.20.

The ground state of nuclear matter is a superfluid liquid with T= 0 and $n_B \approx$ 0.17fm^{-3}. For small temperatures at lower densities, there is a mixed liquid-gas phase (compare figure 11.10), which yields to a Fermi gas at very low density. Since nuclear matter with only neutrons is not bound, the liquid-gas phase transition might go away for isospin-asymmetric matter. Above the ground-state density, the superfluid liquid may turn into a liquid crystal at a few times nuclear density. This liquid crystal would contain standing waves of pions in a nucleon distribution with alternating layers of neutron-rich and proton-rich matter. This "pion condensate" phase would be an extreme consequence of the Kisslinger anomaly, in which the large nucleon density would produce enough attractive interaction energy to overcome the rest mass of short-wavelength pions. Theoretical models disagree on whether the liquid-crystal phase could occur. Even if it existed, its ordered structure would undoubtedly be destroyed by thermal agitation at high temperatures.

At temperatures above the critical temperature T_C for the liquid-gas transition, low- to moderate-density nuclear matter would consist of a uniform gas of nucleons.

At temperatures of a few MeV, this gas would contain many bound clusters of nucleons — deuterons, helium nuclei, etc. At temperatures near $m_\pi c^2$, few clusters would survive but many pions would be found.

Finally, at very large temperature or density, the nucleons would dissolve into their constituent quarks and gluons. At large temperatures, the quarks would be accompanied by many antiquarks. The relative numbers of quarks and antiquarks would be determined by the chemical potential associated with n_B. This deconfined plasma was the original state of all the nuclear matter in the universe.

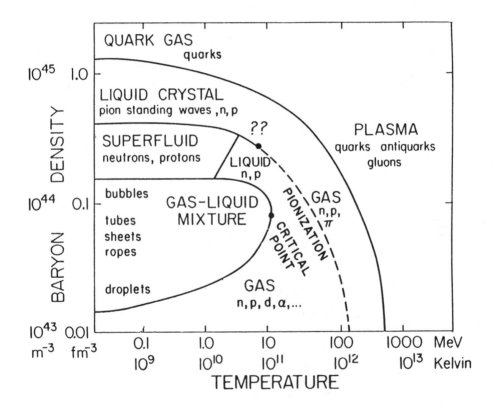

Figure 11.20   Some postulated phases of nuclear matter.

Appendix

Notation

Many experimental and theoretical numbers are used in the text. Sometimes the reference is given explicitly and sometimes only as names which often appear in the reference list at the end of the book. If nothing but numbers are given, they are either obtained from Nuclear Data Sheets, the Particle Data Group Rev. Mod. Phys. publications or they represent our own judgment.

The quantitites used in the book are defined when they are introduced for the first time. To facilitate the reading we have adopted various systematics of notation. Throughout (N, Z) are the number of neutrons and protons, and A = N + Z is the total number of nucleons of a given nucleus. The mass of the particle "i" is denoted m_i, e.g., m_m, m_p, m_π for neutron, proton and pion, respectively. When sufficient accuracy is obtained without distinction between neutrons and protons, we use m_N. These name indices are also used on other quantities, e.g., separation energies. Other types of indices usually are abbreviations of descriptive terms, e.g., B for barrier, MF for mean field, cr for cranking, F for Fermi, etc.

Single-particle quantities are usually denoted by lower-case letters, e.g., spin ($\vec{s}$), orbital ($\vec{\ell}$), and total angular momentum ($\vec{j}$). The corresponding many-particle quantities are then denoted by uppercase letters like $\vec{S}$, $\vec{L}$, and $\vec{J}$. This also applies to isospin denoted by $\vec{t} = (t_1, t_2, t_3)$ where $t_3|n\rangle = -\frac{1}{2}|n\rangle$ and $t_3|p\rangle = \frac{1}{2}|p\rangle$ for the single-nucleon operators.

One-body operators $\mathcal{O}_x = \sum_{i=1}^{A} \mathcal{O}^x(i)$ are denoted by the same letter for both the many-particle ($\mathcal{O}_x$) and the single-particle operator ($\mathcal{O}^x$). The index is moved from subscript to superscript to indicate the difference.

Basic and effective two-particle potentials are distinguished by their respective names V and $\mathfrak{V}$. One-particle or mean-field potentials are called U and W for real and imaginary parts, respectively.

Single-particle and Bogolyubov quasi-particle energies are denoted ε and ϵ while many-particle energies are called E or $\mathcal{E}$.

Spatial single-particle wave functions are called ϕ, the two-component spin-spinor wave function is called χ, and full spinors with both spatial and spin variables are denoted Φ. Four-component spin and isospin- spinors are named $\mathcal{X}$ and the full-spinors are called $\mathbf{\Phi}$. Plane-wave eigenstates of momentum $\vec{p}$ are called $\Upsilon_{\vec{p}}$. We use ψ

for many-particle wave functions and Ψ for particularly complicated many-particle wave functions.

Symbols above Letters		Example
" $\rightarrow$ " — vector quantity	:	$\vec{V}$
" $\cdot$ " — time derivative	:	$\dot{Q}$
" $\sim$ " — smooth part	:	$\tilde{E}$
" $\sim$ " — distorted wave quantity	:	$\tilde{\Upsilon}$
" $\sim$ " — Fourier transform	:	$\tilde{\chi}$
" $_-$ " — anti particle	:	$\bar{\nu}$
" $_-$ " — time reversed	:	$\overline{|jm\rangle}$
" $_-$ " — adjoint Dirac spinor	:	$\bar{\Phi} \equiv \Phi^+ \Upsilon_0$
" $\wedge$ " — parametrized form of potential	:	$\hat{U}$
" $\wedge$ " — energy resulting from parametrized potential	:	$\hat{\varepsilon}$
" $\wedge$ " — unit vector	:	$\hat{n}$

Various Symbols

$[A, B] = AB-BA$ — commutator

$\{A, B\} = AB+BA$ — anticommutator

$\{\vec{\phi}_i\}$ — set of values $(\vec{\phi}_1, \vec{\phi}_2, \ldots)$

$\langle\psi|\theta|\psi\rangle$ — expectation value

$\langle\theta\rangle \equiv \langle\psi|\theta|\psi\rangle/\langle\psi|\psi\rangle$ — mean value

θ^+ — Hermitean conjugate of θ

θ^* — complex conjugate of θ

$\vec{\nabla}$ — gradient operator

$\vec{\nabla}^2 \equiv \vec{\nabla} \cdot \vec{\nabla}$ — Laplace operator

$V = |\vec{V}|$; $\vec{V} = (V_x, V_y, V_z)$

$\vec{V} \cdot \vec{u}$ — vector dot product

$\vec{V} \times \vec{u}$ — vector cross product

P — Cauchy principal value

$|vac\rangle$ — vacuum state

$|gs\rangle$ — ground state

$\langle j_1\mu_1 j_2\mu_2|j\mu\rangle$ — Clebsch-Gordon Coefficient with the phase convention used in Bohr and Mottelson.

Functions and Operators

$Y_\ell^m(\theta, \phi)$ — Spherical harmonic

$P_\ell(x)$ — Legendre polynomial

$j_\ell(x)$ — spherical Bessel function

$\vec{\tilde{\Phi}}_{\ell\lambda}^\mu(\vec{\omega})$ — vector spherical harmonic

$\mathcal{D}_{MK}^{J}(\vec{\omega})$ — $\mathcal{D}$ – function used in Bohr and Mottelson
$\mathfrak{R}_{\vec{\omega}} \equiv e^{-i\psi J_z}e^{-i\theta J_y}e^{-i\phi J_z}$ — rotation operator
a_k^{+}, a_k — creation and annihilation operators for particle in the state $|k\rangle$.
α_k^{+}, α_k — creation and annihilation operators for Bogolyubov quasiparticle in
  the state $|k\rangle$.

$$\{a_i^{+}, a_k\} = \{\alpha_i^{+}, \alpha_k\} = \delta_{ik}, \{a_i, a_k\} = \{\alpha_i, \alpha_k\} = 0$$

$a_{\vec{k}\vec{e}}^{+}$, $a_{\vec{k}\vec{e}}$ — creation and annihilation operators of photons of momentum $\hbar\vec{k}$ and
  polarization $\vec{e}$.

Constants
for rest energies of various particles see Table 1.1
$e^2 = 1.44$ MeV fm
$c = 3 \times 10^{23}$ fm sec^{-1}
$\hbar c = 197$ MeV fm

Relativistic Conventions
Our metric tensor and γ matrices follow the convention of Bjorken and Drell:

$$g_{\mu\nu} = \begin{pmatrix} 1 & 0 & 0 & 0 \\ 0 & -1 & 0 & 0 \\ 0 & 0 & -1 & 0 \\ 0 & 0 & 0 & -1 \end{pmatrix}, \gamma^0 = \begin{pmatrix} 1 & 0 & 0 & 0 \\ 0 & 1 & 0 & 0 \\ 0 & 0 & -1 & 0 \\ 0 & 0 & 0 & -1 \end{pmatrix}, \gamma^5 = \begin{pmatrix} 0 & 0 & 1 & 0 \\ 0 & 0 & 0 & 1 \\ 1 & 0 & 0 & 0 \\ 0 & 1 & 0 & 0 \end{pmatrix}$$

$$\gamma^1 = \begin{pmatrix} 0 & 0 & 0 & 1 \\ 0 & 0 & 1 & 0 \\ 0 & -1 & 0 & 0 \\ -1 & 0 & 0 & 0 \end{pmatrix}, \gamma^2 = \begin{pmatrix} 0 & 0 & 0 & -i \\ 0 & 0 & i & 0 \\ 0 & i & 0 & 0 \\ -i & 0 & 0 & 0 \end{pmatrix}, \gamma^3 = \begin{pmatrix} 0 & 0 & 1 & 0 \\ 0 & 0 & 0 & -1 \\ -1 & 0 & 0 & 0 \\ 0 & 1 & 0 & 0 \end{pmatrix}$$

Bibliography

Agassi, D., C.M. Ko, and H.A. Weidenmüller, Ann. Phys. **117**(1979)237,407.

Alder, K., and Aa. Winther, *Coulomb Excitation* (Academic Press, New York, 1966).

Arima, A. and F. Iachello, in Advances in Nuclear Physics **13** J. Negele and E. Vogt, eds, (Plenum, New York, 1984).

Ashcroft, N.W. and N.D. Mermin, *Solid State Physics* (Holt Rinehart and Winston, New York, 1976).

Bardeen, J., L.N. Cooper, and J. R. Schrieffer, Phys. Rev. **106**(1957)162 and **108**(1957)1175.

Barrett, B.R., in *Nuclear Structure: Proceedings of the International School on Nuclear Structure (Alushta, October 19-22,1985)*, V.G. Soloviev and Yu. P. Popov, eds. (Dubna, USSR, 1985)153.

Bethe, H.A. and R.F. Bacher, Rev. Mod. Phys. **8**(1936)82.

Bethe, H.A. and R.W. Jackiw, *Intermediate Quantum Mechanics* (Benjamin, New York, 1968).

Bijl, A. see N.F. Mott, Phil. Mag. **40**(1949)61.

Bjorken, J.D. and S.D. Drell *Relativistic Quantum Mechanics*, (McGraw-Hill, New York, 1964).

Bjorken, J.D. and S.D. Drell *Relativistic Quantum Fields*, (McGraw-Hill, New York, 1965).

Blaizot, J. and G. Ripka *Quantum Theory of Finite Systems*, (MIT Press, Cambridge Mass., 1985).

Blatt, J.M. and V.F. Weisskopf, *Theoretical Nuclear Physics* (Wiley, New York, 1952).

Blocki, J., Y. Boneh, J.R. Nix, J. Randrup, M. Robel, A.J. Sierk, and W.J. Swiatecki, Prog. Part. and Nucl. Phys. **4**(1980)383.

Bohr, A. and B.R. Mottelson, *Nuclear Structure, Vol 1 and 2* (Benjamin, Reading Massachusetts, 1969,1975).

Bohr, N., Nature **137**(1936)344,351.

Bohr, N. and J.A. Wheeler, Phys. Rev. **56**(1939)426.

Bonche, P., S. Levit, and D. Vautherin, Nucl. Phys. **A427**(1984)278.

Brack, M., *Proc. Symp. Physics and Chemistry of Fission* (Jülich, IAEA Vienna,1980)227

Brink, D.M., Phys. Lett. **B40**(1972)37.

Brown, G.E., *Unified Theory of Nuclear Models and Forces* 2nd edition, (North Holland, Amsterdam, 1967).

Brown, G.E. and A.D. Jackson, *The Nucleon-Nucleon Interaction* (North Holland, Amsterdam, 1976).

de Shalit, A. and H. Feshbach, *Theoretical Nuclear Physics* (Wiley, New York, 1974).

de Shalit, A. and I. Talmi, *Nuclear Shell Theory* (Academic Press, New York, 1963).

Edmonds, A.R., *Angular Momentum in Quantum Mechanics* (Princeton University Press, Princeton New Jersey, 1960).

Ericson, T., Advances in Physics **9**(1960)425.

Fisher, M., Physics **3**(1967)255.

Fröman, N. and P.O. Fröman, *JWKB Approximations* (North Holland, Amsterdam, 1965).

Gasiorowicz, S., *Elementary Particle Physics* (Wiley, New York, 1966).

Gogny, D., Lecture Notes in Physics **108**(1979)88.

Goldhaber, A.S., Phys. Lett. **B53**(1974)306.

Halzen, F. and A.D. Martin, *Quarks and Leptons* (Wiley, New York, 1984).

Hill, D.L. and J.A. Wheeler, Phys. Rev. **89**(1953)1102.

Hofmann, H. and A.S. Jensen, Nucl. Phys. **A428**(1984)1c.

Hofmann, H. and P. Siemens, Nucl. Phys. **A275**(1977)464.

Holinde, K., Phys. Lett. **C68**(1981)121.

Inglis, D.R., Phys. Rev. **96**(1954)1059.

Jackson, J.D., *Classical Electrodynamics* 2nd edition, (Wiley, New York, 1975).

Jeukenne, J.P., A. Lejeune, and C. Mahaux, Phys. Lett. **C25**(1976)83.

Kadanoff, L. and G. Baym, *Quantum Statistical Mechanics* (Benjamin, New York, 1976).

Kisslinger, L.S., Phys. Rev. **98**(1955)761.

Kramers, H.A., Physica **7**(1940)284.

Mayer, M.G. and J.H.D. Jensen, *Elementary Theory of Nuclear Shell Structure* (Wiley, New York, 1955).

Messiah, A., *Quantum Mechanics 1 & 2* (North Holland, Amsterdam, 1962).

Nakayama, K., S. Krewald, and J. Speth, Nucl. Phys. **A451**(1986)243.

Negele, J.W., Rev. Mod. Phys. **54**(1982)913.

Negele, J. and D. Vautherin, Phys. Rev. **C5**(1972)1486.

Oset, E., H. Toki, and W. Weise, Phys. Lett. **C83**(1982)281.

Peierls, R.E. and D.J. Thouless, Nucl. Phys. **38** (1962)154.

Peierls, R.E. and J. Yoccoz, Proc. Phys. Soc. **A70**(1957)381.

Penzias, A.A. and R.W. Wilson, Astrophysics J. **142**(1965)419.

Perey, A.M. and F.G. Perey, Atomic Data and Nuclear Data Tables **17**(1976)1.

Pethick, C.J. and D.G. Ravenhall, *Fragmentation and the Liquid-Gas Phase Transition*, report no. P/86/5/75 (Univ. Illinois, Urbana, 1986).

Preston, M.A., *Physics of the Nucleus* (Addison-Wesley, Reading Massachusetts, 1962).

Preston, M.A. and R.K. Bhaduri, *Structure of the Nucleus* (Addison-Wesley, Reading Massachusetts, 1975).

Randrup, J., Nucl. Phys. **A327**(1979)490.

Reid, R.V., Ann. Phys. **50**(1968)411.

Rose, M.E., *Elementary Theory of Angular Momentum* (Wiley, New York, 1957).

Rosenbluth, M., Phys. Rev. **79**(1950)615.

Rutherford, E., Phil. Mag. **21**(1911)669.

Satchler, G.R., *Direct Nuclear Reactions* (Oxford Press, New York, 1983).

Satz, H., Nucl. Phys. **A400**(1983)541c.

Schiff, L.I., *Quantum Mechanics* (McGraw-Hill, New York, 1954).

Schrieffer, J.R., *Theory of Superconductivity* (Benjamin, New York, 1964).

Siemens, P. and J. Kapusta, Phys. Rev. Lett. **43**(1979)1486.

Stocker, W., Nucl. Phys. **A202**(1973)265.

Thouless, D.J., Nucl. Phys. **22**(1961)78.

Thouless, D.J., *The Quantum Mechanics of Many-Body Systems* (Academic Press, New York, 1961).

Villars, F. *Many-Body Description of Nuclear Structure and Reactions*, Proc. Int. School of Physics, Enrico Fermi **36**(1966)14.

Vinh Mau, R., in *Mesons in Nuclei*, M. Rho and P. Wilkinson, eds. (North Holland, Amsterdam, 1979)179; see also Lacombe et al., Phys. Rev. **C21**(1979)861.

Weinberg, S., *The First Three Minutes* (Basic Books, Inc., New York, 1977).

Wigner, E.P., Proc. Nat. Acad. Sci. **22**(1936)662.

Wilczynski, J., Phys. Lett. **B47**(1973)484.

Wu, C.S. and S.A. Moszkowski, *Beta Decay* (Interscience, Wiley, New York, 1966).

Yukawa, H., Proc. Phys.-Mat. Soc. Japan **17**(1935)48.

Index

For any specific entry only the first occurence in a subsection is given.

The Addison-Wesley **Advanced Book Program** would like to offer you the opportunity to learn about our new physics and scientific computing titles in advance. To be placed on our mailing list and receive pre-publication notices and special offers, just **fill out this card completely** and return to us, postage paid. Thank you.

Title and Author of this book: **Date purchased:**
_____   _____

Name _____
Title _____
School/Company _____
Department _____
Street Address _____
City _____ State _____ Zip _____
Telephone/s(   ) _____ (   ) _____

Where did you buy/obtain this book?

☐ Bookstore   ☐ Mail Order   ☐ School (Required for Class)
☐ Campus Bookstore   ☐ Toll Free # to Publisher   ☐ Professional Meeting
☐ Other _____   ☐ Publisher's Representative

What professional scientific and engineering associations are you an active member of?

☐ AAPT (Amer Assoc of Physics Teachers)   ☐ APS (Amer Physical Society)   ☐ SPS (Society of Physics Students)
☐ AIP (Amer Institute of Physics)   ☐ Sigma Pi Sigma   ☐ AAAS (Amer Assoc for the Advancement of Science)
☐ Other _____

Check your areas of interest.

⑩ ✔**Physics**

11 ☐ Quantum Mechanics   18 ☐ Materials Science   25 ☐ Geophysics
12 ☐ Particle/Astro Physics   19 ☐ Biological Physics   26 ☐ Medical Physics
13 ☐ Condensed Matter   20 ☐ High Polymer Physics   27 ☐ Optics
14 ☐ Mathematical Physics   21 ☐ Chemical Physics   28 ☐ Vacuum Physics
15 ☐ Nuclear Physics   22 ☐ Fluid Dynamics
16 ☐ Electron/Atomic Physics 23 ☐ History of Physics
17 ☐ Plasma/Fusion Physics   24 ☐ Statistical Physics
29 ☐ Other _____

Are you more interested in: ☐ theory ☐ experimentation?

Are you currently writing, or planning to write a textbook, research monograph, reference work, or create software in any of the above areas?
        ☐ Yes   ☐ No
        Area: _____

(If Yes) **Are you interested in discussing your project with us?**
        ☐ Yes   ☐ No

Physics

‖‖‖‖‖‖‖‖‖‖‖‖‖‖‖‖‖‖‖‖‖‖‖‖‖‖‖‖‖‖‖‖

BUSINESS REPLY MAIL
FIRST CLASS PERMIT NO. 828   REDWOOD CITY, CA   94065

Postage will be paid by Addressee:

ADDISON-WESLEY
PUBLISHING COMPANY, INC.®

Advanced Book Program
390 Bridge Parkway, Suite 202
Redwood City, CA 94065-1522